Multiobjective Linear Programming

Dinh The Luc

Multiobjective Linear Programming

An Introduction

 Springer

Dinh The Luc
Avignon University
Avignon
France

ISBN 978-3-319-21090-2   ISBN 978-3-319-21091-9   (eBook)
DOI 10.1007/978-3-319-21091-9

Library of Congress Control Number: 2015943841

Springer Cham Heidelberg New York Dordrecht London

Springer International Publishing AG Switzerland is part of Springer Science+Business Media
(www.springer.com)

To Dieu Huyen,
Liuli and The Duc

Preface

Multiobjective optimization problems arise in decision-making processes in many areas of human activity including economics, engineering, transportation, water resources, and the social sciences. Although most real-life problems involve non-linear objective functions and constraints, solution methods are principally straightforward in problems with a linear structure. Apart from Zeleny's classic 1974 work entitled "Linear Multiobjective Programming" and Steuer's 1986 book "Multiple Criteria Optimization: Theory, Computation and Application," nearly all textbooks and monographs on multiobjective optimization are devoted to non-convex problems in a general setting, sometimes with set-valued data, which are not always accessible to practitioners. The main purpose of this book is to introduce readers to the field of multiobjective optimization using problems with fairly simple structures, namely those in which the objective and constraint functions are linear. By working with linear problems, readers will easily come to grasp the fundamental concepts of vector problems, recognize parallelisms in more complicated problems with scalar linear programming, analyze difficulties related to multi-dimensionality in the outcome space, and develop effective methods for treating multiobjective problems.

Because of the introductory nature of the book, we have sought to present the material in as elementary a fashion as possible, so as to require only a minimum of mathematical background knowledge. The first part of the book consists of two chapters providing the necessary concepts and results on convex polyhedral sets and linear programming to prepare readers for the new area of optimization with several objective functions. The second part of the book begins with an examination of the concept of Pareto optimality, distinguishing it from the classical concept of optimality used in traditional optimization. Two of the most interesting topics in this part of the book involve duality and stability in multiple objective linear programming, both of which are discussed in detail. The third part of the book is devoted to numerical algorithms for solving multiple objective linear programs. This includes the well-known multiple objective simplex method, the outcome space method, and a recent method using normal cone directions.

Although some new research results are incorporated into the book, it is well suited for use in the first part of a course on multiobjective optimization for undergraduates or first-year graduate students in applied mathematics, engineering, computer science, operations research, and economics. Neither integer problems nor fuzzy linear problems are addressed. Further, applications to other domains are not tackled, though students will certainly have no real difficulty in studying them, once the basic results of this book assimilated.

During the preparation of this manuscript I have benefited from the assistance of many people. I am grateful to my Post-Ph.D. and Ph.D. students Anulekha Dhara, Truong Thi Thanh Phuong, Tran Ngoc Thang, and Moslem Zamani for their careful reading of the manuscript. I would also like to thank Moslem Zamani for the illustrative figures he made for this book. I want to take this opportunity to give special thanks to Juan-Enrique Martinez-Legaz (Autonomous University of Barcelona), Boris Mordukhovich (Wayne State University), Nguyen Thi Bach Kim (Hanoi Polytechnical University), Panos Pardalos (University of Florida), Michel Thera (University of Limoges), Majid Soleimani-Damaneh (University of Tehran), Ralph E. Steuer (University of Georgia), Michel Volle (University of Avignon), and Mohammad Yaghoobi (University of Kerman) for their valued support in this endeavor.

Avignon   Dinh The Luc
December 2014

Contents

Part III   Methods

Notations

$\mathbb{N}$	Natural numbers
$\mathbb{R}$	Real numbers
$\mathbb{R}^n$	Euclidean n-dimensional space
$L(\mathbb{R}^n, \mathbb{R}^m)$	Space of $m \times n$ matrices
B_n	Closed unit ball in $\mathbb{R}^n$
S_n	Unit sphere in $\mathbb{R}^n$
$B_{m \times n}$	Closed unit ball in $L(\mathbb{R}^n, \mathbb{R}^m)$
e	Vector of ones
e^i	i-th coordinate unit vector
Δ	Standard simplex
$\|x\|$	Euclidean norm
$\|x\|_\infty$	Max-norm
$\langle x, y \rangle$	Canonical scalar product
$\leqq$	Less than or equal to
$\leq$	Less than but not equal to
$<$	Strictly less than
$\mathrm{aff}(A)$	Affine hull
$\mathrm{cl}(A), \bar{A}$	Closure
$\mathrm{int}(A)$	Interior
$\mathrm{ri}(A)$	Relative interior
$\mathrm{co}(A)$	Convex hull
$\overline{\mathrm{co}}(A)$	Closed convex hull
$\mathrm{cone}(A)$	Conic hull
$\mathrm{pos}(A)$	Positive hull
$\mathrm{Max}(A)$	Set of maximal elements
$\mathrm{WMax}(A)$	Set of weakly maximal elements
$\mathrm{Min}(A)$	Set of minimal elements
$\mathrm{WMin}(A)$	Set of weakly minimal elements
S(MOLP)	Efficient solution set
WS(MOLP)	Weakly efficient solution set

$\sup(A)$	Supremum
$I(x)$	Active index set at x
$A^{\perp}$	Orthogonal
A^{-}	Negative polar cone
A_{∞}	Recession/asymptotic cone
$N_A(x)$	Normal cone
$d(x, C)$	Distance function
$h(A, B)$	Hausdorff distance
$\mathrm{gr}(G)$	Graph
$\mathrm{supp}(x)$	Support

Chapter 1
Introduction

Mathematical optimization studies the problem of finding the best element from a set of feasible alternatives with regard to a criterion or objective function. It is written in the form

$$\text{optimize} \quad f(x)$$
$$\text{subject to} \quad x \in X,$$

where X is a nonempty set, called a feasible set or a set of feasible alternatives, and f is a real function on X, called a criterion or objective function. Here "optimize" stands for either "minimize" or "maximize" which amounts to finding $\bar{x} \in X$ such that either $f(\bar{x}) \leqq f(x)$ for all $x \in X$, or $f(\bar{x}) \geqq f(x)$ for all $x \in X$.

This model offers a general framework for studying a variety of real-world and theoretical problems in the sciences and human activities. However, in many practical situations, we tend to encounter problems that involve not just one criterion, but a number of criteria, which are often in conflict with each other. It then becomes impossible to model such problems in the above-mentioned optimization framework. Here are some instances of such situations.

Automotive design The objective of automotive design is to determine the technical parameters of a vehicle to minimize (1) production costs, (2) fuel consumption, and (3) emissions, while maximizing (4) performance and (5) crash safety. These criteria are not always compatible; for instance a high-performance engine often involves very high production costs, which means that no design can optimally fulfill all criteria.

House purchase Buying property is one of life's weightiest decisions and often requires the help of real estate agencies. An agency suggests a number of houses or apartments which roughly meet the potential buyer's budget and requirements. In order to make a decision, the buyer assesses the available offers on the basis of his or her criteria. The final choice should satisfy the following: minimal cost, minimal maintenance charges, maximal quality and comfort, best environment etc. It is quite

© Springer International Publishing Switzerland 2016
D.T. Luc, *Multiobjective Linear Programming*,
DOI 10.1007/978-3-319-21091-9_1

natural that the higher the quality of the house, the more expensive it is; as such, it is impossible to make the best choice without compromising.

Distributing electrical power In a system of thermal generators the chief problem concerns allocating the output of each generator in the system. The aim is not only to satisfy the demand for electricity, but also to fulfill two main criteria: minimizing the costs of power generation and minimizing emissions. Since the costs and the emissions are measured in different units, we cannot combine the two criteria into one.

Queen Dido's city Queen Dido's famous problem consists of finding a territory bounded by a line which has the maximum area for a given perimeter. According to elementary calculus, the solution is known to be a circle. However, as it is inconceivable to have a city touching the sea without a seashore, Queen Dido set another objective, namely for her territory to have as large a seashore as possible. As a result, a semicircle partly satisfies her two objectives, but fails to maximize either aspect.

As we have seen, even in the simplest situations described above there can be no alternative found that simultaneously satisfies all criteria, which means that the known concepts of optimization do not apply and there is a real need to develop new notions of optimality for problems involving multiple objective functions. Such a concept was introduced by Pareto (1848–1923), an Italian economist who explained the Pareto optimum as follows: "The optimum allocation of the resources of a society is not attained so long as it is possible to make at least one individual better off in his own estimation while keeping others as well off as before in their own estimation." Prior to Pareto, the Irish economist Edgeworth (1845–1926) had defined an optimum for the multiutility problem of two consumers P and Q as "a point (x, y) such that in whatever direction we take an infinitely small step, P and Q do not increase together but that, while one increases, the other decreases." According to the definition put forward by Pareto, among the feasible alternatives, those that can simultaneously be improved with respect to all criteria cannot be optimal. And an alternative is optimal if any alternative better than it with respect to a certain criterion is worse with respect to some other criterion, that is, if a tradeoff takes place when trying to find a better alternative. From the mathematical point of view, if one defines a domination order in the set of feasible alternatives by a set of criteria—an alternative a dominates an alternative b if the value of every criterion function at a is bigger than that at b—then an alternative is optimal in the Pareto sense if it is dominated by no other alternatives. In other words, an alternative is optimal if it is maximal with respect to the above order. This explains the mathematical origin of the theory of multiple objective optimization, which stems from the theory of ordered spaces developed by Cantor (1845–1918) and Hausdorff (1868–1942).

A typical example of ordered spaces, frequently encountered in practice, is the finite dimensional Euclidean space $\mathbb{R}^n$ with $n \geq 2$, in which two vectors a and b are comparable, let's say a is bigger than or equal to b if all coordinates of a are bigger than or equal to the corresponding coordinates of b. A multiple objective optimization problem is then written as

$$\text{Maximize} \quad F(x) := (f_1(x), \ldots, f_k(x))$$
$$\text{subject to} \quad x \in X,$$

where $f_1, \ldots, f_k$ are real objective functions on X and "Maximize" signifies finding an element $\bar{x} \in X$ such that no value $F(x), x \in X$ is bigger than the value $F(\bar{x})$. It is essential to note that the solution $\bar{x}$ is not worse than any other solution, but in no ways it is the best one, that is, the value $F(\bar{x})$ cannot be bigger than or equal to all values $F(x), x \in X$ in general. A direct consequence of this observation is the fact that the set of "optimal values" is not a singleton, which forces practitioners to find a number of "optimal solutions" before making a final decision. Therefore, solving a multiple objective optimization problem is commonly understood as finding the entire set of "optimal solutions" or "optimal values", or at least a representative portion of them. Indeed, this is the point that makes multiple objective optimization a challenging and fascinating field of theoretical research and application.

Part I
Background

Chapter 2
Convex Polyhedra

We begin the chapter by introducing basic concepts of convex sets and linear functions in a Euclidean space. We review some of fundamental facts about convex polyhedral sets determined by systems of linear equations and inequalities, including Farkas' theorem of the alternative which is considered a keystone of the theory of mathematical programming.

2.1 The Space $\mathbb{R}^n$

Throughout this book, $\mathbb{R}^n$ denotes the n-dimensional Euclidean space of real column n-vectors. The norm of a vector x with components $x_1, \cdots, x_n$ is given by

$$\|x\| = \left[\sum_{i=1}^{n} (x_i)^2 \right]^{1/2}.$$

The inner product of two vectors x and y in $\mathbb{R}^n$ is expressed as

$$\langle x, y \rangle = \sum_{i=1}^{n} x_i y_i.$$

The closed unit ball, the open unit ball and the unit sphere of $\mathbb{R}^n$ are respectively defined by

$$B_n := \left\{ x \in \mathbb{R}^n : \|x\| \leqq 1 \right\},$$
$$\text{int}(B_n) := \left\{ x \in \mathbb{R}^n : \|x\| < 1 \right\},$$
$$S_n := \left\{ x \in \mathbb{R}^n : \|x\| = 1 \right\}.$$

© Springer International Publishing Switzerland 2016
D.T. Luc, *Multiobjective Linear Programming*,
DOI 10.1007/978-3-319-21091-9_2

Given a nonempty set $Q \subseteq \mathbb{R}^n$, we denote the *closure* of Q by cl(Q) and its *interior* by int(Q). The *conic hull*, the *positive hull* and the *affine hull* of Q are respectively given by

$$\text{cone}(Q) := \{ta : a \in Q, t \in \mathbb{R}, t \geq 0\},$$

$$\text{pos}(Q) := \left\{ \sum_{i=1}^{k} t_i a^i : a^i \in Q, t_i \in \mathbb{R}, t_i \geq 0, i = 1, \cdots, k \text{ with } k \in \mathbb{N} \right\},$$

$$\text{aff}(Q) := \left\{ \sum_{i=1}^{k} t_i a^i : a^i \in Q, t_i \in \mathbb{R}, i = 1, \cdots, k \text{ and } \sum_{i=1}^{k} t_i = 1 \text{ with } k \in \mathbb{N} \right\},$$

where $\mathbb{N}$ denotes the set of natural numbers (Figs. 2.1, 2.2 and 2.3).

Fig. 2.1 Conic hull (with $Q = Q_1 \cup Q_2$)

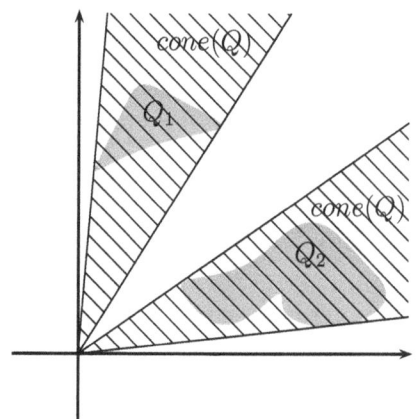

Fig. 2.2 Positive hull (with $Q = Q_1 \cup Q_2$)

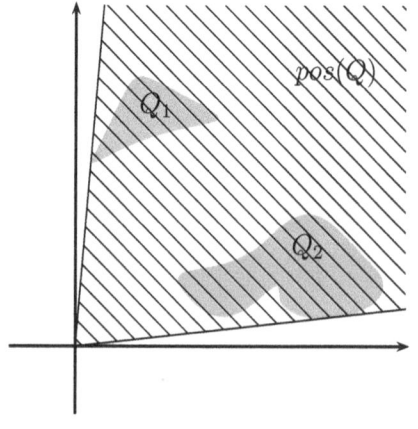

Fig. 2.3 Affine hull (with $Q = Q_1 \cup Q_2$)

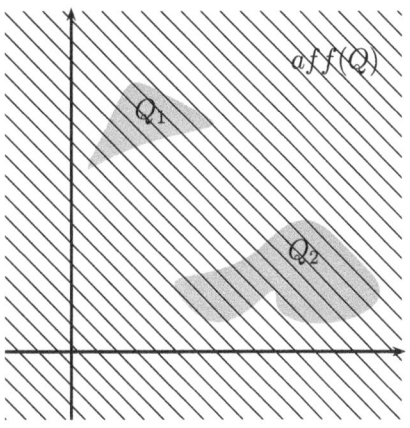

Among the sets described above $\text{cone}(Q)$ and $\text{pos}(Q)$ are cones, that is, they are invariant under multiplication by positive numbers; $\text{pos}(Q)$ is also invariant under addition of its elements; and $\text{aff}(Q)$ is an affine subspace of $\mathbb{R}^n$. For two vectors x and y of $\mathbb{R}^n$, inequalities $x > y$ and $x \geqq y$ mean respectively $x_i > y_i$ and $x_i \geqq y_i$ for all $i = 1, \cdots, n$. When $x \geqq y$ and $x \neq y$, we write $x \geq y$. So a vector x is positive, that is $x \geqq 0$, if its components are non-negative; and it is strictly positive if its components are all strictly positive. The set of all positive vectors of $\mathbb{R}^n$ is the positive orthant $\mathbb{R}^n_+$. Sometimes row vectors are also considered. They are transposes of column vectors. Operations on row vectors are performed in the same manner as on column vectors. Thus, for two row n-vectors c and d, their inner product is expressed by

$$\langle c, d \rangle = \langle c^T, d^T \rangle = \sum_{i=1}^{n} c_i d_i,$$

where the upper index T denotes the transpose. On the other hand, if c is a row vector and x is a column vector, then the product cx is understood as a matrix product which is equal to the inner product $\langle c^T, x \rangle$.

Convex sets

We call a subset Q of $\mathbb{R}^n$ *convex* if the segment joining any two points of Q lies entirely in Q, which means that for every $x, y \in Q$ and for every real number $\lambda \in [0, 1]$, one has $\lambda x + (1 - \lambda)y \in Q$ (Figs. 2.4, 2.5). It follows directly from the definition that the intersection of convex sets, the Cartesian product of convex sets, the image and inverse image of a convex set under a linear transformation, the interior and the closure of a convex set are convex. In particular, the sum $Q_1 + Q_2 := \{x + y : x \in Q_1, y \in Q_2\}$ of two convex sets Q_1 and Q_2 is convex; the conic hull of a convex set is convex. The positive hull and the affine hull of any set are convex.

The *convex hull* of Q, denoted $\text{co}(Q)$ (Fig. 2.6), consists of all convex combinations of elements of Q, that is,

Fig. 2.4 Convex set

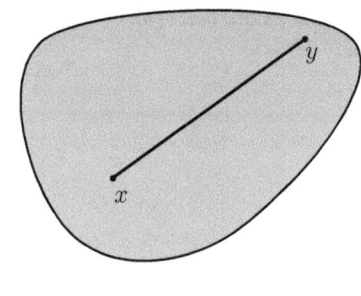

Fig. 2.5 Nonconvex set

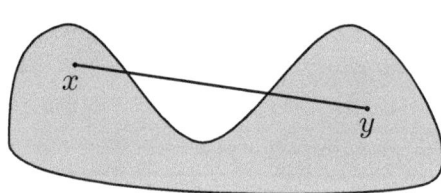

$$\mathrm{co}(Q) := \left\{ \sum_{i=1}^{k} \lambda_i x^i : x^i \in Q, \lambda_i \geq 0, i = 1, \cdots, k \text{ and } \sum_{i=1}^{k} \lambda_i = 1 \text{ with } k \in \mathbb{N} \right\}.$$

It is the intersection of all convex sets containing Q. The closure of the convex hull of Q will be denoted by $\overline{\mathrm{co}}(Q)$, which is exactly the intersection of all closed convex sets containing Q. The positive hull of a set is the conic hull of its convex hull. A convex combination $\sum_{i=1}^{k} \lambda_i x^i$ is strict if all coefficients λ_i are strictly positive.

Given a nonempty convex subset Q of $\mathbb{R}^n$, the *relative interior* of Q, denoted ri(Q), is its interior relative to its affine hull, that is,

$$\mathrm{ri}(Q) := \left\{ x \in Q : (x + \varepsilon B_n) \cap \mathrm{aff}(Q) \subseteq Q \text{ for some } \varepsilon > 0 \right\}.$$

Equivalently, a point x in Q is a relative interior point if and only if for any point y in Q there is a positive number δ such that the segment joining the points $x - \delta(x - y)$ and $x + \delta(x - y)$ entirely lies in Q. As a consequence, any strict convex combination of a finite collection $\{x^1, \cdots, x^k\}$ belongs to the relative interior of its convex hull (see also Lemma 6.4.8). It is important to note also that every nonempty convex set in $\mathbb{R}^n$ has a nonempty relative interior. Moreover, if two convex sets Q_1 and Q_2 have at least one relative interior point in common, then $\mathrm{ri}(Q_1 \cap Q_2) = \mathrm{ri}(Q_1) \cap \mathrm{ri}(Q_2)$.

Fig. 2.6 Convex hull of Q

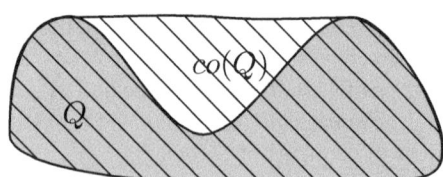

Fig. 2.7 The standard
simplex in $\mathbb{R}^3$

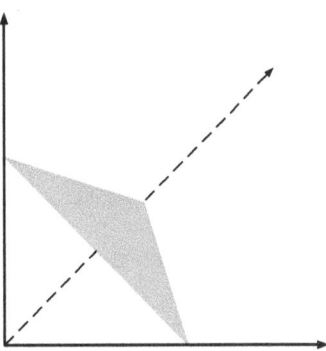

Example 2.1.1 (*Standard simplex*) Let e^i be the ith coordinate unit vector of $\mathbb{R}^n$, that is its components are all zero except for the ith component equal to one. Let Δ denote the convex hull of $e^1, \cdots, e^n$. Then a vector x with components $x_1, \cdots, x_n$ is an element of Δ if and only if $x_i \geq 0, i = 1, \cdots, n$ and $\sum_{i=1}^n x_i = 1$. This set has no interior point. However, its relative interior consists of x with $x_i > 0, i = 1, \cdots, n$ and $\sum_{i=1}^n x_i = 1$. The set Δ is called the *standard simplex* of $\mathbb{R}^n$ (Fig. 2.7).

Caratheodory's theorem

It turns out that the convex hull of a set Q in the space $\mathbb{R}^n$ can be obtained by convex combinations of at most $n + 1$ elements of Q. First we see this for positive hull.

Theorem 2.1.2 *Let $\{a^1, \cdots, a^k\}$ be a collection of vectors in $\mathbb{R}^n$. Then for every nonzero vector x from the positive hull* $\mathrm{pos}\{a^1, \cdots, a^k\}$ *there exists an index set $I \subseteq \{1, \cdots, k\}$ such that*

(i) *the vectors $a^i, i \in I$ are linearly independent;*
(ii) *x belongs to the positive hull* $\mathrm{pos}\{a^i, i \in I\}$.

Proof Since the collection $\{a^1, \cdots, a^k\}$ is finite, we may choose an index set I of minimum cardinality such that $x \in \mathrm{pos}\{a^i, i \in I\}$. It is evident that there are strictly positive numbers $t_i, i \in I$ such that $x = \sum_{i \in I} t_i a^i$. We prove that (i) holds for this I. Indeed, if not, one can find an index $j \in I$ and real numbers s_i such that

$$a^j - \sum_{i \in I \setminus \{j\}} s_i a^i = 0.$$

Set

$$\varepsilon = \min \left\{ t_j \text{ and } -\frac{t_i}{s_i} : i \in I \text{ with } s_i < 0 \right\}$$

and express

$$x = \sum_{i \in I} t_i a^i - \varepsilon\left(a^j - \sum_{i \in I \setminus \{j\}} s_i a^i\right)$$

$$= (t_j - \varepsilon)a^j + \sum_{i \in I \setminus \{j\}} (t_i + \varepsilon s_i)a^i.$$

It is clear that in the latter sum those coefficients corresponding to the indices that realize the minimum in the definition of ε are equal to zero. By this, x lies in the positive hull of less than $|I|$ vectors of the collection. This contradiction completes the proof.   □

A collection of vectors $\{a^1, \cdots, a^k\}$ in $\mathbb{R}^n$ is said to be *affinely independent* if the dimension of the subspace $\mathrm{aff}\{a^1, \cdots, a^k\}$ is equal to $k - 1$. By convention a set consisting of a solitary vector is affinely independent. The next result is a version of Caratheodory's theorem and well-known in convex analysis.

Corollary 2.1.3 *Let $\{a^1, \cdots, a^k\}$ be a collection of vectors in $\mathbb{R}^n$. Then for every $x \in \mathrm{co}\{a^1, \cdots, a^k\}$ there exists an index set $I \subseteq \{1, \cdots, k\}$ such that*

(i) *the vectors $a^i, i \in I$ are affinely independent*
(ii) *x belongs to the convex hull of $a^i, i \in I$.*

Proof We consider the collection of vectors $v^i = (a^i, 1), i = 1, \cdots, k$ in the space $\mathbb{R}^n \times \mathbb{R}$. It is easy to verify that x belongs to the convex hull $\mathrm{co}\{a^1, \cdots, a^k\}$ if and only if the vector $(x, 1)$ belongs to the positive hull $\mathrm{pos}\{v^1, \cdots, v^k\}$. Applying Theorem 2.1.2 to the latter positive hull we deduce the existence of an index set $I \subseteq \{1, \cdots, k\}$ such that the vector $(x, 1)$ belongs to the positive hull $\mathrm{pos}\{v^i, i \in I\}$ and the collection $\{v^i, i \in I\}$ is linearly independent. Then x belongs to the convex hull $\mathrm{co}\{a^i, i \in I\}$ and the collection $\{a^i, i \in I\}$ is affinely independent.   □

Linear operators and matrices

A mapping $\phi : \mathbb{R}^n \to \mathbb{R}^k$ is called a linear operator between $\mathbb{R}^n$ and $\mathbb{R}^k$ if

(i) $\phi(x + y) = \phi(x) + \phi(y)$,
(ii) $\phi(tx) = t\phi(x)$

for every $x, y \in \mathbb{R}^n$ and $t \in \mathbb{R}$. The kernel and the image of ϕ are the sets

$$\mathrm{Ker}\phi = \{x \in \mathbb{R}^n : \phi(x) = 0\},$$
$$\mathrm{Im}\phi = \{y \in \mathbb{R}^k : y = \phi(x) \text{ for some } x \in \mathbb{R}^n\}.$$

These sets are linear subspaces of $\mathbb{R}^n$ and $\mathbb{R}^k$ respectively.

We denote the $k \times n$-matrix whose columns are $c_1, \cdots, c_n$ by C, where c_i is the vector image of the ith coordinate unit vector e^i by ϕ. Then for every vector x of $\mathbb{R}^n$ one has

$$\phi(x) = Cx.$$

The mapping $x \mapsto Cx$ is clearly a linear operator from $\mathbb{R}^n$ to $\mathbb{R}^k$. This explains why one can identify a linear operator with a matrix. The space of $k \times n$ matrices is denoted by $L(\mathbb{R}^n, \mathbb{R}^k)$. The *transpose* of a matrix C is denoted by C^T. The norm and the inner product in the space of matrices are given by

$$\|C\| = \left(\sum_{i=1,\cdots,n} \sum_{j=1,\cdots,n} |c_{ij}|^2 \right)^{1/2},$$

$$\langle C, B \rangle = \sum_{i=1,\cdots,n} \sum_{j=1,\cdots,n} c_{ij} b_{ij}.$$

The norm $\|C\|$ is called also the *Frobenius norm*.

The inner product $\langle C, B \rangle$ is nothing but the trace of the matrix $C B^T$. Sometimes the space $L(\mathbb{R}^n, \mathbb{R}^k)$ is identified with the $n \times k$-dimensional Euclidean space $\mathbb{R}^{n \times k}$.

Linear functionals

A particular case of linear operators is when the value space is one-dimensional. This is the space of linear functionals on $\mathbb{R}^n$ and often identified with the space $\mathbb{R}^n$ itself. Thus, each linear functional ϕ is given by a vector d_ϕ by the formula

$$\phi(x) = \langle d_\phi, x \rangle.$$

When $d_\phi \neq 0$, the kernel of ϕ is called a *hyperplane*; the vector d_ϕ is a *normal vector* to this hyperplane. Geometrically, d_ϕ is orthogonal to the hyperplane $\text{Ker}\phi$. The sets

$$\left\{ x \in \mathbb{R}^n : \langle d_\phi, x \rangle \geq 0 \right\},$$
$$\left\{ x \in \mathbb{R}^n : \langle d_\phi, x \rangle \leq 0 \right\}$$

are closed halfspaces and the sets

$$\left\{ x \in \mathbb{R}^n : \langle d_\phi, x \rangle > 0 \right\},$$
$$\left\{ x \in \mathbb{R}^n : \langle d_\phi, x \rangle < 0 \right\}$$

are open halfspaces bounded by the hyperplane $\text{Ker}\phi$. Given a real number α and a nonzero vector d of $\mathbb{R}^n$, one also understands a hyperplane of type

$$H(d, \alpha) = \left\{ x \in \mathbb{R}^n : \langle d, x \rangle = \alpha \right\}.$$

The sets

$$H_+(d, \alpha) = \left\{ x \in \mathbb{R}^n : \langle d, x \rangle \geq \alpha \right\},$$
$$H_-(d, \alpha) = \left\{ x \in \mathbb{R}^n : \langle d, x \rangle \leq \alpha \right\}$$

are positive and negative halfspaces and the sets

$$\text{int}\big(H_+(d, \alpha)\big) = \big\{x \in \mathbb{R}^n : \langle d, x \rangle > \alpha\big\},$$
$$\text{int}\big(H_-(d, \alpha)\big) = \big\{x \in \mathbb{R}^n : \langle d, x \rangle < \alpha\big\}$$

are positive and negative open halfspaces.

Theorem 2.1.4 *Let Q be a nonempty convex set in $\mathbb{R}^n$ and let $\langle d, . \rangle$ be a positive functional on Q, that is $\langle d, x \rangle \geq 0$ for every $x \in Q$. If $\langle d, x \rangle = 0$ for some relative interior point x of Q, then $\langle d, . \rangle$ is zero on Q.*

Proof Let y be any point in Q. Since x is a relative interior point, there exists a positive number δ such that $x + t(y - x) \in Q$ for $|t| \leq \delta$. Applying $\langle d, . \rangle$ to this point we obtain

$$\langle d, x + t(y - x) \rangle = t \langle d, y \rangle \geq 0$$

for all $t \in [-\delta, \delta]$. This implies that $\langle d, y \rangle = 0$ as requested.   □

2.2 System of Linear Inequalities

We shall mainly deal with two kinds of systems of linear equations and inequalities. The first system consists of k inequalities

$$\langle a^i, x \rangle \leq b_i, \quad i = 1, \cdots, k, \tag{2.1}$$

where $a^1, \cdots, a^k$ are n-dimensional column vectors and $b_1, \cdots, b_k$ are real numbers; and the second system consists of k equations which involves positive vectors only

$$\langle a^i, x \rangle = b_i, \quad i = 1, \cdots, k \tag{2.2}$$
$$x \geq 0.$$

Denoting by A the $k \times n$-matrix whose rows are the transposes of $a^1, \cdots, a^k$ and by b the column k-vector of components $b_1, \cdots, b_k$, we can write the systems (2.1) and (2.2) in matrix form

$$Ax \leq b \tag{2.3}$$

and

$$Ax = b \tag{2.4}$$
$$x \geq 0.$$

Notice that any system of linear equations and inequalities can be converted to the two matrix forms described above. To this end it suffices to perform three operations:

(a) Express each variable x_i as difference of two non-negative variables $x_i = x_i^+ - x_i^-$ where

$$x_i^+ = \max\{x_i; 0\},$$
$$x_i^- = \max\{-x_i; 0\}.$$

(b) Introduce a non-negative *slack variable* y_i in order to obtain equivalence between inequality $\langle a^i, x \rangle \leq b_i$ and equality $\langle a^i, x \rangle + y_i = b_i$. Similarly, with a non-negative *surplus variable* z_i one may express inequality $\langle a^i, x \rangle \geq b_i$ as equality $\langle a^i, x \rangle - z_i = b_i$.
(c) Express equality $\langle a^i, x \rangle = b_i$ by two inequalities $\langle a^i, x \rangle \leq b_i$ and $\langle a^i, x \rangle \geq b_i$.

Example 2.2.1 Consider the following system

$$x_1 + 2x_2 = 1,$$
$$-x_1 - x_2 \geq 0.$$

It is written in form (2.3) as

$$\begin{pmatrix} 1 & 2 \\ -1 & -2 \\ 1 & 1 \end{pmatrix} \begin{pmatrix} x_1 \\ x_2 \end{pmatrix} \leq \begin{pmatrix} 1 \\ -1 \\ 0 \end{pmatrix}$$

and in form (2.4) with a surplus variable y as

$$\begin{pmatrix} 1 & -1 & 2 & -2 & 0 \\ -1 & 1 & -1 & 1 & -1 \end{pmatrix} (x_1^+, x_1^-, x_2^+, x_2^-, y)^T = \begin{pmatrix} 1 \\ 0 \end{pmatrix},$$
$$(x_1^+, x_1^-, x_2^+, x_2^-, y)^T \geq 0.$$

Redundant equation

Given a system (2.4) we say it is *redundant* if at least one of the equations (called *redundant equation*) can be expressed as a linear combination of the others. In other words, it is redundant if there is a nonzero k-dimensional vector λ such that

$$A^T \lambda = 0,$$
$$\langle b, \lambda \rangle = 0.$$

Moreover, redundant equations can be dropped from the system without changing its solution set. Similarly, an inequation of (2.1) is called redundant if its removal from the system does not change the solution set.

Proposition 2.2.2 *Assume that $k \leq n$ and that the system (2.4) is consistent. Then it is not redundant if and only if the matrix A has full rank.*

Proof If one of equations, say $\langle a^1, x \rangle = b_1$, is redundant, then a^1 is a linear combination of $a^2, \cdots, a^k$. Hence the rank of A is not maximal, it is less than k. Conversely, when the rank of A is maximal (equal to k), no row of A is a linear combination of the others. Hence no equation of the system can be expressed as a linear combination of the others.   □

Farkas' theorem

One of the theorems of the alternative that are pillars of the theory of linear and nonlinear programming is Farkas' theorem or Farkas' lemma. There are a variety of ways to prove it, the one we present here is elementary.

Theorem 2.2.3 (Farkas' theorem) *Exactly one of the following systems has a solution:*

(i) $Ax = b$ *and* $x \geq 0$;
(ii) $A^T y \geq 0$ *and* $\langle b, y \rangle < 0$.

Proof If the first system has a solution x, then for every y with $A^T y \geq 0$ one has

$$\langle b, y \rangle = \langle Ax, y \rangle = \langle x, A^T y \rangle \geq 0,$$

which shows that the second system has no solution.

  Now suppose the first system has no solution. Then either the system

$$Ax = b$$

has no solution, or it does have a solution, but every solution of it is not positive. In the first case, choose m linearly independent columns of A, say $a_1, \cdots, a_m$, where m is the rank of A. Then the vectors $a_1, \cdots, a_m, b$ are linearly independent too (because b does not lie in the space spanned by $a_1, \cdots, a_m$). Consequently, the system

$$\langle a_i, y \rangle = 0, \quad i = 1, \cdots, m,$$
$$\langle b, y \rangle = -1$$

admits a solution. This implies that the system (ii) has solutions too. It remains to prove the solvability of (ii) when $Ax = b$ has solutions and they are all non-positive. We do it by induction on the dimension of x. Assume $n = 1$. If the system $a_{i1}x_1 = b_i, i = 1, \cdots, k$ has a negative solution x_1, then $y = -(b_1, \cdots, b_k)^T$ is a solution of (ii) because $A^T y = -(a_{11}^2 + \cdots + a_{k1}^2)x_1 > 0$ and $\langle b, y \rangle = -(b_1^2 + \cdots + b_k^2) < 0$. Now assume $n > 1$ and that the result is true for the case of dimension $n-1$. Given an n-vector x, denote by $\bar{x}$ the $(n-1)$-vector consisting of the first $(n-1)$ components of x. Let $\bar{A}$ be the matrix composed of the first $(n-1)$ columns of A. It is clear that the system

$$\bar{A}\bar{x} = b \text{ and } \bar{x} \geq 0$$

has no solution. By induction there is some $\bar{y}$ such that

$$\bar{A}^T \bar{y} \geq 0,$$
$$\langle b, \bar{y} \rangle < 0.$$

If $\langle a_n, \bar{y} \rangle \geq 0$, we are done. If $\langle a_n, \bar{y} \rangle < 0$, define new vectors

$$\hat{a}_i = \langle a_i, \bar{y} \rangle a_n - \langle a_n, \bar{y} \rangle a_i, \quad i = 1, \cdots, n-1,$$
$$\hat{b} = \langle b, \bar{y} \rangle a_n - \langle a_n, \bar{y} \rangle b$$

and consider a new system

$$\hat{a}_1 \xi_1 + \cdots + \hat{a}_{n-1} \xi_{n-1} = \hat{b}. \tag{2.5}$$

We claim that this system of k equations has no positive solution. Indeed, if not, say $\xi_1, \cdots, \xi_{n-1}$ were non-negative solutions, then the vector x with

$$x_i = \xi_i, \quad i = 1, \cdots, n-1,$$
$$x_n = -\frac{1}{\langle a_n, \bar{y} \rangle} \Big(\langle a_1 \xi_1 + \cdots + a_{n-1} \xi_{n-1}, \bar{y} \rangle - \langle b, \bar{y} \rangle \Big)$$

should be a positive solution of (i) because $-\langle b, \bar{y} \rangle > 0$ and $\langle \bar{A}\xi, \bar{y} \rangle = \langle \xi, \bar{A}^T \bar{y} \rangle \geq 0$ for $\xi = (\xi_1, \cdots, \xi_{n-1})^T \geq 0$, implying $x_n \geq 0$. Applying the induction hypothesis to (2.5) we deduce the existence of a k-vector $\hat{y}$ with

$$\langle \hat{a}_i, \hat{y} \rangle \geq 0, \quad i = 1, \cdots, n-1,$$
$$\langle \hat{b}, \hat{y} \rangle < 0.$$

Then the vector $y = \langle a_n, \hat{y} \rangle \bar{y} - \langle a_n, \bar{y} \rangle \hat{y}$ satisfies the system (ii). The proof is complete.   □

A number of consequences can be derived from Farkas' theorem which are useful in the study of linear systems and linear programming problems.

Corollary 2.2.4 *Exactly one of the following systems has a solution:*

(i) $Ax = 0$, $\langle c, x \rangle = 1$ and $x \geq 0$;
(ii) $A^T y \geq c$.

Proof If (ii) has a solution y, then for a positive vector x with $Ax = 0$ one has

$$0 = \langle y, Ax \rangle = \langle A^T y, x \rangle \geq \langle c, x \rangle.$$

So (i) is not solvable. Conversely, if (i) has no solution, then applying Farkas' theorem to the inconsistent system

$$\begin{pmatrix} A \\ c^T \end{pmatrix} x = \begin{pmatrix} 0 \\ 1 \end{pmatrix} \quad \text{and } x \geq 0$$

yields the existence of a vector y and of a real number t such that

$$\left(A^T c \right) \begin{pmatrix} y \\ t \end{pmatrix} \geq 0 \quad \text{and} \quad \left\langle \begin{pmatrix} 0 \\ 1 \end{pmatrix}, \begin{pmatrix} y \\ t \end{pmatrix} \right\rangle < 0.$$

Hence $t < 0$ and $-y/t$ is a solution of (ii).   □

Corollary 2.2.5 *Exactly one of the following systems has a solution:*

(i) $Ax \geq 0$ *and* $x \geq 0$;
(ii) $A^T y \leq 0$ *and* $y > 0$.

Proof By introducing a surplus variable $z \in \mathbb{R}^k$ we convert (i) to an equivalent system

$$Ax - Iz = 0,$$
$$\begin{pmatrix} x \\ z \end{pmatrix} \geq 0,$$
$$\left\langle c, \begin{pmatrix} x \\ z \end{pmatrix} \right\rangle = 1,$$

where c is an $(n+k)$-vector whose n first components are all zero and the remaining components are one. According to Corollary 2.2.4 it has no solution if and only if the following system has a solution

$$\begin{pmatrix} A^T \\ -I \end{pmatrix} y \geq c.$$

It is clear that the latter system is equivalent to (ii).   □

The next corollary is known as Motzkin's theorem of the alternative.

Corollary 2.2.6 (Motzkin's theorem) *Let A and B be two matrices having the same number of columns. Exactly one of the following systems has a solution:*

(i) $Ax > 0$ *and* $Bx \geq 0$;
(ii) $A^T y + B^T z = 0$, $y \geq 0$ *and* $z \geq 0$.

Proof The system (ii) is evidently equivalent to the following one

$$
\begin{pmatrix} A^T & B^T \\ e^T & 0 \end{pmatrix} \begin{pmatrix} y \\ z \end{pmatrix} = \begin{pmatrix} 0 \\ 1 \end{pmatrix}
$$

$$
\begin{pmatrix} y \\ z \end{pmatrix} \geq 0.
$$

By Farkas' theorem it is compatible (has a solution) if and only if the following system is incompatible:

$$
\begin{pmatrix} A & e \\ B & 0 \end{pmatrix} \begin{pmatrix} x \\ t \end{pmatrix} \geq \begin{pmatrix} 0 \\ 0 \end{pmatrix}
$$

$$
\left\langle \begin{pmatrix} x \\ t \end{pmatrix}, \begin{pmatrix} 0 \\ 1 \end{pmatrix} \right\rangle < 0.
$$

The latter system is evidently equivalent to the system of (i).   □

Some classical theorems of alternatives are immediate from Corollary 2.2.6.

- *Gordan's theorem* (B is the zero matrix):
  Exactly one of the following systems has a solution

  (1) $Ax > 0$;
  (2) $A^T y = 0$ and $y \geq 0$.

- *Ville's theorem* (B is the identity matrix):
  Exactly one of the following systems has a solution

  (3) $Ax > 0$ and $x \geq 0$;
  (4) $A^T y \leq 0$ and $y \geq 0$.

- *Stiemke's theorem* (A is the identity matrix and B is replaced by $\begin{pmatrix} B \\ -B \end{pmatrix}$):
  Exactly one of the following systems has a solution

  (5) $Bx = 0$ and $x > 0$;
  (6) $B^T y \geq 0$.

2.3 Convex Polyhedra

A set that can be expressed as the intersection of a finite number of closed half-spaces is called a *convex polyhedron*. A convex bounded polyhedron is called a *polytope*. According to the definition of closed half-spaces, a convex polyhedron is the solution set to a finite system of inequalities

$$\langle a^i, x \rangle \leqq b_i, \quad i = 1, \cdots, k \qquad (2.6)$$

where $a^1, \cdots, a^k$ are n-dimensional column vectors and $b_1, \cdots, b_k$ are real numbers. When $b_i = 0, i = 1, \cdots, k$, the solution set to (2.6) is a cone and called a *convex polyhedral cone*. We assume throughout this section that the system is not redundant and solvable.

Supporting hyperplanes and faces

Let P be a convex polyhedron and let

$$H = \{x \in \mathbb{R}^n : \langle v, x \rangle = \alpha\}$$

be a hyperplane with v nonzero. We say H is a *supporting hyperplane* of P at a point $x \in P$ if the intersection of H with P contains x and P is contained in one of the closed half-spaces bounded by H (Fig. 2.8). In this case, the nonempty set $H \cap P$ is called a *face* of P. Thus, a nonempty subset F of P is a face if there is a nonzero vector $v \in \mathbb{R}^n$ such that

$$\langle v, y \rangle \leqq \langle v, x \rangle \quad \text{for all } x \in F, \ y \in P.$$

When a face is zero-dimensional, it is called a *vertex*. A nonempty polyhedron may have no vertex. By convention P is a face of itself; other faces are called *proper faces*. One-dimensional faces are called *edges*. Two vertices are said to be adjacent if they are end-points of an edge.

Example 2.3.1 Consider a system of three inequalities in $\mathbb{R}^2$:

$$x_1 + x_2 \leqq 1 \qquad (2.7)$$
$$-x_1 - x_2 \leqq 0 \qquad (2.8)$$
$$-x_1 \leqq 0. \qquad (2.9)$$

Fig. 2.8 Supporting hyperplane

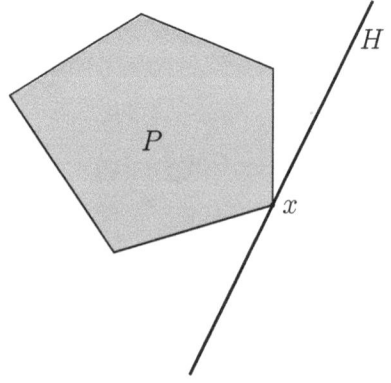

The polyhedron defined by (2.7) and (2.8) has no vertex. It has two one-dimensional faces determined respectively by $x_1 + x_2 = 1$ and $x_1 + x_2 = 0$, and one two-dimensional face, the polyhedron itself. The polyhedron defined by (2.7)–(2.9) has two vertices (zero-dimensional faces) determined respectively by

$$\begin{cases} x_1 = 0 \\ x_2 = 0 \end{cases} \text{ and } \begin{cases} x_1 = 0 \\ x_2 = 1 \end{cases},$$

three one-dimensional faces given by

$$\begin{cases} x_1 + x_2 \leqq 1 \\ -x_1 - x_2 \leqq 0 \\ x_1 = 0 \end{cases}, \begin{cases} x_1 + x_2 = 1 \\ -x_1 \leqq 0 \end{cases} \text{ and } \begin{cases} -x_1 - x_2 = 0 \\ -x_1 \leqq 0 \end{cases},$$

and one two-dimensional face, the polyhedron itself.

Proposition 2.3.2 *Let P be a convex polyhedron. The following properties hold.*

  (i) *The intersection of any two faces is a face if it is nonempty.*
 (ii) *Two different faces have no relative interior point in common.*

Proof We prove (i) first. Assume F_1 and F_2 are two faces with nonempty intersection. If they coincide, there is nothing to prove. If not, let H_1 and H_2 be two supporting hyperplanes that generate these faces, say

$$H_1 = \{x \in \mathbb{R}^n : \langle v^1, x \rangle = \alpha_1\},$$
$$H_2 = \{x \in \mathbb{R}^n : \langle v^2, x \rangle = \alpha_2\}.$$

Since these hyperplanes contain the intersection of distinct faces F_1 and F_2, the vector $v = v^1 + v^2$ is not zero. Consider the hyperplane

$$H = \{x \in \mathbb{R}^n : \langle v, x \rangle = \alpha_1 + \alpha_2\}.$$

It is a supporting hyperplane of P because it evidently contains the intersection of the faces F_1 and F_2, and for every point x in P, one has

$$\langle v, x \rangle = \langle v^1, x \rangle + \langle v^2, x \rangle \leqq \alpha_1 + \alpha_2. \tag{2.10}$$

It remains to show that the intersection of H and P coincides with the intersection $F_1 \cap F_2$. The inclusion

$$F_1 \cap F_2 \subseteq H \cap P$$

being clear, we show the converse. Let x be in $H \cap P$. Then (2.10) becomes equality for this x. But $\langle v^1, x \rangle \leqq \alpha_1$ and $\langle v^2, x \rangle \leqq \alpha_2$, so that equality of (2.10) is possible only when the two latter inequalities are equalities. This proves that x belongs to both F_1 and F_2.

For the second assertion notice that if F_1 and F_2 have a relative interior point in common, then in view of Theorem 2.1.4, the functional $\langle v^1, . \rangle$ is constant on F_2. It follows that $F_2 \subseteq H_1 \cap P \subseteq F_1$. Similarly, one has $F_1 \subseteq F_2$, and hence equality holds.   □

Let x be a solution of the system (2.6). Define the *active index set* at x to be the set

$$I(x) = \left\{ i \in \{1, \cdots, k\} : \langle a^i, x \rangle = b_i \right\}.$$

The remaining indices are called *inactive indices*.

Theorem 2.3.3 *Assume that P is a convex polyhedron given by (2.6). A nonempty proper convex subset F of P is a face if and only if there is a nonempty maximal index set $I \subseteq \{1, \cdots, k\}$ such that F is the solution set to the system*

$$\langle a^i, x \rangle = b_i, \ i \in I \tag{2.11}$$

$$\langle a^j, x \rangle \leqq b_j, \ j \in \{1, \cdots, k\} \backslash I, \tag{2.12}$$

in which case the dimension of F is equal to $n - rank\{a^i : i \in I\}$.

Proof Denote the solution set to the system (2.11, 2.12) by F' that we suppose nonempty. To prove that it is a face, we set

$$v = \sum_{i \in I} a^i \ \text{ and } \ \alpha = \sum_{i \in I} b_i.$$

Notice that v is nonzero because F' is not empty and the system (2.6) is not redundant. It is clear that the negative half-space $H_-(v, \alpha)$ contains P. Moreover, if x is a solution to the system, then, of course, x belongs to P and to H at the same time, which implies $F' \subseteq H \cap P$. Conversely, any point x of the latter intersection satisfies

$$\langle a^i, x \rangle \leqq b_i, \ \ i = 1, \cdots, k,$$

$$\sum_{i \in I} \langle a^i, x \rangle = \sum_{i \in I} b_i.$$

The latter equality is possible only when those inequalities with indices from I are equalities. In other words, x belongs to F'.

Now, let F be a proper face of P. Pick a relative interior point $\overline{x}$ of F and consider the system (2.11, 2.12) with $I = I(\overline{x})$ the active index set of $\overline{x}$. Being a proper face of P, F has no interior point, and so the set I is nonempty. As before, F' is the solution set to that system. By the first part, it is a face. We wish to show that it coincides with F. For this, in view of Proposition 2.3.2 it suffices to show that $\overline{x}$ is also a relative interior point of F'. Let x be another point in F'. We have to prove that there is a positive number δ such that the segment $[\overline{x}, \overline{x} + \delta(x - \overline{x})]$ lies in F'. Indeed, note that for indices j outside the set I, inequalities $\langle a^j, \overline{x} \rangle \leqq b_j$ are strict. Therefore, there is $\delta > 0$ such that

$$\langle a^j, \overline{x} \rangle + \delta \langle a^j, x - \overline{x} \rangle \leqq b_j$$

for all $j \notin I$. Moreover, being a linear combination of $\overline{x}$ and x, the endpoint $\overline{x} + \delta(x - \overline{x})$ satisfies the equalities (2.11) too. Consequently, this point belongs to F', and hence so does the whole segment. Since F and F' are two faces with a relative interior point in common, they must be the same. $\qquad\square$

In general, for a given face F of P, there may exist several index sets I for which F is the solution set to the system (2.11, 2.12). We shall, however, understand that no inequality can be equality without changing the solution set when saying that the system (2.11, 2.12) determines the face F. So, if two inequalities combined yields equality, their indices will be counted in I.

Corollary 2.3.4 *If an m-dimensional convex polyhedron has a vertex, then it has faces of any dimension less than m.*

Proof The corollary is evident for a zero-dimensional polyhedron. Suppose P is a polyhedron of dimension $m > 0$. By Theorem 2.3.3 without loss of generality we may assume that P is given by the system (2.11, 2.12) with $|I| = n - m$ and that the family $\{a^i, i \in I\}$ is linearly independent. Since P has a vertex, there is some $i_0 \in \{1, \cdots, k\} \backslash I$ such that the vectors $a^i, i \in I \cup \{i_0\}$ are linearly independent. Then the system

$$\langle a^i, x \rangle = b_i, \ i \in I \cup \{i_0\},$$
$$\langle a^j, x \rangle \leqq b_j, \ j \in \{1, \cdots, k\} \backslash (I \cup \{i_0\})$$

generates an $(m-1)$-dimensional face of P. Notice that this system has a solution because P is generated by the non-redundant system (2.11, 2.12). Continuing the above process we are able to construct a face of any dimension less than m. $\qquad\square$

Corollary 2.3.5 *Let F be a face of the polyhedron P determined by the system (2.11, 2.12). Then for every $x \in F$ one has*

$$I(x) \supseteq I.$$

Equality holds if and only if x is a relative interior point of F.

Proof The inclusion $I \subseteq I(x)$ is evident because $x \in F$. For the second part, we first assume $I(x) = I$, that is

$$\langle a^i, x \rangle = b_i, \ i \in I,$$
$$\langle a^j, x \rangle < b_j, \ j \in \{1, \cdots, k\} \backslash I.$$

It is clear that if $y \in \text{aff}(F)$, then $\langle a^i, y \rangle = b_i, \ i \in I$, and if $y \in x + \varepsilon B_k$ with $\varepsilon > 0$ sufficiently small, then $\langle a^j, y \rangle < b_j, \ j \in \{1, \cdots, k\} \backslash I$. We deduce that

$\mathrm{aff}(F) \cap (x + \varepsilon B_k) \subseteq F$, which shows that x is a relative interior point of F. Conversely, let x be a relative interior point of F. Using the argument in the proof of Theorem 2.3.3 we know that F is also a solution set to the system

$$\langle a^i, y \rangle = b_i, \ i \in I(x),$$
$$\langle a^j, y \rangle \leqq b_j, \ j \in \{1, \cdots, k\} \backslash I(x).$$

Since the system (2.11, 2.12) determines F, we have $I(x) \subseteq I$, and hence equality follows.   □

Corollary 2.3.6 *Let F be a face of the polyhedron P determined by the system (2.11, 2.12). Then a point $v \in F$ is a vertex of F if and only if it is a vertex of P.*

Proof It is clear that every vertex of P is a vertex of F if it belongs to F. To prove the converse, let us deduce a system of inequalities from (2.11, 2.12) by expressing equalities $\langle a^i, x \rangle = b_i$ as two inequalities $\langle a^i, x \rangle \leqq b_i$ and $\langle -a^i, x \rangle \leqq -b_i$. If v is a vertex of F, then the active constraints at v consists of the vectors $a^i, -a^i, i \in I$ and some $a^j, j \in J \subseteq \{1, \cdots, k\} \backslash I$, so that the rank of the family $\{a^i, -a^i, a^j : i \in I, j \in J\}$ is equal to n. It follows that the family $\{a^i, a^j : i \in I, j \in J\}$ has rank equal to n too. In view of Theorem 2.3.3 the point v is a vertex of P.   □

Given a face F of a polyhedron, according to the preceding corollary the active index set $I(x)$ is constant for every relative interior point x of F. Therefore, we call it active index set of F and denote it by I_F.

A collection of subsets of a polyhedron is said to be a *partition* of it if the elements of the collection are disjoint and their union contains the entire polyhedron.

Corollary 2.3.7 *The collection of all relative interiors of faces of a polyhedron forms a partition of the polyhedron.*

Proof It is clear from Proposition 2.3.2(ii) that relative interiors of different faces are disjoint. Moreover, given a point x in P, consider the active index set $I(x)$. If it is empty, then the point belongs to the interior of P and we are done. If it is not empty, by Corollary 2.3.5, the face determined by system (2.11, 2.12) with $I = I(x)$ contains x in its relative interior. The proof is complete.   □

We now deduce a first result on representation of elements of polyhedra by vertices.

Corollary 2.3.8 *A convex polytope is the convex hull of its vertices.*

Proof The corollary is evident when the dimension of a polytope is less or equal to one in which case it is a point or a segment with two end-points. We make the induction hypothesis that the corollary is true when a polytope has dimension less than m with $1 < m < n$ and prove it for the case when P is a polytope determined by the system (2.6) and has dimension equal to m. Since P is a convex set, the convex hull of its vertices is included in P itself. Conversely, let x be a point of P. In view of

Corollary 2.3.7 it is a relative interior point of some face F of P. If F is a proper face of P, its dimension is less than m, and so we are done. It remains to treat the case where x is a relative interior point of P. Pick any point $y \neq x$ in P and consider the line passing through x and y. Since P is bounded, the intersection of this line with P is a segment, say with end-points c and d. Let F_c and F_d be faces of P that contain c and d in their relative interiors. As c and d are not relative interior points of P, the faces F_c and F_d are proper faces of P, and hence they have dimension strictly less than m. By induction c and d belong to the convex hulls of the vertices of F_c and F_d respectively. By Corollary 2.3.6 they belong to the convex hull of the vertices of P, and hence so does x because x belongs to the convex hull of c and d. $\qquad\square$

A similar result is true for polyhedral cones. It explains why one-dimensional faces of a polyhedral cone are called extreme rays.

Corollary 2.3.9 *A nontrivial polyhedral cone with vertex is the convex hull of its one-dimensional faces.*

Proof By definition a polyhedral cone P is defined by a homogeneous system

$$\langle a^i, x \rangle \leqq 0, \ i = 1, \cdots, k. \tag{2.13}$$

Choose any nonzero point y in P and consider the hyperplane H given by

$$\langle a^1 + \cdots + a^k, x - y \rangle = 0. \tag{2.14}$$

We claim that the vector $a^1 + \cdots + a^k$ is nonzero. Indeed, if not, the inequalities (2.13) would become equalities for all $x \in P$, and P would be either a trivial cone, or a cone without vertex. Moreover, $P \cap H$ is a bounded polyhedron, because otherwise one should find a nonzero vector u satisfying $\langle a^i, u \rangle = 0, i = 1, \cdots, k$ and P could not have vertices. In view of Corollary 2.3.8, $P \cap H$ is the convex hull of its vertices. To complete the proof it remains to show that a vertex v of $P \cap H$ is the intersection of a one-dimensional face of P with H. Indeed, the polytope $P \cap H$ being determined by the system (2.13) and (2.14), there is a set $J \subset \{1, \cdots, k\}$ with $|J| = n - 1$ such that the vectors a^j, $j \in J$ and $a^1 + \cdots + a^k$ are linearly independent and v is given by system

$$\langle a^j, x \rangle = 0, \ j \in J \tag{2.15}$$
$$\langle a^1 + \cdots + a^k, x \rangle = \langle a^1 + \cdots + a^k, y \rangle,$$
$$\langle a^i, x \rangle \leqq 0, \ i \in \{1, \cdots, k\} \backslash J. \tag{2.16}$$

It is clear that (2.15) and (2.16) determine a one-dimensional face of P whose intersection with H is v. $\qquad\square$

Separation of convex polyhedra

Given two convex polyhedra P and Q in $\mathbb{R}^n$, we say that a nonzero vector v separates them if

$$\langle v, x \rangle \geq \langle v, y \rangle \quad \text{for all vectors } x \in P, y \in Q$$

and strict inequality is true for some of them (Fig. 2.9). The following result can be considered as a version of Farkas' theorem or Gordan's theorem.

Theorem 2.3.10 *If P and Q are convex polyhedra without relative interior points in common, then there is a nonzero vector separating them.*

Proof We provide a proof for the case where both P and Q have interior points only. Without loss of generality we may assume that P is determined by the system (2.6) and Q is determined by the system

$$\langle d^j, x \rangle \leq c_j, \quad j = 1, \cdots, m.$$

Thus, the following system:

$$\langle a^i, x \rangle < b_i, \quad i = 1, \cdots, k$$
$$\langle d^j, x \rangle < c_j, \quad j = 1, \cdots, m$$

has no solution because the first k inequalities determine the interior of P and the last m inequalities determine the interior of Q. This system is equivalent to the following one:

$$\begin{pmatrix} -A & b \\ -D & c \\ 0 & 1 \end{pmatrix} \begin{pmatrix} x \\ t \end{pmatrix} > \begin{pmatrix} 0 \\ 0 \\ 0 \end{pmatrix},$$

Fig. 2.9 Separation

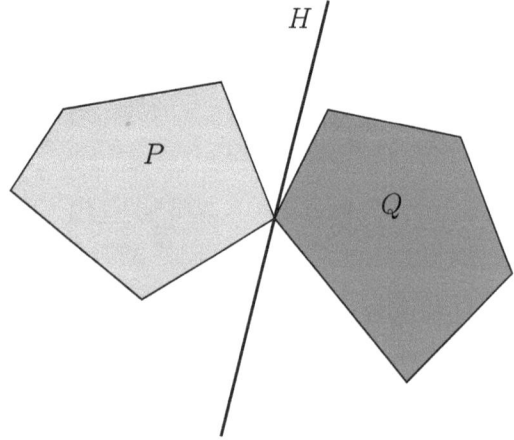

where A is the $k \times n$-matrix whose rows are transposes of $a^1, \cdots, a^k$, D is the $m \times n$-matrix whose rows are transposes of $d^1, \cdots, d^k$, b is the k-vector with the components $b_1, \cdots, b_k$ and c is the m-vector with the components $c_1, \cdots, c_m$. According to Gordan's theorem, there exist positive vectors $\lambda \in \mathbb{R}^k$ and $\mu \in \mathbb{R}^m$ and a real number $s \geq 0$, not all zero, such that

$$A^T \lambda + D^T \mu = 0,$$
$$\langle b, \lambda \rangle + \langle c, \mu \rangle + s = 0.$$

It follows from the latter equality that (λ, μ) is nonzero. We may assume without loss of generality that $\lambda \neq 0$. We claim that $A^T \lambda \neq 0$. Indeed, if not, choose x an interior point of P and y an interior point of Q. Then $D^T \mu = 0$ and hence

$$\langle b, \lambda \rangle > \langle Ax, \lambda \rangle = \langle x, A^T \lambda \rangle = 0$$

and

$$\langle c, \mu \rangle \geq \langle Dy, \mu \rangle = \langle y, D^T \mu \rangle = 0,$$

which is in contradiction with the aforesaid equality. Defining v to be the nonzero vector $-A^T \lambda$, we deduce for every $x \in P$ and $y \in Q$ that

$$\langle v, x \rangle = \langle -A^T \lambda, x \rangle = \langle \lambda, -Ax \rangle \geq \langle \lambda, -b \rangle \geq \langle \mu, c \rangle \geq \langle \mu, Dy \rangle = \langle v, y \rangle.$$

Of course inequality is strict when x and y are interior points. By this v separates P and Q as requested.   □

Asymptotic cones

Given a nonempty convex and closed subset C of $\mathbb{R}^n$, we say that a vector v is an *asymptotic* or a *recession direction* of C if

$$x + tx \in C \quad \text{for all } x \in C, \ t \geq 0.$$

The set of all asymptotic directions of C is denoted by C_∞ (Fig. 2.10). It is a convex cone. It can be seen that a closed convex set is bounded if and only if its *asymptotic cone* is trivial. The set $C_\infty \cap (-C_\infty)$ is a linear subspace and called the *lineality space* of C.

An equivalent definition of asymptotic directions is given next.

Theorem 2.3.11 *A vector v is an asymptotic direction of a convex and closed set C if and only if there exist a sequence of elements $x^s \in C$ and a sequence of positive numbers t_s converging to zero such that $v = \lim_{s \to \infty} t_s x^s$.*

Proof If $v \in C_\infty$ and $x \in C$, then $x^s = x + sv \in C$ for all $s \in \mathbb{N} \backslash \{0\}$. Setting $t_s = 1/s$ we obtain $v = \lim_{s \to \infty} t_s x^s$ with $\lim_{s \to \infty} t_s = 0$. Conversely, assume that

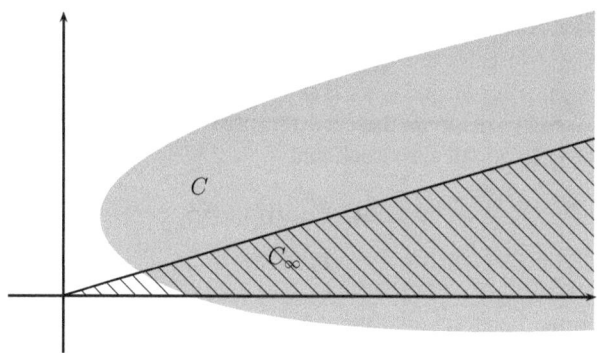

Fig. 2.10 Asymptotic cone

$v = \lim_{s \to \infty} t_s x^s$ for $x^s \in C$ and $t_s > 0$ converging to zero as s tends to ∞. Let $x \in C$ and $t > 0$ be given. Then $t t_s$ converges to zero as $s \to \infty$ and $0 \le t t_s \le 1$ for s sufficiently large. Hence,

$$x + tv = \lim_{s \to \infty} (x + t t_s x^s)$$
$$= \lim_{s \to \infty} \left((1 - t t_s)x + t t_s x^s + t t_s x \right)$$
$$= \lim_{s \to \infty} \left((1 - t t_s)x + t t_s x^s \right).$$

The set C being closed and convex, the points under the latter limit belong to the set C, and therefore their limit $x + tv$ belongs to C too. Since x and $t > 0$ were chosen arbitrarily we conclude that $v \in C_\infty$.   □

Below is a formula to compute the asymptotic cone of a polyhedron.

Theorem 2.3.12 *The asymptotic cone of the polyhedron P determined by the system (2.6) is the solution set to system*

$$\langle a^i, v \rangle \le 0, \ i = 1, \cdots, k. \tag{2.17}$$

Proof Let v be an asymptotic direction of P. Then for every positive number t one has

$$\langle a^i, \bar{x} + tv \rangle \le b_i, \ i = 1, \cdots, k,$$

where $\bar{x}$ is any point in P. By dividing both sides of the above inequalities by $t > 0$ and letting this t tend to ∞ we derive (2.17). For the converse, if v is a solution of (2.17), then for every point x in P one has

$$\langle a^i, x + tv \rangle = \langle a^i, x \rangle + t \langle a^i, v \rangle \le b_i, \ i = 1, \cdots, k$$

for all $t \geq 0$. Thus, the points $x + tv$ with $t \geq 0$, belong to P and v is an asymptotic direction. $\qquad\square$

Example 2.3.13 Consider a (nonempty) polyhedron in $\mathbb{R}^3$ defined by the system:

$$-x_1 - x_2 - x_3 \leq -1,$$
$$x_3 \leq 1,$$
$$x_1, x_2, x_3 \geq 0.$$

The asymptotic cone is given by the system

$$-x_1 - x_2 - x_3 \leq 0,$$
$$x_3 \leq 0,$$
$$x_1, x_2, x_3 \geq 0,$$

in which the first inequality is redundant, and hence it is simply given by $x_1 \geq 0$, $x_2 \geq 0$ and $x_3 = 0$.

Using asymptotic directions we are also able to tell whether a convex polyhedron has a vertex or not. A cone is called *pointed* if it contains no straight line. When a cone C is not pointed, it contains a nontrivial linear subspace $C \cap (-C)$, called also the lineality space of C.

Corollary 2.3.14 *A convex polyhedron has vertices if and only if its asymptotic cone is pointed. Consequently, if a convex polyhedron has a vertex, then so does any of its faces.*

Proof It is easy to see that when a polyhedron has a vertex, it contains no straight line, and hence its asymptotic cone is pointed. We prove the converse by induction on the dimension of the polyhedron. The case where a polyhedron is of dimension less or equal to one is evident because a polyhedron with a pointed asymptotic cone is either a point or a segment or a ray, hence it has a vertex. Assume the induction hypothesis that the conclusion is true for all polyhedra of dimension less than m with $1 < m < n$. Let P be m-dimensional with a pointed asymptotic cone. If P has no proper face, then the inequalities (2.6) are strict, which implies that P is closed and open at the same time. This is possible only when P coincides with the space $\mathbb{R}^n$ which contradicts the hypothesis that P_∞ is pointed. Now, let F be a proper face of P. Its asymptotic cone, being a subset of the asymptotic cone of P is pointed too. By induction, it has a vertex, which in view of Corollary 2.3.6 is also a vertex of P.

To prove the second part of the corollary it suffices to notice that if a face of P has no vertex, by the first part of the corollary, its asymptotic cone contains a straight line, hence so does the set P itself. $\qquad\square$

A second representation result for elements of a convex polyhedron is now formulated in a more general situation.

Corollary 2.3.15 *A convex polyhedron with vertex is the convex hull of its vertices and extreme directions.*

Proof We conduct the proof by induction on the dimension of the polyhedron. The corollary is evident when a polyhedron is zero or one-dimensional. We assume that it is true for all convex polyhedra of dimension less than m with $1 < m < n$ and prove it for an m-dimensional polyhedron P determined by system

$$\langle a^i, x \rangle = b_i, \ i \in I$$
$$\langle a^j, x \rangle \leqq b_j, \ j \in \{1, \cdots, k\} \setminus I,$$

in which $|I| = n - m$ and the vectors $a^i, i \in I$ are linearly independent. Let y be an arbitrary element of P. If it belongs to a proper face of P, then by induction we can express it as a convex combination of vertices and extreme directions. If it is a relative interior point of P, then setting $a = a^1 + \cdots + a^k$ that is nonzero because the asymptotic cone of P is pointed, P itself having a vertex, and considering the intersection of P with the hyperplane H determined by equality $\langle a, x \rangle = \langle a, y \rangle$ we obtain a bounded polyhedron $P \cap H$. By Corollary 2.3.8, the point y belongs to the convex hull of vertices of $P \cap H$. The vertices of $P \cap H$ belong to proper faces of P, by induction, they also belong to the convex hull of vertices and extreme directions of P, hence so does y. $\qquad \square$

When a polyhedron has a non-pointed asymptotic cone, it has no vertex. However, it is possible to express it as a sum of its asymptotic cone and a bounded polyhedron as well.

Corollary 2.3.16 *Every convex polyhedron is the sum of a bounded polyhedron and its asymptotic cone.*

Proof Denote the lineality space of the asymptotic cone of a polyhedron P by M. If M is trivial, we are done in view of Corollary 2.3.15 (the convex hull of all vertices of P serves as a bounded polyhedron). If M is not trivial, we decompose the space $\mathbb{R}^n$ into the direct sum of M and its orthogonal $M^\perp$. Denote by $P_\perp$ the projection of P on $M^\perp$. Then $P = P_\perp + M$. Indeed, let x be an element in P and let $x = x^1 + x^2$ with $x^1 \in M$ and $x^2 \in M^\perp$. Then $x^2 \in P_\perp$ and one has $x \in P_\perp + M$. Conversely, let $x^1 \in M$ and $x^2 \in P_\perp$. By definition there is some $y \in P$, say $y = y^1 + y^2$ with $y^1 \in M$ and $y^2 \in M^\perp$ such that $y^2 = x^2$. Since M is a part of the asymptotic cone of P, one deduces that

$$x = x^1 + x^2 = y^1 + (x^1 - y^1) + y^2 = y + (x^1 - y^1) \in y + M \subseteq P,$$

showing that x belongs to P. Further, we claim that the asymptotic cone of $P_\perp$ is pointed. In fact, if not, say it contains a straight line d. Then the convexity of P implies that the space $M + d$ belongs to the asymptotic cone of P. This is a contradiction because d lies in $M^\perp$ and M is already the biggest linear subspace contained in P. Let Q denote the convex hull of the set of all vertices of $P_\perp$ which is nonempty by Corollary 2.3.14. It follows from Corollary 2.3.18 below that the asymptotic cone of P is the sum of the asymptotic cone of $P_\perp$ and M. We deduce $P = Q + (P_\perp)_\infty + M = Q + P_\infty$ as requested. □

The following calculus rule for asymptotic directions under linear transformations is useful.

Corollary 2.3.17 *Let P be the polyhedron determined by the system (2.6) and let L be a linear operator from $\mathbb{R}^n$ to $\mathbb{R}^m$. Then*

$$L(P_\infty) = [L(P)]_\infty.$$

Proof The inclusion $L(P_\infty) \subseteq [L(P)]_\infty$ is true for any closed convex set. Indeed, if u is an asymptotic direction of P, then for every x in P and for every positive number t one has $x + tu \in P$. Consequently, $L(x) + tL(u)$ belongs to $L(P)$ for all $t \geq 0$. This means that $L(u)$ is an asymptotic direction of $L(P)$. For the converse inclusion, let v be a nonzero asymptotic direction of $L(P)$. By definition, for a fixed x of P, vectors $L(x) + tv$ belong to $L(P)$ for any $t \geq 0$. Thus, there are $x^1, x^2, \cdots$ in P such that $L(x^\nu) = L(x) + \nu v$, or equivalently $v = L(\frac{x^\nu - x}{\nu})$ for all $\nu = 1, 2, \cdots$ Without loss of generality we may assume that the vectors $\frac{x^\nu - x}{\nu}$ converge to some nonzero vector u as ν tends to ∞. Then

$$\langle a^i, u \rangle = \lim_{\nu \to \infty} \left(\langle a^i, \frac{x^\nu}{\nu} \rangle - \langle a^i, \frac{x}{\nu} \rangle \right) \leq 0$$

for all $i = 1, \cdots, k$. In view of Theorem 2.3.12 the vector u is an asymptotic direction of P and $v = L(u) \in L(P_\infty)$ as requested. □

Corollary 2.3.18 *Let P, P_1 and P_2 be polyhedra in $\mathbb{R}^n$ with $P \subseteq P_1$. Then*

$$(P)_\infty \subseteq (P_1)_\infty$$
$$(P_1 \times P_2)_\infty = (P_1)_\infty \times (P_2)_\infty$$
$$(P_1 + P_2)_\infty = (P_1)_\infty + (P_2)_\infty.$$

Proof The first two expressions are direct from the definition of asymptotic directions. For the third expression consider the linear transformation L from $\mathbb{R}^n \times \mathbb{R}^n$ to $\mathbb{R}^n$ defined by $L(x, y) = x + y$, and apply Corollary 2.3.17 and the second expression to conclude. □

Polar cones

Given a cone C in $\mathbb{R}^n$, the (negative) *polar cone* of C (Fig. 2.11) is the set

$$C^\circ := \left\{ v \in \mathbb{R}^n : \langle v, x \rangle \leq 0 \text{ for all } x \in C \right\}.$$

The polar cone of C° is called the bipolar cone of C. Here is a formula to compute the polar cone of a polyhedral cone.

Theorem 2.3.19 *The polar cone of the polyhedral cone determined by the system*

$$\langle a^i, x \rangle \leqq 0, i = 1, \cdots, k,$$

is the positive hull of the vectors $a^1, \cdots, a^k$.

Proof It is clear that any positive combination of vectors $a^1, \cdots, a^k$ belongs to the polar cone of the polyhedral cone. Let v be a nonzero vector in the polar cone. Then the following system has no solution

$$\langle a^i, x \rangle \leqq 0, i = 1, \cdots, k,$$
$$\langle v, x \rangle > 0.$$

According to Farkas' theorem the system

$$y_1 a^1 + \cdots + y_k a^k = v,$$
$$y_1, \cdots, y_k \geqq 0$$

has a solution, which completes the proof.   □

Example 2.3.20 Let C be a polyhedral cone in $\mathbb{R}^3$ defined by the system:

$$x_1 - x_2 \leqq 0,$$
$$x_3 = 0.$$

By expressing the latter equality as two inequalities $x_3 \leqq 0$ and $-x_3 \leqq 0$, we deduce that the polar cone of C is the positive hull of the three vectors $(1, -1, 0)^T, (0, 0, -1)^T$ and $(0, 0, 1)^T$. In other words, the polar cone C° consists of vectors $(t, -t, s)^T$ with $t \in \mathbb{R}_+$ and $s \in \mathbb{R}$.

Corollary 2.3.21 *Let C_1 and C_2 be polyhedral cones in $\mathbb{R}^n$. Then the following calculus rules hold*

Fig. 2.11 Polar cone

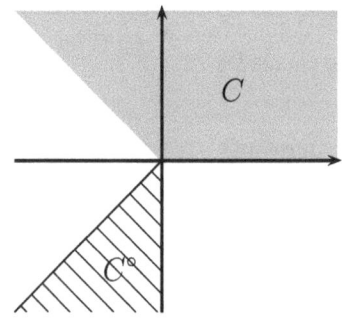

$$(C_1 + C_2)^\circ = C_1^\circ \cap C_2^\circ$$
$$(C_1 \cap C_2)^\circ = C_1^\circ + C_2^\circ.$$

Proof Let $v \in (C_1 + C_2)^\circ$. We have

$$\langle v, x + y \rangle \leqq 0 \text{ for all } x \in C_1, y \in C_2.$$

By setting $y = 0$ in this inequality we deduce $v \in C_1^\circ$. Similarly, by setting $x = 0$ we obtain $v \in C_2^\circ$, and hence $v \in C_1^\circ \cap C_2^\circ$. Conversely, if v belongs to both C_1° and C_2°, then $\langle v, \cdot \rangle$ is negative on C_1 and C_2. Consequently, it is negative on the sum $C_1 + C_2$ by linearity, which shows that $v \in (C_1 + C_2)^\circ$.

For the second equality we observe that the inclusion $C_1^\circ + C_2^\circ \subseteq (C_1 \cap C_2)^\circ$ follows from the definition. To prove the opposite inclusion we assume that C_1 is determined by the system described in Theorem 2.3.19 with $i = 1, \cdots, k_1$ and C_2 is determined by that system with $i = k_1 + 1, \cdots, k_1 + k_2$. Then the polyhedral cone $C_1 \cap C_2$ is determined by that system with $i = 1, \cdots, k_1 + k_2$. In view of Theorem 2.3.19, the polar cone of $C_1 \cap C_2$ is the positive hull of the vectors $a^1, \cdots, a^{k_1+k_2}$, which is evidently the sum of the positive hulls $\mathrm{pos}\{a^1, \cdots, a^{k_1}\}$ and $\mathrm{pos}\{a^{k_1+1}, \cdots, a^{k_1+k_2}\}$, that is the sum of the polar cones C_1° and C_2°. $\square$

Corollary 2.3.22 *The bipolar cone of a polyhedral cone C coincides with the cone C itself.*

Proof According to Theorem 2.3.19 a vector v belongs to the bipolar cone $C^{\circ\circ}$ if and only if

$$\left\langle v, \sum_{i=1}^{k} \lambda_i a^i \right\rangle \leqq 0 \text{ for all } \lambda_i \geqq 0, i = 1, \cdots, k.$$

The latter system is equivalent to

$$\langle a^i, v \rangle \leqq 0, i = 1, \cdots, k,$$

which is exactly the system determining the cone C. $\square$

Corollary 2.3.23 *A vector v belongs to the polar cone of the asymptotic cone of a convex polyhedron if and only if the linear functional $\langle v, . \rangle$ attains its maximum on the polyhedron.*

Proof It suffices to consider the case where v is nonzero. Assume v belongs to the polar cone of the asymptotic cone P_∞. In virtue of Theorems 2.3.12 and 2.3.19, it is a positive combination of the vectors $a^1, \cdots, a^k$. Then the linear functional $\langle v, . \rangle$ is majorized by the same combination of real numbers $b_1, \cdots, b_k$ on P. Let α be its supremum on P. Our aim is to show that this value is realizable, or equivalently, the system

$$\langle a^i, x \rangle \leqq b_i, \ i = 1, \cdots, k$$
$$\langle v, x \rangle \geqq \alpha$$

is solvable. Suppose to the contrary that the system has no solution. In view of Corollary 2.2.4, there are a positive vector y and a real number $t \geq 0$ such that

$$tv = A^T y,$$
$$t\alpha = \langle b, y \rangle + 1.$$

We claim that t is strictly positive. Indeed, if $t = 0$, then $A^T y = 0$ and $\langle b, y \rangle = -1$ and for a vector x in P we would deduce

$$0 = \langle A^T y, x \rangle = \langle y, Ax \rangle \leq \langle y, b \rangle = -1,$$

a contradiction. We obtain expressions for v and α as follows

$$v = \frac{1}{t} A^T y \ \text{ and } \ \alpha = \frac{1}{t} (\langle b, y \rangle + 1).$$

Let $\{x^r\}_{r \geq 1}$ be a maximizing sequence of the functional $\langle v, . \rangle$ on P, which means $\lim_{r \to \infty} \langle v, x^r \rangle = \alpha$. Then, for every r one has

$$\langle v, x^r \rangle = \frac{1}{t} \langle A^T y, x^r \rangle$$
$$\leqq \frac{1}{t} \langle y, b \rangle$$
$$\leqq \alpha - \frac{1}{t},$$

which is a contradiction when r is sufficiently large.

For the converse part, let $\bar{x}$ be a point in P where the functional $\langle v, . \rangle$ achieves its maximum. Then

$$\langle v, x - \bar{x} \rangle \leqq 0 \ \text{ for all } x \in P.$$

In particular,

$$\langle v, u \rangle \leqq 0 \text{ for all } u \in P_\infty,$$

and hence v belongs to the polar cone of P_∞.   □

Normal cones

Given a convex polyhedron P determined by the system (2.6) and a point x in P, we say that a vector v is a *normal vector* to P at x if

$$\langle v, y - x \rangle \leqq 0 \ \text{ for all } \ y \in P.$$

The set of all normal vectors to P at x forms a convex cone called the *normal cone* to P at x and denoted $N_P(x)$ (Fig. 2.12). When x is an interior point of P, the normal cone at that point is zero. When x is a boundary point, the normal cone is computed by the next result.

Theorem 2.3.24 *The normal cone to the polyhedron P at a boundary point x of P is the positive hull of the vectors a^i with i being active indices at the point x.*

Proof Let $\bar{x}$ be a boundary point in P. Then the active index set $I(\bar{x})$ is nonempty. Let v be an element of the positive hull of the vectors a^i, $i \in I(\bar{x})$, say

$$v = \sum_{i \in I(\bar{x})} \lambda_i a^i \quad \text{with} \quad \lambda_i \geq 0, i \in I(\bar{x}).$$

Then for every point x in P and every active index $i \in I(\bar{x})$, one has

$$\langle a^i, x - \bar{x} \rangle = \langle a^i, x \rangle - b_i \leq 0,$$

which yields

$$\langle v, x - \bar{x} \rangle = \sum_{i \in I(\bar{x})} \lambda_i \langle a^i, x - \bar{x} \rangle \leq 0.$$

Hence v is normal to P at $\bar{x}$. For the converse, assume that v is a nonzero vector satisfying

$$\langle v, x - \bar{x} \rangle \leq 0 \quad \text{for all } x \in P. \tag{2.18}$$

We wish to establish that v is a normal vector at 0 to the polyhedron, denoted Q, that is determined by the system

$$\langle a^i, y \rangle \leq 0, \quad i \in I(\bar{x}).$$

This will certainly complete the proof because the normal cone to that polyhedron is exactly its polar cone, the formula of which was already given in Theorem 2.3.19. Observe that normality condition (2.18) can be written as

$$\langle v, y \rangle \leq 0 \quad \text{for all } y \in \mathrm{cone}(P - \bar{x}).$$

Therefore, v will be a normal vector to Q at zero if Q coincides with $\mathrm{cone}(P - \bar{x})$. Indeed, let y be a vector of $\mathrm{cone}(P - \bar{x})$, say $y = t(x - \bar{x})$ for some x in P and some positive number t. Then

$$\langle a^i, y \rangle = t \langle a^i, x - \bar{x} \rangle \leq 0,$$

which yields $y \in Q$. Thus, $\mathrm{cone}(P - \bar{x})$ is a subset of Q. For the reverse inclusion we notice that inequalities with inactive indices are strict at $\bar{x}$. Therefore, given a vector y in Q, one can find a small positive number t such that

Fig. 2.12 Normal cone

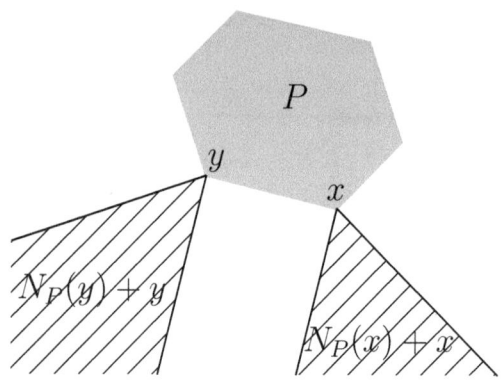

$$\langle a^j, \overline{x} \rangle + t \langle a^j, y \rangle \le b_j$$

for all j inactive. Of course, when i is active, it is true that

$$\langle a^i, \overline{x} + ty \rangle = \langle a^i, \overline{x} \rangle + t \langle a^i, y \rangle \le b_i.$$

Hence, $\overline{x} + ty$ belongs to P, or equivalently y belongs to cone$(P - \overline{x})$. This achieves the proof.   □

Example 2.3.25 Consider the polyhedron in $\mathbb{R}^3$ defined by the system:

$$\begin{array}{rrrl} x_1 + & x_2 + & x_3 \le & 1, \\ -2x_1 - & 3x_2 & \le & -1, \\ & x_1, x_2, x_3 \ge & & 0. \end{array}$$

This is a convex polytope with six vertices

$$v_1 = \begin{pmatrix} 1 \\ 0 \\ 0 \end{pmatrix}, v_2 = \begin{pmatrix} 0 \\ 1 \\ 0 \end{pmatrix}, v_3 = \begin{pmatrix} 0 \\ 1/3 \\ 0 \end{pmatrix},$$

$$v_4 = \begin{pmatrix} 1/2 \\ 0 \\ 0 \end{pmatrix}, v_5 = \begin{pmatrix} 1/2 \\ 0 \\ 1/2 \end{pmatrix}, v_6 = \begin{pmatrix} 0 \\ 1/3 \\ 2/3 \end{pmatrix}$$

and five two-dimensional faces

$$\text{co}\{v_1, v_2, v_5, v_6\}, \text{co}\{v_1, v_2, v_3, v_4\}, \text{co}\{v_1, v_4, v_5\},$$
$$\text{co}\{v_3, v_4, v_5, v_6\}, \text{co}\{v_2, v_3, v_6\}.$$

At the vertex v_1 there are three active constraints:

$$
\begin{aligned}
x_1 + x_2 + x_3 &= 1, \\
x_2 &= 0, \\
x_3 &= 0,
\end{aligned}
$$

and two non-active constraints

$$
\begin{aligned}
-2x_1 - 3x_2 &\leqq -1, \\
-x_1 &\leqq 0.
\end{aligned}
$$

Hence the normal cone at the vertex v_1 is the positive hull of the vectors $u_1 = (1, 1, 1)^T, u_2 = (0, -1, 0)^T$ and $u_3 = (0, 0, -1)^T$. Notice that u_1 generates the normal cone at the point $(1/3, 1/3, 1/3)^T$ on the two-dimensional face $F_1 = \mathrm{co}\{v_1, v_2, v_6, v_5\}$, u_2 generates the normal cone at the point $(2/3, 0, 1/4)^T$ on the two-dimensional face $F_2 = \mathrm{co}\{v_1, v_4, v_5\}$, and the positive hull of u_1 and u_2 is the normal cone at the point $(3/4, 0, 1/4)^T$ on the one-dimensional face $[v_1, v_5]$ that is the intersection of the two-dimensional faces F_1 and F_2.

As a direct consequence of Theorem 2.3.24, we observe that the normal cone is the same at any relative interior point of a face. We refer to this cone as the normal cone to a face. In view of Corollary 2.3.7 we obtain a collection of all normal cones of faces, whose union is called the normal cone of P and denoted by N_P. Thus, if $\mathcal{F} := \{F_1, \cdots, F_q\}$ is the collection of all faces of P, then

$$
N_P = \bigcup_{i=1}^{q} N(F_i).
$$

It is to point out a distinction between this cone and the cone $N(P)$, the normal cone to P when P is considered as a face of itself. We shall see now that the collection $\mathcal{N}$ of all normal cones $N(F_i)$, $i = 1, \cdots, q$, is a nice dual object of the collection $\mathcal{F}$.

Theorem 2.3.26 *Assume that P is a convex polyhedron given by the system*

$$
\langle a^i, x \rangle \leqq b_i, \quad i = 1, \cdots, k.
$$

Then the following assertions hold.

(i) *The normal cone of P is composed of all normal cones to P at its points, that is*

$$
N_P = \bigcup_{x \in P} N_P(x)
$$

and coincides with the polar cone of the asymptotic cone of P. In particular, it is a polyhedral cone, and it is the whole space if and only if P is a polytope (bounded polyhedron).

(ii) *In the collection $\mathcal{F}$, if F_i is a face of F_j, then $N(F_j)$ is a face of $N(F_i)$. Moreover, if $i \neq j$, then the normal cones $N(F_i)$ and $N(F_j)$ have no relative interior point in common.*

(iii) *In the collection $\mathcal{N}$, if N is a face of $N(F_i)$, then there is a face F_ℓ containing the face F_i such that $N = N(F_\ell)$.*

Proof For the first property it is evident that N_P is contained in the union of the right hand side. Let $x \in P$. There exists an index $i \in \{1, \cdots, k\}$ such that $x \in \mathrm{ri}(F_i)$. Then $N_P(x) = N(F_i)$ and equality of (i) is satisfied. To prove that N_P coincides with $(P_\infty)^\circ$, let v be a vector of the normal cone $N(F_i)$ for some i. Choose a relative interior point x_0 of the face F_i. Then, by definition,

$$\langle v, x - x_0 \rangle \leqq 0 \text{ for all } x \in P.$$

By Corollary 2.3.23 the vector v belongs to the polar cone of the cone P_∞. Conversely, let v be in $(P_\infty)^\circ$. In view of the same corollary, the linear functional $\langle v, . \rangle$ attains its maximum on P at some point $\overline{x}$, which means that

$$\langle v, x - \overline{x} \rangle \leqq 0 \text{ for all } x \in P.$$

By definition, v is a normal vector to P at $\overline{x}$.

For (ii), assume that F_i is a face of F_j with $i \neq j$, which implies that the active index set I_{F_i} of F_i contains the active index set I_{F_j} of F_j. Let x_j be a relative interior point of F_j. Then one has

$$N(F_j) = N_P(x_j) \subset N(F_i).$$

Suppose that $N(F_j)$ is not a face of $N(F_i)$. There exists a face

$$N_0 = \mathrm{pos}\{a^\ell : \ell \in I_0\} \subseteq N(F_i)$$

for some $I_0 \subseteq I_{F_i}$, which contains $N(F_j)$ as a proper subset and such that its relative interior meets $N(F_j)$ at some point, say v_0. Let F_0 be the solution set to the system

$$\begin{aligned}
\langle a^\ell, x \rangle &= b_\ell , & \ell \in I_0, \\
\langle a^\ell, x \rangle &\leqq b_\ell , & \ell \in \{1, \cdots, p\} \backslash I_0 .
\end{aligned}$$

We see that $I_{F_j} \subseteq I_0 \subseteq I_{F_i}$, hence $F_i \subseteq F_0 \subseteq F_j$. In particular $F_0 \neq \emptyset$, hence it is a face of P. Let x_0 be a relative interior point of F_0. We claim that

$$\langle v, x_j - x_0 \rangle = 0 \text{ for all } v \in N_0.$$

Indeed, consider the linear functional $v \mapsto \langle v, x_j - x_0 \rangle$ on N_0. On the one hand, $\langle v, x_j - x_0 \rangle \leqq 0$ for all $v \in N_0$ because $x_0 \in \mathrm{ri}(F_0)$. On the other hand, for $v_0 \in \mathrm{ri}(N_0) \cap N(F_j)$ above, one has $\langle v_0, x_0 - x_j \rangle \leqq 0$, hence $\langle v_0, x_j - x_0 \rangle = 0$.

Consequently, $\langle v, x_j - x_0 \rangle = 0$ on N_0. Using this fact we derive for every $v \in N_0$ that

$$\langle v, x - x_j \rangle = \langle v, x - x_0 \rangle + \langle v, x_0 - x_j \rangle \leqq 0,$$

for all $x \in M$, which implies $v \in N(F_j)$ and arrive at the contradiction $N(F_j) = N_0$. To prove the second part of assertion (ii), suppose to the contrary that the normal cones $N(F_i)$ and $N(F_j)$ have a relative interior point v in common. Then for each $x \in F_i$ and $y \in F_j$ one has

$$\langle v, x - y \rangle = 0.$$

Since $\langle u, y - x \rangle \leqq 0$ for all $u \in N(F_i)$ and v is a relative interior point of $N(F_i)$, one deduces

$$\langle u, x - y \rangle = 0 \text{ for all } u \in N(F_i).$$

Consequently, for $u \in N(F_i)$ it is true that

$$\langle u, z - y \rangle = \langle u, z - x \rangle + \langle u, x - y \rangle \leqq 0 \text{ for all } z \in P,$$

which shows $u \in N(F_j)$. In other words $N(F_i) \subseteq N(F_j)$. The same argument with i and j interchanging the roles, leads to equality $N(F_i) = N(F_j)$. In view of the first part we arrive at the contradiction $F_i = F_j$.

We proceed to (iii). Let N be a face of $N(F_i)$ for some $i : 1 \leq i \leq k$. The case $N = N(F_i)$ being trivial, we may assume $N \neq N(F_i)$. Let $I \subseteq I_{F_i}$ be a subset of indices such that

$$N = \text{cone}\{a^\ell : \ell \in I\} \subseteq N(F_i) = \text{cone}\{a^\ell : \ell \in I_{F_i}\} .$$

Let F be the solution set to the system

$$\langle a^\ell, x \rangle = b_\ell , \ \ell \in I,$$
$$\langle a^\ell, x \rangle \leqq b_\ell , \ \ell \in \{1, \cdots, p\} \backslash I .$$

Since $I \subseteq I_{F_i}$, we have $F_i \subseteq F$. In particular $F \neq \emptyset$ and F is a face of P. Now we show that $N(F) = N$ and F_i is a proper face of F. Indeed, as N is a proper face of $N(F_i)$, I is a proper subset of I_{F_i} and there is a nonzero vector $u \in R^n$ such that

$$\langle a^\ell, u \rangle = 0 \text{ for } \ell \in I,$$
$$\langle a^\ell, u \rangle < 0 \text{ for } \ell \in I_{F_i} \backslash I .$$

Take $x \in \text{ri}(F_i)$ and consider the point $x + tu$ with $t > 0$. One obtains

$$\langle a^\ell, x + tu \rangle = b_\ell , \ \ell \in I,$$

$$\langle a^\ell, x + tu \rangle = \langle a^\ell, x \rangle + t \langle a^\ell, u \rangle < b_\ell , \ \ell \in I_{F_i} \backslash I .$$

Moreover, since $\langle a^\ell, x \rangle < b_\ell$ for $\ell \in \{1, \cdots, p\}\backslash I_{F_i}$, when t is sufficiently small, one also has

$$\langle a^\ell, x + tu \rangle < b_\ell , \quad \ell \in \{1, \cdots p\}\backslash I_{F_i} .$$

Consequently,

$$N(F) = N_P(x + tu) = \text{pos}\{a^\ell : \ell \in I\} = N$$

when t is sufficiently small. It is evident that $F \neq F_i$. The proof is complete.   □

Example 2.3.27 Consider the polyhedron P in $\mathbb{R}^2$ defined by the system:

$$-x_1 - x_2 \leqq -1,$$
$$-x_1 + x_2 \leqq 1,$$
$$-x_2 \leqq 0.$$

It has two vertices F_1 and F_2 determined respectively by

$$\begin{cases} -x_1 -x_2 = -1 \\ -x_1 +x_2 \leqq 1 \\ -x_2 = 0 \end{cases} \text{ and } \begin{cases} -x_1 -x_2 = -1 \\ -x_1 +x_2 = 1 \\ -x_2 \leqq 0 \end{cases}$$

three one-dimensional faces F_3, F_4 and F_5 determined respectively by

$$\begin{cases} -x_1 -x_2 \leqq -1 \\ -x_1 +x_2 \leqq 1 \\ -x_2 = 0 \end{cases}, \begin{cases} -x_1 -x_2 = -1 \\ -x_1 +x_2 \leqq 1 \\ -x_2 \leqq 0 \end{cases} \text{ and } \begin{cases} -x_1 -x_2 \leqq -1 \\ -x_1 +x_2 = 1 \\ -x_2 \leqq 0 \end{cases}$$

and P itself is the unique two-dimensional face. Denote by $v_1 = (-1, -1)^T$, $v_2 = (-1, 1)^T$ and $v_3 = (0, -1)^T$. Then the normal cones of the faces $F_1, \cdots, F_5$ are respectively the positive hulls of the families $\{v_1, v_3\}$, $\{v_1, v_2\}$, $\{v_3\}$, $\{v_1\}$ and $\{v_2\}$. The normal cone of P is zero. Moreover, the union N_P of these normal cones is the positive hull of the vectors v_2 and v_3. It is the polar cone of the asymptotic cone of P which is defined by the system

$$-x_1 - x_2 \leqq 0,$$
$$-x_1 + x_2 \leqq 0,$$
$$-x_2 \leqq 0,$$

in which the first inequality is redundant and hence it is reduced to $x_1 \geqq x_2 \geqq 0$.

Next we prove that the normal cone of a face is obtained from the normal cones of its vertices.

Corollary 2.3.28 *Assume that a face F of P is the convex hull of its vertices* $v^1, \cdots, v^q$. *Then*

$$N(F) = \bigcap_{i=1}^{q} N_P(v^i).$$

Proof The inclusion $N(F) \subseteq \bigcap_{i=1}^{q} N_P(v^i)$ is clear from (ii) of Theorem 2.3.26. We prove the converse inclusion. Let u be a nonzero vector of the intersection $\bigcap_{i=1}^{q} N_P(v^i)$. Let x be a relative interior point of F. Then x is a convex combination of the vertices $v^1, \cdots, v^q$:

$$x = \sum_{i=1}^{q} \lambda_i v^i$$

with $\lambda_i \geq 0, i = 1, \cdots, q$ and $\lambda_1 + \cdots + \lambda_q = 1$. We have then

$$\langle u, x' - v^i \rangle \leq 0 \text{ for all } x' \in P, i = 1, \cdots, q.$$

This implies

$$\langle u, x' - x \rangle = \langle u, \sum_{i=1}^{q} \lambda_i x' - \sum_{i=1}^{q} \lambda_i v^i \rangle$$

$$= \sum_{i=1}^{q} \lambda_i \langle u, x' - v^i \rangle \leq 0$$

By this, u is a normal vector to P at x, and $u \in N(F)$. $\qquad\square$

Combining this corollary with Corollary 2.3.8 we conclude that the normal cone of a bounded face is the intersection of the normal cones of all proper faces of that bounded face. This is not true for unbounded faces, for instance when a face has no proper face.

2.4 Basis and Vertices

In this section we consider a polyhedron P given by the system

$$Ax = b \qquad\qquad (2.19)$$
$$x \geq 0.$$

We assume throughout that the matrix A has n columns denoted $a_1, \cdots, a_n$ and k rows that are transposes of $a^1, \cdots, a^k$ and linearly independent, and that the components $b_1, \cdots, b_k$ of the vector b are non-negative numbers. A point x in P is said to be an *extreme point* of P if it cannot be expressed as a convex combination $x = ta + (1-t)a'$ for some $0 < t < 1$ and $a, a' \in P$ with $a \neq a'$. It can be seen that extreme points correspond to vertices we have defined in the previous section. Certain results we have obtained for polyhedra given in a general form (by inequalities) will be recaptured here, but our emphasis will be laid on computing issues which are much simplified under equality form (2.19).

A $k \times k$-submatrix B composed of columns of A is said to be a *basis* if it is invertible.

Let B be a basis. By using a permutation one may assume that B is composed of the first k columns of A, and the remaining columns form a $k \times (n-k)$-submatrix N, called a *non-basic part* of A. Let x be a vector with components x_B and x_N, where x_B is a k-dimensional vector and x_N is an $(n-k)$-dimensional vector satisfying

$$Bx_B = b,$$
$$x_N = 0.$$

If x_B is a positive vector, then x is a solution to (2.19) and called a *feasible basic solution* (associated with the basis B). If in addition x_B has no zero component, it is called *non-degenerate*; otherwise it is *degenerate*.

Example 2.4.1 Consider the polyhedron in $\mathbb{R}^3$ defined by the system:

$$\begin{aligned}
x_1 + x_2 + x_3 &= 1, \\
3x_1 + 2x_2 \quad\quad &= 1, \\
x_1, x_2, x_3 &\geq 0.
\end{aligned}$$

The vectors $a_1 = (1, 1, 1)^T$ and $a_2 = (3, 2, 0)^T$ are linear independent. There are three bases

$$B_1 = \begin{pmatrix} 1 & 1 \\ 3 & 2 \end{pmatrix}, B_2 = \begin{pmatrix} 1 & 1 \\ 3 & 0 \end{pmatrix} \text{ and } B_3 = \begin{pmatrix} 1 & 1 \\ 2 & 0 \end{pmatrix}.$$

The basic solutions corresponding to B_1, B_2 and B_3 are respectively $(-1, 2, 0)^T$, $(1/3, 0, 2/3)^T$ and $(0, 1/2, 1/2)^T$. The first solution is unfeasible, while the two last ones are feasible and non-degenerate.

Given a vector $x \in \mathbb{R}^n$, its *support*, denoted $\text{supp}(x)$, consists of indices i for which the component x_i is nonzero. The support of a nonzero vector is always nonempty.

Theorem 2.4.2 *A vector x is a vertex of the polyhedron P if and only if it is a feasible basic solution of the system (2.19).*

Proof Let x be a feasible basic solution. Assume that it is a convex combination of two solutions y and z of the system (2.19), say $x = ty + (1-t)z$ with $t \in (0, 1)$. Then for any nonbasic index j, the component x_j is zero, so that $ty_j + (1-t)z_j = 0$. Remembering that y and z are positive vectors, we derive $y_j = z_j = 0$. Moreover, the basic components of solutions to (2.19) satisfy equation

$$Bx_B = b$$

with B nonsingular. Therefore, they are unique, that is $x_B = y_B = z_B$. Consequently, the three solutions x, y and z are the same.

Conversely, let x be an extreme point of the polyhedron. Our aim is to show that the columns a^i, $i \in \mathrm{supp}(x)$ are linearly independent. It is then easy to find a basis B such that x is the basic solution associated with that basis. To this end, we prove first that $\mathrm{supp}(x)$ is minimal by inclusion among solutions of the system (2.19). In fact, if not, one can find another solution, say y, with minimal support such that $\mathrm{supp}(y)$ is a proper subset of $\mathrm{supp}(x)$. Choose an index j from the support of y such that

$$\frac{x_j}{y_j} = \min\{\frac{x_i}{y_i} : i \in \mathrm{supp}(y)\}.$$

Let $t > 0$ be that quotient. Then

$$A(x - ty) = (1-t)b \quad \text{and} \quad x - ty \geqq 0.$$

If $t \geqq 1$, then by setting $z = x - y$ we can express

$$x = \frac{1}{2}(y + \frac{2}{3}z) + \frac{1}{2}(y + \frac{4}{3}z),$$

a convex combination of two distinct solutions of (2.19), which is a contradiction. If $t < 1$, then take

$$z = \frac{1}{1-t}(x - ty).$$

We see that z is a solution to (2.19) and different from x because its support is strictly contained in the support of x. It is also different from y because the component y_j is not zero while the component z_j is zero. We derive from the definition of z that x is a strict convex combination of y and z, which is again a contradiction.

Now we prove that the columns a^i, $i \in \mathrm{supp}(x)$ are linearly independent. Suppose the contrary: there is a vector y different from x (if not take $2y$ instead) with

$$Ay = 0 \quad \text{and} \quad \mathrm{supp}(y) \subseteq \mathrm{supp}(x).$$

By setting

$$
t = \begin{cases} -\min\{\frac{x_i}{y_i} : i \in \text{supp}(y)\} & \text{if } y \geq 0 \\ \min\{-\frac{x_i}{y_i} : i \in \text{supp}(y), \, y_i < 0\} & \text{else} \end{cases}
$$

we obtain that $z = x + ty$ is a solution to (2.19) whose support is strictly contained in the support of x and arrive at a contradiction with the minimality of the support of x. It remains to complete the vectors a^i, $i \in \text{supp}(x)$ to a basis to see that x is indeed a basic solution. $\qquad\square$

Corollary 2.4.3 *The number of vertices of the polyhedron P does not exceed the binomial coefficient* $\binom{n}{k}$.

Proof This follows from Theorem 2.4.2 and the fact that the number of bases of the matrix A is at most $\binom{n}{k}$. Notice that not every basic solution has positive components. $\qquad\square$

We deduce again Corollary 2.3.8 about the description of polytopes in terms of extreme points (vertices), but this time for a polytope determined by the system (2.19).

Corollary 2.4.4 *If P is a polytope, then any point in it can be expressed as a convex combination of vertices.*

Proof Let x be any solution of (2.19). If the support of x is minimal, then in view of Theorem 2.4.2 that point is a vertex. If not, then there is a solution y^1 different from x, with minimal support and $\text{supp}(y^1) \subset \text{supp}(x)$. Set

$$
t_1 = \min\left\{\frac{x_j}{y_j^1} : j \in \text{supp}(y^1)\right\}.
$$

This number is positive and strictly smaller than one, because otherwise the nonzero vector $x - y^1$ should be an asymptotic direction of the polyhedron and P should be unbounded. Consider the vector

$$
z^1 = \frac{1}{1 - t_1}(x - t_1 y^1).
$$

It is clear that this vector is a solution to (2.19) and its support is strictly smaller than the support of x. If the support of z^1 is minimal, then z^1 is a vertex and we obtain a convex combination

$$
x = t_1 y^1 + (1 - t_1)z^1,
$$

in which y^1 and z^1 are vertices. If not, we continue the process to find a vertex y^2 whose support is strictly contained in the support of z^1 and so on. In view of Corollary 2.4.3 after a finite number of steps one finds vertices $y^1, \cdots, y^p$ such that x is a convex combination of them. $\qquad\square$

Extreme rays

Extreme direction of a convex polyhedron P in $\mathbb{R}^n$ can be defined to be a direction that cannot be expressed as a strictly positive combination of two linearly independent asymptotic vectors of P. As the case when a polyhedron is given by a system of linear inequalities (Corollary 2.3.15), we shall see that a polyhedron determined by (2.19) is completely determined by its vertices and extreme directions.

Theorem 2.4.5 *Assume that the convex polyhedron P is given by the system (2.19). Then*

(i) *A nonzero vector v is an asymptotic direction of P if and only if it is a solution to the associated homogenous system*

$$Ax = 0,$$
$$x \geq 0.$$

(ii) *A nonzero vector v is an extreme asymptotic direction of P if and only if it is a positive multiple of a vertex of the polyhedron determined by the system*

$$Ay = 0 \tag{2.20}$$
$$y_1 + \cdots + y_n = 1,$$
$$y \geq 0.$$

Consequently P_∞ consists of all positive combinations of the vertices of this latter polyhedron.

Proof The first assertion is proven as in Theorem 2.3.12. For the second assertion, let v be a nonzero extreme direction. Then $Av = 0$ by (i) and $t := v_1 + \cdots + v_n > 0$. The vector v/t is in the polyhedron of (ii), denoted Q. Since each point of that polyhedron is an asymptotic direction of P, if v/t were a convex combination of two distinct points y^1 and y^2 in Q, then v would be a convex combination of two linearly independent asymptotic directions ty^1 and ty^2 of P, which is a contradiction. Conversely, let v be a vertex of Q. It is clear that v is nonzero. If $v = tx + (1-t)y$ for some nonzero asymptotic directions x and y of P and some $t \in (0, 1)$, then with

$$t' = \frac{t \sum_{i=1}^n x_i}{t \sum_{i=1}^n x_i + (1-t) \sum_{i=1}^n y_i} = t \sum_{i=1}^n x_i,$$

$$x' = \frac{1}{\sum_{i=1}^n x_i} x,$$

$$y' = \frac{1}{\sum_{i=1}^n y_i} y,$$

we express v as a convex combination $t'x' + (1-t')y'$ of two points of Q. Note that $t' > 0$. By hypothesis, $x' = y'$ which means that x and y are linearly dependent. The proof is complete. $\qquad\square$

Corollary 2.4.6 *A nonzero vector is an extreme asymptotic direction of P if and only if it is a basic feasible solution of the system (2.20). Consequently, the number of extreme asymptotic directions of P does not exceed the binomial coefficient*
$$\binom{n}{k+1}.$$

Proof This is obtained from Theorems 2.4.2 and 2.4.5. $\qquad\square$

Example 2.4.7 Consider the polyhedron in $\mathbb{R}^3$ defined by the system:
$$x_1 - x_2 = 1$$
$$x_1, x_2, x_3 \geqq 0.$$

The asymptotic cone of this polyhedron is the solution set to the system
$$x_1 - x_2 = 0$$
$$x_1, x_2, x_3 \geqq 0.$$

Any vector $(t, t, s)^T$ with $t \geqq 0$ and $s \geqq 0$ is an asymptotic direction. To obtain extreme asymptotic directions we solve the system
$$\begin{aligned} y_1 - \ y_2 \qquad\quad &= 0 \\ y_1 + \ y_2 + \ y_3 &= 1 \\ y_1, \ y_2, \ y_3 &\geqq 0. \end{aligned}$$

There are three bases corresponding to basic variables $\{y_1, y_2\}, \{y_1, y_3\}$ and $\{y_2, y_3\}$:
$$B_1 = \begin{pmatrix} 1 & -1 \\ 1 & 1 \end{pmatrix}, \quad B_2 = \begin{pmatrix} 1 & 0 \\ 1 & 1 \end{pmatrix} \text{ and } B_3 = \begin{pmatrix} -1 & 0 \\ 1 & 1 \end{pmatrix}.$$

The basic solution $y = (1/2, 1/2, 0)^T$ is associated with B_1 and the basic solution $y = (0, 0, 1)^T$ is associated with B_2 and B_3. Both of them are feasible, and hence they are extreme asymptotic directions.

In the following we describe a practical way to compute extreme rays of the polyhedron P.

Corollary 2.4.8 *Assume that B is a basis of the matrix A and a_s is a non-basic column of A such that the system*
$$By = -a_s$$

has a positive solution $\bar{y} \geq 0$. Then the vector $\bar{x}$ whose basic components are equal to $\bar{y}$, the sth component is equal to 1 and the other non-basic components are all zero, is an extreme ray of the polyhedron P.

Proof It is easy to check that the submatrix corresponding to the variables of $\bar{y}$ and the variable y_s is a feasible basis of the system (2.20). It remains to apply Corollary 2.4.6 to conclude.   □

In Example 2.4.7 we have $A = (1, -1, 0)$. For the basis $B = (1)$ corresponding to the basic variable x_1 and the second non-basic column, the system $By = -a_s$ takes the form $y = 1$ and has a positive solution $y = 1$. In view of Corollary 2.4.8 the vector $(1, 1, 0)^T$ is an extreme asymptotic direction. Note that using the same basis B and the non-basic column $a_3 = (0)$ we obtain the system $y = 0$ which has a positive (null) solution. Hence the vector $(0, 0, 1)^T$ is also an extreme asymptotic direction.

Representation of Elements of a Polyhedron

A finitely generated convex set is defined to be a set which is the convex hull of a finite set of points and directions, that is, each element of it is the sum of a convex combination of a finite set of points and a positive combination of a finite set of directions. The next theorem states that convex polyhedra are finitely generated, which is Corollary 2.3.15 for a polyhedron determined by the system (2.19).

Theorem 2.4.9 *Every point of a convex polyhedron given by the system (2.19) can be expressed as a convex combination of its vertices, possibly added to a positive combination of the extreme asymptotic directions.*

Proof Let x be any point in P. If its support is minimal, then, according to the proof of Theorem 2.4.2 that point is a vertex. If not, there is a vertex v^1 whose support is minimal and strictly contained in the support of x. Set

$$t = \min\left\{\frac{x_j}{v_j^1} : j \in \text{supp}(v^1)\right\}$$

and consider the vector $x - tv^1$. If $t \geq 1$, then the vector $z = x - v^1$ is an asymptotic direction of the polyhedron and then x is the sum of the vertex v^1 and an asymptotic direction. The direction z, in its turn, is expressed as a convex combination of extreme asymptotic directions. So the corollary follows. If $t < 1$, the technique of proof of Theorem 2.4.2 can be applied. Expressly, setting $z = (x - tv^1)/(1 - t)$ we deduce that $z \geq 0$ and

$$Az = \frac{1}{1-t}b - \frac{t}{1-t}b = b.$$

Moreover, the support of z is a proper subset of the support of x because the components j of z with j realizing the value of $t = x_j/v_j^1$ are zero. Then $x = tv^1 + (1-t)z$ with strict inclusion $\text{supp}(z) \subset \text{supp}(x)$. Continuing this process we arrive at finding a finite number of vertices $v^1, \cdots, v^p$ and an asymptotic direction z such that x is the sum of a convex combination of $v^1, \cdots, v^p$ and z. Then expressing z as a convex combination of asymptotic extreme directions we obtain the conclusion.   □

In view of Corollaries 2.4.3 and 2.4.6 the numbers of vertices and extreme asymptotic directions of a polyhedron P are finite. Denote them respectively by $v^1, \cdots, v^p$ and $z^1, \cdots, z^q$. Then each element x of P is expressed as

$$x = \sum_{i=1}^{p} \lambda_i v^i + \sum_{j=1}^{q} \mu_j z^j$$

with

$$\sum_{i=1}^{p} \lambda_i = 1, \lambda_i \geq 0, i = 1, \cdots, p \text{ and } \mu_j \geq 0, j = 1, \cdots, q.$$

Notice that the above representation is not unique, that is, an element x of P can be written as several combinations of $v^i, i = 1, \cdots, p$ and $z^j, j = 1, \cdots, q$ with different coefficients λ_i and μ_j. An easy example can be observed for the center x of the square with vertices

$$v^1 = \begin{pmatrix} 0 \\ 0 \end{pmatrix}, v^2 = \begin{pmatrix} 1 \\ 0 \end{pmatrix}, v^3 = \begin{pmatrix} 0 \\ 1 \end{pmatrix} \text{ and } v^4 = \begin{pmatrix} 1 \\ 1 \end{pmatrix}.$$

It is clear that x can be seen as the middle point of v^1 and v^4, and as the middle point of v^2 and v^3 too.

Another point that should be made clear is the fact that the results of this section are related to polyhedra given by the system (2.19) and they might be false under systems of different type. For instance, in view of Theorem 2.4.9 a polyhedron determined by (2.19) has at least a vertex. This is no longer true if a polyhedron is given by another system. Take a hyperplane determined by equation $\langle d, x \rangle = 0$ for some nonzero vector $d \in \mathbb{R}^2$. It is a polyhedron without vertices. An equivalent system is given in form of (2.19) as follows

$$\langle d, x^+ \rangle - \langle d, x^- \rangle = 0,$$
$$x^+, x^- \geq 0.$$

The latter system generates a polyhedron in $\mathbb{R}^4$ that does have vertices. However, a vertex $(x^+, x^-)^T$ of this polyhedron gives an element $x = x^+ - x^-$ of the former polyhedron, but not a vertex of it.

Chapter 3
Linear Programming

A linear mathematical programming problem is a problem of finding a maximum or minimum of a linear functional over a convex polyhedron. The functional to optimize is called an objective or cost function, and the linear equalities and linear inequalities that define the polyhedron are called constraints.

3.1 Optimal Solutions

We consider the following linear programming problem, denoted (LP):

$$\text{maximize} \quad \langle c, x \rangle$$
$$\text{subject to} \quad Ax = b \tag{3.1}$$
$$x \geqq 0, \tag{3.2}$$

where c is an n-vector, A is an $m \times n$-matrix and b is an m-vector. Under these constraints we say (LP) is given in standard form. It is given in canonical form when the constraints (3.1) and (3.2) are substituted by inequalities $Ax \leqq b$. As we have already discussed in Sect. 2.2, linear equalities can be converted to linear inequalities and vice versa, any linear programming problem may be set in form as (LP) above. We denote the feasible set of the problem (LP) by X, that is, X is the solution set to the system (3.1–3.2). A feasible solution $\bar{x} \in X$ is *optimal* if $\langle c, \bar{x} \rangle \geqq \langle c, x \rangle$ for all $x \in X$. The linear function $x \mapsto \langle c, x \rangle$ is called the cost function of the problem. A fundamental theorem of linear programming is given next.

Theorem 3.1.1 *Assume that X is nonempty. Then the four conditions below are equivalent.*

(i) *(LP) admits an optimal solution.*
(ii) *(LP) admits an optimal vertex solution.*

© Springer International Publishing Switzerland 2016
D.T. Luc, *Multiobjective Linear Programming*,
DOI 10.1007/978-3-319-21091-9_3

(iii) *The cost function is non-positive on every asymptotic direction of X.*
(iv) *The cost function is bounded above on X.*

Proof The scheme of our proof is as follows: (i)$\Rightarrow$ (iv) $\Rightarrow$ (iii) $\Rightarrow$ (ii) $\Rightarrow$ (i). The first and the last implications are immediate. We proceed to the second implication. Let α be an upper bound of the cost function on X and let u be a nonzero asymptotic direction of X if it exists. Pick any point x in X which is nonempty by hypothesis. Then for every positive number t, the point $x + tu$ belongs to X. Hence

$$\langle c, x + tu \rangle = \langle c, x \rangle + t \langle c, u \rangle \leq \alpha.$$

This inequality being true for all t positive, we must have $\langle c, u \rangle \leq 0$.

To establish the third implication let $\{v^1, \cdots, v^p\}$ be the collection of all vertices and let $\{u^1, \cdots, u^q\}$ be the collection of all extreme rays of the polyhedron X. The collection of extreme rays may be empty. Choose a vertex v^{i_0} such that

$$\langle c, v^{i_0} \rangle = \max\{\langle c, v^1 \rangle, \cdots, \langle c, v^p \rangle\}.$$

Let x be any point in X. In view of Theorem 2.4.9, there are non-negative numbers t_i and s_j with $\sum_{i=1}^{p} t_i = 1$ such that

$$x = \sum_{i=1}^{p} t_i v^i + \sum_{j=1}^{q} s_j u^j.$$

We deduce

$$\langle c, x \rangle = \sum_{i=1}^{p} t_i \langle c, v^i \rangle + \sum_{j=1}^{q} s_j \langle c, u^j \rangle$$
$$\leq \langle c, v^{i_0} \rangle,$$

which shows that the vertex v^{i_0} is an optimal solution. The proof is complete. □

Existence of optimal solutions is always guaranteed when the feasible solution set is bounded as the next corollary shows.

Corollary 3.1.2 *If the problem (LP) has a bounded feasible set, then it has optimal solutions.*

Proof When the set X is bounded, it has no nonzero asymptotic direction. Hence condition (iii) of the previous theorem is fulfilled and the problem (LP) has optimal solutions. □

The result below expresses a necessary and sufficient condition for optimal solutions in terms of normal directions.

Theorem 3.1.3 *Assume that X is nonempty. Then the following statements are equivalent.*

(i) $\bar{x}$ *is an optimal solution of (LP).*
(ii) *The vector c belongs to the normal cone to the set X at $\bar{x}$.*
(iii) *The whole face of X which contains $\bar{x}$ as a relative interior point is an optimal solution face.*

Consequently, if (LP) has an optimal solution, then the optimal solution set is a face of the feasible polyhedron.

Proof The implication (iii) $\Rightarrow$ (i) is evident, so we have to show the implications (i) $\Rightarrow$ (ii) and (ii) $\Rightarrow$ (iii). For the first implication we observe that if $\bar{x}$ is an optimal solution, then

$$\langle c, x - \bar{x} \rangle \leqq 0 \text{ for all } x \in X.$$

By definition, c is a normal vector to X at $\bar{x}$ which yields (ii). Now, assume (ii) and let x be any point in the face that has $\bar{x}$ as a relative interior point. There is a positive number δ such that the points $\bar{x} + \delta(\bar{x} - x)$ and $\bar{x} - \delta(\bar{x} - x)$ belong to X. We have then

$$\langle c, \bar{x} + \delta(\bar{x} - x) - \bar{x} \rangle \leqq 0,$$
$$\langle c, \bar{x} - \delta(\bar{x} - x) - \bar{x} \rangle \leqq 0.$$

This produces

$$\langle c, \bar{x} - x \rangle \leqq 0,$$
$$\langle c, -(\bar{x} - x) \rangle \leqq 0.$$

Consequently, $\langle c, x \rangle = \langle c, \bar{x} \rangle$, which together with the normality of c at $\bar{x}$ shows that x is an optimal solution too.

For the second part of the theorem, set $\alpha = \langle c, \bar{x} \rangle$ where $\bar{x}$ is an optimal solution of (LP). Then the intersection of the hyperplane

$$H = \left\{ x \in \mathbb{R}^n : \langle c, x \rangle = \alpha \right\}$$

with the feasible set X is a face of X and contains all optimal solutions of the problem. $\square$

Given a feasible basis B, we call it an *optimal basis* if the associated basic solution is an optimal solution of (LP). We shall decompose the cost vector c into the basic component vector c_B and non-basic component vector c_N. The vector

$$\bar{c}_N = c_N - (B^{-1}N)^T c_B$$

is called the *reduced cost vector*.

Theorem 3.1.4 *Let B be a feasible basis and $\overline{x}$ the feasible basic solution associated with B. The following statements hold.*

(i) *If the reduced cost vector $\overline{c}_N$ is negative, then B is optimal.*
(ii) *When B is non-degenerate, it is optimal if and only if the reduced cost vector $\overline{c}_N$ is negative.*

Proof Up to a suitable permutation we may assume that the matrix A is decomposed by $(B \ N)$, the basic index set is $\{1, \cdots, m\}$ and the non-basic index set is $\{m + 1, \cdots, n\}$. To prove (i) let x be any feasible solution of the problem. Since x is a solution to the system $Ax = b$, the basic part x_B of x corresponding to the basic columns of B is expressed by its non-basic components via

$$x_B = B^{-1}b - B^{-1}Nx_N = \overline{x}_B - B^{-1}Nx_N. \tag{3.3}$$

The cost function at x is then given by

$$\begin{aligned}
\langle c, x \rangle &= \langle c_B, x_B \rangle + \langle c_N, x_N \rangle \\
&= \langle c_B, B^{-1}b - B^{-1}Nx_N \rangle + \langle c_N, x_N \rangle \\
&= \langle c_B, \overline{x}_B \rangle + \langle \overline{c}_N, x_N \rangle \\
&= \langle c, \overline{x} \rangle + \langle \overline{c}_N, x_N \rangle.
\end{aligned}$$

Since x_N is positive and by hypothesis $\overline{c}_N$ is negative, we deduce

$$\langle c, x \rangle \leqq \langle c, \overline{x} \rangle.$$

As x was an arbitrary feasible solution of the problem, we deduce that $\overline{x}$ is an optimal solution and B is an optimal basis.

For (ii) we need to prove the "only if" part. Suppose to the contrary that the reduced cost vector is not negative, that is, $\overline{c}_j > 0$ for some non-basic index j. Our aim is to find a new feasible solution $\hat{x}$ with

$$\langle c, \overline{x} \rangle < \langle c, \hat{x} \rangle \tag{3.4}$$

which yields a contradiction. We look for a solution $\hat{x}$ in special form:

$$\hat{x} = \begin{pmatrix} \hat{x}_B \\ \hat{x}_N \end{pmatrix} \text{ with } \hat{x}_N = \overline{x}_N + te^j = te^j$$

where e^j is the non-basic part of the jth coordinate unit vector in $\mathbb{R}^n$ and t is a positive number to be chosen such that $\hat{x}$ be feasible. Since $\hat{x}_N$ is positive, in view of (3.3) the feasibility of $\hat{x}$ means that

$$\hat{x}_B = \overline{x}_B - tB^{-1}Ne^j \geqq 0.$$

The basis B being non-degenerate, the vector $\overline{x}_B$ is strictly positive, hence $\hat{x}_B$ is positive whenever $t > 0$ is sufficiently small. We fix such a value of t and calculate the cost function at that point by using (3.3):

$$
\begin{aligned}
\langle c, \hat{x} \rangle &= \langle c_B, \hat{x}_B \rangle + \langle c_N, \hat{x}_N \rangle \\
&= \langle c_B, \overline{x}_B \rangle - t \langle c_B, B^{-1} N e^j \rangle + t \langle c_N, ej \rangle \\
&= \langle c_B, \overline{x}_B \rangle + t \overline{c}_j \\
&> \langle c_B, \overline{x}_B \rangle,
\end{aligned}
$$

which contradicts the optimality of $\overline{x}$. □

We note that degeneracy is caused by redundancy of equality constraints that define a vertex under consideration. When some of the equality constraints are redundant, that vertex is the solution of at least two different sets of equality constraints and may be associated with several bases. When a feasible basis B is degenerate, three things may happen. First, a strictly positive solution t to determine a new basic feasible solution in the proof of the preceding theorem does not necessarily exist. In such a situation one must look for another basis that defines the same feasible solution as B does and search for t with this new basis. Second, even if a new feasible solution can be found, it is not necessary that the value of the objective function increases when moving to the new solution. In principle, when a feasible vertex is not optimal, there always exists a basis associated with it, which allows to find a new feasible solution where the value of the objective function is strictly bigger than its value at the current vertex. Third, the current vertex may be optimal even if the reduced cost vector is not negative. In other words, the second conclusion of Theorem 3.1.4 is not true if the basis under consideration is degenerate.

Example 3.1.5 Consider the following linear programming problem

$$
\begin{array}{llll}
\text{maximize} & x_1 - 3x_2 & & \\
\text{subject to} & -x_1 + x_2 + x_3 & = 0 \\
& x_1 - 2x_2 & + x_4 = 0 \\
& x_1, x_2, x_3, x_4 \geq 0.
\end{array}
$$

A tangible basis corresponding to basic variables x_3 and x_4 is given by

$$
B_{3,4} = \begin{pmatrix} 1 & 0 \\ 0 & 1 \end{pmatrix}.
$$

Its associated solution is the vector $\overline{x} = (0, 0, 0, 0)^T$, which is feasible and degenerate. The reduced cost vector at this basis is given by

$$
\overline{c}_N = \begin{pmatrix} 1 \\ -3 \end{pmatrix} - \left[\begin{pmatrix} 1 & 0 \\ 0 & 1 \end{pmatrix} \begin{pmatrix} -1 & 1 \\ 1 & -2 \end{pmatrix} \right]^T \begin{pmatrix} 0 \\ 0 \end{pmatrix} = \begin{pmatrix} 1 \\ -3 \end{pmatrix}.
$$

It has a strictly positive component. However the basic solution $\bar{x}$ obtained above is optimal. Indeed, let us examine another basis with which the solution $\bar{x}$ is associated, namely the basis corresponding to the basic variables x_1 and x_3:

$$B_{1,3} = \begin{pmatrix} -1 & 1 \\ 1 & 0 \end{pmatrix}.$$

Of course, like the preceding one, this basis is degenerate. Its reduced cost vector is computed by

$$\bar{c}_N = \begin{pmatrix} -3 \\ 0 \end{pmatrix} - \left[\begin{pmatrix} 0 & 1 \\ 1 & 1 \end{pmatrix} \begin{pmatrix} 1 & 0 \\ -2 & 1 \end{pmatrix} \right]^T \begin{pmatrix} 1 \\ 0 \end{pmatrix} = \begin{pmatrix} -1 \\ 0 \end{pmatrix}.$$

It is a negative vector. In view of Theorem 3.1.4 (i), the solution $\bar{x}$ is optimal.

Example 3.1.6 Consider the following linear programming problem

$$
\begin{aligned}
\text{maximize } & x_1 + x_2 + x_3 \\
\text{subject to } & x_1 + x_2 \qquad\quad +x_4 \qquad\qquad = 8 \\
& \quad - x_2 + x_3 \qquad\qquad + x_5 = 0 \\
& \qquad\qquad\qquad\qquad x_1, \cdots, x_5 \geq 0.
\end{aligned}
$$

A visible basis corresponding to basic variables x_4 and x_5 is given by

$$B_{4,5} = \begin{pmatrix} 1 & 0 \\ 0 & 1 \end{pmatrix}.$$

Its associated solution is the vector $\bar{x} = (0, 0, 0, 8, 0)^T$, which is feasible and degenerate. The reduced cost vector is given by

$$\bar{c}_N = \begin{pmatrix} 1 \\ 1 \\ 1 \end{pmatrix} - \left[\begin{pmatrix} 1 & 0 \\ 0 & 1 \end{pmatrix} \begin{pmatrix} 1 & 1 & 0 \\ 0 & -1 & 1 \end{pmatrix} \right]^T \begin{pmatrix} 0 \\ 0 \end{pmatrix} = \begin{pmatrix} 1 \\ 1 \\ 1 \end{pmatrix}.$$

At this stage Theorem 3.1.4 is not applicable as the basic solution is degenerate. Let us try to find a new feasible solution with a bigger cost value. To this end, we observe that the reduced cost vector has three strictly positive components, and so any of the coordinate unit vectors e_N^j, $j = 1, 2, 3$ is a suitable choice to determine a new feasible solution $\hat{x}$ as described in the proof of Theorem 3.1.4. We set for instance

$$\hat{x}_N = t \begin{pmatrix} 1 \\ 0 \\ 0 \end{pmatrix}$$

$$\hat{x}_B = \begin{pmatrix} 8 \\ 0 \end{pmatrix} - t \begin{pmatrix} 1 & 0 \\ 0 & 1 \end{pmatrix} \begin{pmatrix} 1 & 1 & 0 \\ 0 & -1 & 1 \end{pmatrix} \begin{pmatrix} 1 \\ 0 \\ 0 \end{pmatrix} = \begin{pmatrix} 8-t \\ 0 \end{pmatrix}.$$

Since $\hat{x}$ is positive, the largest t for which $\hat{x}_N$ and $\hat{x}_B$ are positive is the value $t = 8$. We obtain then a new feasible solution $\hat{x} = (8, 0, 0, 0, 0)^T$. This solution is degenerate. A basis associated to it is $B_{1,5}$ which is identical with $B_{4,5}$. The reduced cost vector at $\hat{x}$ using the basis $B_{1,5}$ is given by

$$\bar{c}_N = c_N - (B_{1,5}^{-1} N)^T c_B$$

$$= \begin{pmatrix} 1 \\ 1 \\ 0 \end{pmatrix} - \left[\begin{pmatrix} 1 & 0 \\ 0 & 1 \end{pmatrix} \begin{pmatrix} 1 & 0 & 1 \\ -1 & 1 & 0 \end{pmatrix} \right]^T \begin{pmatrix} 1 \\ 0 \end{pmatrix}$$

$$= \begin{pmatrix} 0 \\ 1 \\ -1 \end{pmatrix}.$$

As before, the solution $\hat{x}$ being degenerate, Theorem 3.1.4 is not applicable. We try again to find a better feasible solution. As the second component of the reduced cost vector is strictly positive, we choose a solution y by help of the vector $e_N^3 = (0, 1, 0)^T$ (basic components being y_1 and y_5):

$$y_N = t \begin{pmatrix} 0 \\ 1 \\ 0 \end{pmatrix}$$

$$y_B = \begin{pmatrix} 8 \\ 0 \end{pmatrix} - t \begin{pmatrix} 1 & 0 \\ 0 & 1 \end{pmatrix} \begin{pmatrix} 1 & 0 & 1 \\ -1 & 1 & 0 \end{pmatrix} \begin{pmatrix} 0 \\ 1 \\ 0 \end{pmatrix} = \begin{pmatrix} 8 \\ -t \end{pmatrix}.$$

The feasibility of y requires $y_N \geq 0$ and $y_B \geq 0$, which enforce $t = 0$. Thus, with the basis $B_{1,5}$ we move nowhere and remain at the same solution $\hat{x}$. One notes nevertheless two more bases associated with the solution $\hat{x}$. They are given below

$$B_{1,2} = \begin{pmatrix} 1 & 1 \\ 0 & -1 \end{pmatrix} \text{ and } B_{1,3} = \begin{pmatrix} 1 & 0 \\ 0 & 1 \end{pmatrix},$$

which correspond to the pairs of basic variables x_1, x_2 and x_1, x_3 respectively. The reduced cost vectors at $\hat{x}$ related to these bases are then

$$\hat{c}_{2,4,5} = \begin{pmatrix} 1 \\ -1 \\ -1 \end{pmatrix} \text{ and } \hat{c}_{3,4,5} = \begin{pmatrix} 1 \\ -1 \\ 0 \end{pmatrix}.$$

Both of these reduced cost vectors suggest to find a new feasible solution z that may increase the cost value, with help of the coordinate vector $e_N^1 = (1, 0, 0)^T$. Thus, by picking for instance the basis $B_{1,3}$ we obtain a system that determines z as follows

$$z_N = t \begin{pmatrix} 1 \\ 0 \\ 0 \end{pmatrix}$$

$$z_B = \begin{pmatrix} 8 \\ 0 \end{pmatrix} - t \begin{pmatrix} 1 & 0 \\ 0 & 1 \end{pmatrix} \begin{pmatrix} 1 & 1 & 0 \\ -1 & 0 & 1 \end{pmatrix} \begin{pmatrix} 1 \\ 0 \\ 0 \end{pmatrix} = \begin{pmatrix} 8 - t \\ t \end{pmatrix}.$$

The biggest t that makes z feasible is the value $t = 8$. The new feasible solution is then $z = (0, 8, 8, 0, 0)^T$. Its associated basis is

$$B_{2,3} = \begin{pmatrix} 1 & 0 \\ -1 & 1 \end{pmatrix}.$$

It is feasible and non-degenerate. A direct calculation gives the negative reduced cost vector $\hat{c}_N = (-1, -2, -1)^T$. By Theorem 3.1.4 the basis $B_{2,3}$ is optimal.

The following corollary is useful in computing optimal solutions.

Corollary 3.1.7 *Let B be a feasible non-degenerate basis and let $\overline{x}$ be the associated basic feasible solution. If there is a non-basic variable x_s for which the sth component $\overline{c}_s$ is strictly positive, then*

(i) *either the variable x_s can take any value bigger than $\overline{x}_s$ without getting out of the feasible region X, in which case the optimal value of (LP) is unbounded;*
(ii) *or another feasible basis $\overline{B}$ can be obtained to which the associated feasible basic solution $\hat{x}$ satisfies*

$$\langle c, \hat{x} \rangle > \langle c, \overline{x} \rangle.$$

Proof Under the hypothesis of the corollary the basic solution $\overline{x}$ is not optimal by the preceding theorem. Our aim is to find another feasible solution that produces a bigger value of the objective function. To this purpose choose $\hat{x}$ with

$$\hat{x}_N = t e^s$$
$$\hat{x}_B = \overline{b} - t \overline{a}_s$$

where $\bar{b}$ denotes the vector $B^{-1}b$ which is the same as $\bar{x}_B$ and $\bar{a}_s = B^{-1}Ne^s$. If the vector $\bar{a}_s$ is negative, then $\hat{x}_B$ is positive and hence $\hat{x}$ is feasible for every positive value of t. Moreover,

$$\langle c, \hat{x} \rangle = \langle c_B, \bar{x}_B \rangle + t\bar{c}_s$$

which diverges to ∞ as t tends to ∞. The optimal value of (LP) is then unbounded.

If the vector $\bar{a}_s$ is not negative, say $\bar{a}_{is} > 0$ for some index i, then $\hat{x}$ cannot be feasible (positive) when t is large. A necessary and sufficient condition for that vector to be feasible is that t be smaller than the value

$$\hat{t} := \min\left\{\frac{\bar{b}_i}{\bar{a}_{is}} : i \in \{1, \cdots, m\}, \bar{a}_{is} > 0\right\}.$$

Let r be a basic index for which the above minimum is reached. Then the solution

$$\hat{x} = \begin{pmatrix} \bar{b} - \hat{t}\bar{a}_s \\ \hat{t}e^s \end{pmatrix}$$

is feasible. The rth component $\bar{x}_r > 0$ becomes $\hat{x}_r = 0$, while the sth component $\bar{x}_s = 0$ becomes $\hat{x}_s > 0$. Denote by $\hat{B}$ the matrix deduced from B by using the column a_s instead of a_r, so that they differ from each other by one column only. We show that this new matrix is a basis. Remember that $a_s = B\bar{a}_s$ which means that a_s is a vector in the column space of the matrix B. The coefficient corresponding to the vector a_r in that linear combination is $\bar{a}_{rs} > 0$. Consequently, the vector a_r can be expressed as a linear combination of the remaining column vectors of B and the vector a_s. Since the columns of B are linearly independent, we deduce the same property for the columns of the matrix $\hat{B}$. Thus $\hat{B}$ is a basis. It is clear that $\hat{x}$ is the basic feasible solution associated with this basis. Moreover, the value of the objective function at this solution is given by $\langle c, \hat{x} \rangle = \langle c, \bar{x} \rangle + \hat{t}\bar{c}_s$ that is strictly bigger than the value $\langle c, \bar{x} \rangle$. The proof is complete.   □

Example 3.1.8 Consider the following linear programming problem

$$\begin{aligned} \text{maximize} \quad & x_1 - x_2 + x_3 \\ \text{subject to} \quad & x_1 + x_2 + x_3 = 1 \\ & 3x_1 + 2x_2 \quad\quad = 1 \\ & x_1, x_2, x_3 \geq 0. \end{aligned}$$

There are three bases

$$B_{1,2} = \begin{pmatrix} 1 & 1 \\ 3 & 2 \end{pmatrix}, \ B_{1,3} = \begin{pmatrix} 1 & 1 \\ 3 & 0 \end{pmatrix} \ \text{and} \ B_{2,3} = \begin{pmatrix} 1 & 1 \\ 2 & 0 \end{pmatrix}.$$

The first basis is not feasible because the associated solution $(-1, 2, 0)^T$ is not feasible, having one negative component. The second and the third bases are feasible and non-degenerate. Their associated basic solutions are respectively $u = (1/3, 0, 2/3)^T$ and $v = (0, 1/2, 1/2)^T$. For the basis $B_{2,3}$, the non-basic variable is x_1, the basic component cost vector $c_{2,3} = (-1, 1)^T$ and the non-basic part of the constraint matrix is $N_1 = (1, 3)^T$. By definition, the reduced cost one-dimensional vector is computed by

$$\bar{c}_1 = 1 - (-1, 1) \begin{pmatrix} 0 & 1/2 \\ 1 & -1/2 \end{pmatrix} \begin{pmatrix} 1 \\ 3 \end{pmatrix} = 3.$$

In view of Theorem 3.1.4 the vertex v is not optimal. Let us follow the method of Corollary 3.1.7 to get a better solution. Remembering that x_1 is the non-basic variable with the corresponding reduced cost $\bar{c}_1 = 3 > 0$ we compute $\bar{b}$ and $\bar{a}_1$ by

$$\bar{b} = (B_3)^{-1}b = \begin{pmatrix} 0 & 1/2 \\ 1 & -1/2 \end{pmatrix} \begin{pmatrix} 1 \\ 1 \end{pmatrix} = \begin{pmatrix} 1/2 \\ 1/2 \end{pmatrix}$$

and

$$\bar{a}_1 = (B_3)^{-1}a_1 = \begin{pmatrix} 0 & 1/2 \\ 1 & -1/2 \end{pmatrix} \begin{pmatrix} 1 \\ 3 \end{pmatrix} = \begin{pmatrix} 3/2 \\ -1/2 \end{pmatrix}.$$

The positive number $\hat{t}$ in the proof of Corollary 3.1.7 expressing the length to move from v to a new vertex is $\hat{t} = \bar{b}_1/\bar{a}_{11} = (1/2)/(3/2) = 1/3$. Hence the new feasible solution $\hat{x}$ is given by

$$\hat{x} = \begin{pmatrix} \hat{t}e_1 \\ \bar{b} - \hat{t}\bar{a}_1 \end{pmatrix} = \begin{pmatrix} 1/3 \\ 0 \\ 2/3 \end{pmatrix}.$$

This is exactly the feasible basic solution u. For this solution the non-basic variable is x_2 and the corresponding reduced cost $\bar{c}_2$ is given by

$$\bar{c}_2 = -1 - (1, 1) \begin{pmatrix} 0 & 1/3 \\ 1 & -1/3 \end{pmatrix} \begin{pmatrix} 1 \\ 2 \end{pmatrix} = -2 < 0.$$

By Theorem 3.1.4 the solution u is optimal and the optimal value of the problem is equal to $\langle c, v \rangle = 1$.

3.2 Dual Problems

Associated with the linear problem (LP) we define a new linear problem, denoted (LD) and called the *dual* of (LP). We display both (LP) and (LD) below

$$
\begin{array}{ll}
\text{maximize} \quad \langle c, x \rangle & \qquad \text{minimize} \quad \langle b, y \rangle \\
\text{subject to } Ax = b & \qquad \text{subject to } A^T y \geq c. \\
\qquad x \geq 0,
\end{array}
$$

In this dual formulation the problem (LP) is called the *primal* problem. Using the method of converting linear inequalities to linear equalities, one may obtain from the pair (LP) and (LD) above the dual of a linear problem given in canonical form. In fact, suppose we are given the problem

$$
\begin{array}{l}
\text{maximize} \quad \langle c, x \rangle \\
\text{subject to } Ax \leq b.
\end{array}
$$

It is equivalent to the following problem

$$
\text{maximize } (c, -c, 0) \begin{pmatrix} x^+ \\ x^- \\ z \end{pmatrix}
$$

$$
\text{subject to } (A, -A, I) \begin{pmatrix} x^+ \\ x^- \\ z \end{pmatrix} = b
$$

$$
x^+, x^-, z \geq 0,
$$

in which I is the identity $m \times m$ matrix and z is an m-vector variable. It follows from the scheme (LP)-(LD) that the dual of the latter problem is given by

$$
\text{minimize} \qquad \langle b, y \rangle
$$

$$
\text{subject to } \begin{pmatrix} A^T \\ -A^T \\ I \end{pmatrix} y \geq \begin{pmatrix} c^T \\ -c^T \\ 0 \end{pmatrix},
$$

which is equivalent to

$$
\begin{array}{l}
\text{minimize} \quad \langle b, y \rangle \\
\text{subject to } A^T y = c \\
\qquad\qquad y \geq 0.
\end{array}
$$

On the other hand, by putting the minimization of the function $\langle b, y \rangle$ as minus of the maximization of the function $\langle -b, y \rangle$ and applying the primal-dual scheme above we come to conclusion that the dual of (LD) is exactly (LP). In other words the dual of the dual is the primal. In this sense the pair (LP) and (LD) is known as a *symmetric form* of duality. This symmetry can also be seen when we write down a primal and dual pair in a general mixed form in which both equality and inequality constraints as well as positive, negative variables and free (unrestricted) variables are present:

$$
\begin{array}{ll}
\text{maximize } \sum_{i=1}^{3} \langle c^i, x^i \rangle & \text{minimize } \sum_{i=1}^{3} \langle b^i, y^i \rangle \\
\text{subject to} \quad A^1 x \le b^1 & \text{subject to} \qquad y^1 \ge 0 \\
\qquad\qquad A^2 x \ge b^2 & \qquad\qquad\qquad y^2 \le 0 \\
\qquad\qquad A^3 x = b^3 & \qquad\qquad\qquad y^3 \text{ free} \\
\qquad\qquad x^1 \ge 0 & \qquad\qquad\qquad A_1^T y \ge c^1 \\
\qquad\qquad x^2 \le 0 & \qquad\qquad\qquad A_2^T y \le c^2 \\
\qquad\qquad x^3 \text{ free} & \qquad\qquad\qquad A_3^T y = c^3
\end{array}
$$

in which the variables x and y are decomposed into three parts: positive part, negative part and unrestricted part, and the dimensions of the vectors c^i, b^i and the matrices A^i, $i = 1, 2, 3$ are in concordance. The matrices A_1, A_2 and A_3 are composed

of the columns of the matrix $\begin{pmatrix} A^1 \\ A^2 \\ A^3 \end{pmatrix}$ corresponding to the variables x^1, x^2 and x^3

respectively. The aforementioned scheme of duality is clearly seen in the following example.

Example 3.2.1 Consider the following problem of three variables:

$$
\begin{array}{ll}
\text{maximize} & c_1 x_1 + c_2 x_2 + c_3 x_3 \\
\text{subject to} & a_{11} x_1 + a_{12} x_2 + a_{13} x_3 \le b_1 \\
& a_{21} x_1 + a_{22} x_2 + a_{23} x_3 \ge b_2 \\
& a_{31} x_1 + a_{32} x_2 + a_{33} x_3 = b_3 \\
& x_1 \ge 0 \\
& x_2 \le 0 \\
& x_3 \text{ free.}
\end{array}
$$

By using the duality scheme we obtain the dual problem

$$
\begin{array}{ll}
\text{minimize} & b_1 y_1 + b_2 y_2 + b_3 y_3 \\
\text{subject to} & y_1 \ge 0 \\
& y_2 \le 0 \\
& y_3 \text{ free} \\
& a_{11} y_1 + a_{21} y_2 + a_{31} y_3 \ge c_1 \\
& a_{12} y_1 + a_{22} y_2 + a_{32} y_3 \le c_2 \\
& a_{13} y_1 + a_{23} y_2 + a_{33} Y_3 = c_3.
\end{array}
$$

Lagrangian functions

There are several ways to obtain the dual problem (LD) from the primal problem. Here is a method by Lagrangian functions. Let us define a function of two variables x and y in the product space $\mathbb{R}^n \times \mathbb{R}^m$ with values in the extended real line $\mathbb{R} \cup \{\pm\infty\}$:

$$L(x, y) = \begin{cases} \langle c, x \rangle + \langle y, b - Ax \rangle & \text{if} \quad x \geqq 0 \\ -\infty & \text{else.} \end{cases}$$

The inner product $\langle y, b - Ax \rangle$ can be interpreted as a measure of violation of the equality constraint $Ax = b$, which is added to the objective function as a penalty. The function $L(x, y)$ is called the *Lagrangian function* of (LP). We see now that both problems (LP) and (LD) are obtained from this function.

Proposition 3.2.2 *For every fixed vector x in $\mathbb{R}^n$ and every fixed vector y in $\mathbb{R}^m$ one has*

$$\inf_{y' \in \mathbb{R}^m} L(x, y') = \begin{cases} \langle c, x \rangle & \text{if} \quad Ax = b, x \geqq 0 \\ -\infty & \text{else} \end{cases}$$

$$\sup_{x' \in \mathbb{R}^n} L(x', y) = \begin{cases} \langle b, y \rangle & \text{if} \quad A^T y \geqq c \\ +\infty & \text{else.} \end{cases}$$

Proof To compute the first formula of the proposition we distinguish three possible cases of x: (a) $x \ngeqq 0$; (b) $x \geqq 0$ with $Ax \neq b$, and (c) $x \geqq 0$ and $Ax = b$. In the case (a), the function $L(x, y')$ takes the value $-\infty$ for any y'. In the case (b) we put $y_t = t(Ax - b)$ for $t > 0$. Then

$$\lim_{t \to \infty} L(x, y_t) = \lim_{t \to \infty} \langle c, x \rangle - t \|Ax - b\|^2 = -\infty.$$

In the last case, it is obvious that $L(x, y') = \langle c, x \rangle$ for every y'. By this the first formula is proven.

As to the second formula we have

$$\sup_{x' \in \mathbb{R}^n} L(x', y) = \sup_{x' \in \mathbb{R}^n_+} L(x', y)$$

$$= \sup_{x' \in \mathbb{R}^n_+} \left(\langle c, x' \rangle + \langle y, b - Ax' \rangle \right)$$

$$= \sup_{x' \in \mathbb{R}^n_+} \left(\langle c - A^T y, x' \rangle + \langle b, y \rangle \right).$$

Given y in $\mathbb{R}^m$, if $c - A^T y \leqq 0$, then the maximum on the right hand side is attained at $x' = 0$ and equal to $\langle b, y \rangle$. If $c - A^T y \nleqq 0$, there is a strictly positive component, say

$c_i - (A^T x)_i > 0$. Then by setting $x_t = te^i$ for $t > 0$, where e^i is the ith coordinate unit vector of $\mathbb{R}^n$ we obtain

$$\langle c - A^T y, x_t \rangle + \langle b, y \rangle = t(c_i - (A^T y)_i) + \langle b, y \rangle,$$

which tends to ∞ as t tends to ∞. This completes the proof.   □

As a consequence of Proposition 3.2.2 the primal problem is written in form

$$\text{maximize} \left\{ \inf_{y \in \mathbb{R}^m} L(x, y) \right\}$$
$$\text{subject to} \quad x \in \mathbb{R}^n$$

and the dual problem is written in form

$$\text{minimize} \left\{ \sup_{x \in \mathbb{R}^n} L(x, y) \right\}$$
$$\text{subject to} \quad y \in \mathbb{R}^m.$$

The utility of dual problems will be clear in the sequel.

Duality relations

Intimate and mutual ties between the primal and dual problems stimulate our insight of existence and sensibility of optimal solutions as well as solving methods of a linear problem and economic interpretation of the models where that came from. The theorem below describes a complete relationship between (LP) and (LD), their values and variables.

Theorem 3.2.3 *For the couple of the primal and dual problems (LP) and (LD) the following statements hold.*

(i) *(Weak duality) For each feasible solution x of (LP) and each feasible solution y of (LD),*
$$\langle c, x \rangle \leqq \langle b, y \rangle.$$

In particular, if inequality becomes equality, then x and y are optimal solutions. Moreover, two feasible solutions x and y are optimal if and only if the complementary slackness holds:

$$\langle A^T y - c, x \rangle = 0.$$

(ii) *(Strong duality) If the primal and the dual problems have feasible solutions, then they have optimal solutions and the two optimal values are equal.*

(iii) *If either problem has unbounded optimal value, then the other problem has no feasible solution.*

Proof For the weak duality relation we have $b = Ax$ and $A^T y \geq c$. As x is positive, we deduce

$$\langle b, y \rangle = \langle Ax, y \rangle = \langle x, A^T y \rangle \geq \langle c, x \rangle.$$

Assume now equality holds for some feasible solutions x^0 and y^0 of (LP) and (LD). Then for every feasible solutions x and y the weak duality yields respectively

$$\langle c, x^0 \rangle = \langle b, y^0 \rangle \geq \langle c, x \rangle,$$
$$\langle b, y^0 \rangle = \langle c, x^0 \rangle \leq \langle b, y \rangle,$$

which prove that x^0 and y^0 are optimal.

Furthermore, if the complementary slackness holds for feasible solutions x and y, then the weak duality relation becomes equality, and hence they are optimal. Conversely, let x and y be primal and dual optimal solutions. We wish to establish equality in the weak duality relation. Suppose to the contrary that for these optimal solutions, the weak duality relation is strict, which means that the system

$$A^T y \geq c$$
$$\langle b, y \rangle \leq \langle c, x \rangle$$

is not solvable. In view of Corollary 2.2.4 there exist a positive vector x' and a real number $t \geq 0$ such that

$$(A - b)\begin{pmatrix} x' \\ t \end{pmatrix} = 0$$

$$\left\langle \begin{pmatrix} c \\ -\langle c, x \rangle \end{pmatrix}, \begin{pmatrix} x' \\ t \end{pmatrix} \right\rangle = 1.$$

If $t = 0$, then $Ax' = 0$ and $\langle c, x' \rangle = 1$. This means that the objective function is strictly positive on an asymptotic direction of X and shows that the problem has unbounded optimal value. Thus, t is strictly positive and the vector $\frac{1}{t} x'$ is a feasible solution at which the value of the objective function is

$$\langle c, \frac{1}{t} x' \rangle = \langle c, x \rangle + \frac{1}{t} > \langle c, x \rangle.$$

This is a contradiction and hence the two optimal values are equal.

We proceed to (ii). Assume that (LP) and (LD) have feasible solutions. By the weak duality the cost function $\langle c, . \rangle$ of (LP) is bounded above on the feasible set. In view of Theorem 3.1.1, (LP) has an optimal solution. The same argument shows that the dual (LD) possesses an optimal solution too. Let x be an optimal solution of (LP) and y an optimal solution of (LD). By the first part, they satisfy the complementary slackness condition. Taking into account the fact that $Ax = b$ we deduce

$$\langle c, x \rangle = \langle A^T y, x \rangle = \langle Ax, y \rangle = \langle b, y \rangle,$$

and so the two optimal values are equal.

The last statement is an immediate consequence of the weak duality relation because if the primal has a feasible solution x, then the dual is bounded below by $\langle c, y \rangle$. Likewise, if the dual has a feasible solution, then the primal is bounded above. □

So far the preceding theorem describes almost all possible situations of the primal and dual pair, it remains to notice the last one when both of them are infeasible. Here is an example of such a situation.

Example 3.2.4 Consider the problem

$$
\begin{aligned}
\text{maximize} \quad & x_1 + x_2 \\
\text{subject to} \quad & x_1 - x_2 \leqq 1 \\
& -x_1 + x_2 \leqq -2 \\
& x_1, x_2 \geqq 0,
\end{aligned}
$$

which is infeasible. Its dual takes the form

$$
\begin{aligned}
\text{minimize} \quad & y_1 - 2y_2 \\
\text{subject to} \quad & y_1 - y_2 \geqq 1 \\
& -y_1 + y_2 \geqq 1 \\
& y_1, \ y_2 \geqq 0.
\end{aligned}
$$

It is evident that the dual problem has no feasible solution.

The next corollary shows how to obtain an optimal dual vertex from an optimal primal basis.

Corollary 3.2.5 *If x is an optimal basic solution of (LP) corresponding to a basis B and the reduced cost vector $\bar{c}_N$ is negative, then the vector $y = (B^{-1})^T c_B$ is an optimal basic solution of (LD).*

Proof We show that $y = (B^{-1})^T c_B$ is feasible. In fact,

$$A^T y = A^T (B^{-1})^T c_B = (B \ N)^T (B^{-1})^T c_B = \begin{pmatrix} c_B \\ (B^{-1}N)^T c_B \end{pmatrix}.$$

By hypothesis $x = \begin{pmatrix} x_B \\ 0 \end{pmatrix}$ and the vector $(B^{-1}N)^T c_B - c_N$ (the opposite of the reduced cost vector) is positive. Hence $A^T y \geqq c$ and y is feasible. To prove that y is optimal, we calculate

$$\langle A^T y, x \rangle = \langle c_B, x_B \rangle + \langle (B^{-1}N)^T c_B, x_N \rangle = \langle c_B, x_B \rangle = \langle c, x \rangle.$$

By Theorem 3.2.3, y is optimal.   □

The hypothesis of the preceding corollary is satisfied if the basis is optimal and non-degenerate. When it is degenerate, the reduced cost vector is not necessarily negative, and so there is no guarantee that the vector y defined by the formula of the corollary is dual feasible.

Example 3.2.6 Consider the problem

$$\begin{array}{ll}
\text{maximize } & x_1 + x_2 \\
\text{subject to } & x_1 + x_2 + x_3 \quad\quad = 1 \\
& x_1 + 2x_2 \quad\quad + x_4 = 2 \\
& x_1, x_2, x_3, x_4 \geq 0.
\end{array}$$

The dual is written as follows

$$\begin{array}{ll}
\text{minimize } & y_1 + 2y_2 \\
\text{subject to } & y_1 + y_2 \geq 1 \\
& y_1 + 2y_2 \geq 1 \\
& y_1, \; y_2 \geq 0.
\end{array}$$

We choose the basis $B_{2,3} = \begin{pmatrix} 1 & 1 \\ 2 & 0 \end{pmatrix}$ that corresponds to the basic variables x_2 and x_3. The associated basic solution $\bar{x} = (0, 1, 0, 0)^T$ is degenerate. The reduced cost vector at this basis is

$$\bar{c}_N = c_N - \left[B_{2,3}^{-1} N \right]^T c_B$$

$$= \begin{pmatrix} 1 \\ 0 \end{pmatrix} - \left[\begin{pmatrix} 0 & 1/2 \\ 1 & -1/2 \end{pmatrix} \begin{pmatrix} 1 & 0 \\ 1 & 1 \end{pmatrix} \right]^T \begin{pmatrix} 1 \\ 0 \end{pmatrix}$$

$$= \begin{pmatrix} 1/2 \\ -1/2 \end{pmatrix}.$$

This vector is not negative, and the dual vector

$$y = \left[B_{2,3}^{-1} \right]^T c_B = \begin{pmatrix} 0 & 1/2 \\ 1 & -1/2 \end{pmatrix}^T \begin{pmatrix} 1 \\ 0 \end{pmatrix} = \begin{pmatrix} 0 \\ 1/2 \end{pmatrix}$$

is not feasible for the dual problem.

On the other hand, if we choose the basis $B_{2,4} = \begin{pmatrix} 1 & 0 \\ 2 & 1 \end{pmatrix}$ that corresponds to the basic variables x_2 and x_4, then its associated basic solution is exactly the same as that of the basis $B_{2,3}$, that is $\bar{x} = (0, 1, 0, 0)^T$. The reduced cost vector at this basis is different, however, and given below

$$\bar{c}_N = c_N - \left[B_{2,4}^{-1}N\right]^T c_B$$

$$= \begin{pmatrix} 1 \\ 0 \end{pmatrix} - \left[\begin{pmatrix} 1 & 0 \\ -2 & 1 \end{pmatrix}\begin{pmatrix} 1 & 1 \\ 1 & 0 \end{pmatrix}\right]^T \begin{pmatrix} 1 \\ 0 \end{pmatrix}$$

$$= \begin{pmatrix} 0 \\ -1 \end{pmatrix}.$$

This time, the reduced cost vector is negative. The dual vector defined by the formula of the corollary

$$y = \left[B_{2,4}^{-1}\right]^T c_B = \begin{pmatrix} 1 \\ 0 \end{pmatrix}$$

is feasible for the dual problem.

Sensitivity

The duality result of Corollary 3.2.5 yields a nice estimate for the change of the optimal value when the constraint vector b undergoes small perturbations.

Corollary 3.2.7 *Assume that* $\bar{x} = \begin{pmatrix} \bar{x}_B \\ 0 \end{pmatrix}$ *is an optimal non-degenerate basic solution of (LP) and B is the corresponding basis. Then for a small perturbation* $b + \Delta b$ *of the vector b, the vector*

$$\hat{x} = \begin{pmatrix} \bar{x}_B + B^{-1}\Delta b \\ 0 \end{pmatrix}$$

is an optimal basic solution of the perturbed problem (LP$_{\Delta b}$):

$$\begin{array}{ll} \text{maximize} & \langle c, x \rangle \\ \text{subject to } Ax = b + \Delta b \\ & x \geq 0 \end{array}$$

and the change in the optimal value is given by

$$\langle c, \hat{x} \rangle - \langle c, \bar{x} \rangle = (c_B)^T B^{-1}\Delta b.$$

In particular, as a function of b, the optimal value of (LP) is differentiable at b and its derivative is the optimal dual solution $\bar{y} = (B^{-1})^T c_B$.

Proof Let Δb be a small increment of b. Consider a perturbed problem $(LP_{\Delta b})$ described in the corollary. We show that $\hat{x}$ is feasible. Indeed,

$$A\hat{x} = (B\ N) \begin{pmatrix} \overline{x}_B + B^{-1}\Delta b \\ 0 \end{pmatrix}$$
$$= B\overline{x}_B + \Delta b$$
$$= b + \Delta b.$$

Since $\overline{x}_B$ is strictly positive, when Δb is sufficiently small, the vector $\overline{x}_B + B^{-1}\Delta b$ is positive. Hence $\hat{x}$ is feasible. We deduce also that B remains a feasible basis of the perturbed problem. According to Corollary 3.2.5 the vector $y = (B^{-1})^T c_B$ is an optimal solution of the dual problem of (LP). It is also a feasible solution of the dual of the perturbed problem as they share the same feasible set. Moreover,

$$\langle b + \Delta b, y \rangle = \langle b + \Delta b, (B^{-1})^T c_B \rangle$$
$$= \langle B^{-1}b + B^{-1}\Delta b, c_B \rangle$$
$$= \langle \hat{x}_B, c_B \rangle = \langle c, \hat{x} \rangle.$$

By duality, $\hat{x}$ is an optimal solution of the perturbed problem and y is an optimal solution of its dual. We deduce also the change of the optimal value:

$$\langle c, \hat{x} \rangle - \langle c, \overline{x} \rangle = \langle c_B, \hat{x}_B \rangle - \langle c_B, \overline{x}_B \rangle$$
$$= \langle b + \Delta b, y \rangle - \langle b, y \rangle$$
$$= \langle \Delta b, (B^{-1})^T c_B \rangle$$
$$= (c_B)^T B^{-1}\Delta b,$$

which yields the requested formula. The last assertion is immediate from the formula of the change of the optimal value.   □

3.3 The Simplex Method

The simplex method is aimed at solving the linear problem (LP). Its strategy is to start with a feasible vertex and search an adjacent vertex that increases the value of the objective function until either a ray on which the objective function is unbounded is identified or an optimal vertex is found.

Description of the method

Let us assume for a moment that we have a feasible basis B_0 in our disposition. Here is the algorithm.

Step 1: Compute the associated feasible vertex x^0 whose components are $x_B^0 = (B_0)^{-1}b$ and $x_N^0 = 0$. Iteration $k = 0$.
Step 2. Set $k := k + 1$. Let B_k be the current feasible basis and its associated basic vertex x^k with two components x_B^k and x_N^k. Compute

$$\bar{b} = B_k^{-1}b$$
$$\bar{c}_N = c_N - [B_k^{-1}N]^T c_B.$$

Step 3. If $\bar{c}_N \leq 0$, then stop. The current vertex x^k is optimal.
Otherwise go to the next step.
Step 4. Let s be an index for which $\bar{c}_s > 0$. Pick the column a_s of the matrix A and compute

$$\bar{a}_s = B_k^{-1}a_s.$$

If this vector is negative, then stop. The problem is unbounded.
Otherwise find an index ℓ such that

$$\hat{x}_s := \frac{\bar{b}_\ell}{\bar{a}_{\ell s}} = \min\left\{\frac{\bar{b}_i}{\bar{a}_{is}} : \bar{a}_{is} > 0\right\}.$$

Step 5. Form a new feasible basis B_{k+1} from B_k by deleting the column a_ℓ and entering the column a_s instead. The new associated vertex x^{k+1} is obtained from x^k by setting the variable $x_s = \hat{x}_s > 0$ and the variable $x_\ell = 0$.
Step 6. Compute the inverse matrix B_{k+1}^{-1} of the new basis B_{k+1} and return to Step 2.

The element $\bar{a}_{\ell s}$ obtained above from the matrix A is called a *pivot*, the column $a_s = (\bar{a}_{1s}, \cdots, \bar{a}_{ms})^T$ and the row $\bar{a}^\ell = (\bar{a}_{\ell 1}, \cdots, \bar{a}_{\ell n})$ are called the *pivotal column* and the *pivotal row* of the algorithm.

Theorem 3.3.1 *If all feasible bases of the matrix A are non-degenerate, then the simplex algorithm terminates at a finite number of iterations.*

Proof Note that the number of the vertices of the polyhedron X is finite, say equal to p. Moreover, in the simplex method, the objective function increases its value each time when passes from a vertex to a vertex. Thus, after a finite number of iterations (at most p), one obtains a vertex which is either optimal, or the objective function increases along a ray starting from it.   □

Finding a feasible basis

We are considering the constraints of the problem (LP)

$$Ax = b \text{ and } x \geq 0.$$

By multiplying both side of an equality by (-1) one may assume that the right hand side vector b has non-negative components only. Moreover, as we saw in the first chapter when the constraints have a solution, one may remove some of them so that the constraints remain without redundant equations and producing the same solution set. From now on, we suppose the two conditions on the constraints: (1) the vector b is positive, and (2) no equality is redundant.

Since the choice of a feasible basis for starting the simplex algorithm is not evident, one introduces a vector of artificial variables $y = (y_1, \cdots, y_m)^T$ and consider the linear problem

$$\text{minimize} \qquad y_1 + \cdots + y_m \qquad\qquad (3.5)$$
$$\text{subject to } Ax + y = b \text{ and } x \geq 0, y \geq 0.$$

Proposition 3.3.2 *The problem (LP) has a feasible solution if and only if the problem (3.5) has a minimum value equal to zero with $y = 0$.*

Proof Assume that x is a feasible solution of (LP). Then (x, y) with $y = 0$ is a feasible solution of (3.5), and the minimum value is zero. Conversely, if the optimal value of (3.5) is zero, then at an optimal solution (x, y) one has $y = 0$, which implies that x is a feasible solution of (LP). □

Notice that the artificial problem (3.5) has optimal solutions. This is because the feasible set is nonempty, it contains, for instance, the solution with $x = 0$ and $y = b$ which is a basic feasible solution associated with the feasible basis $B = I$ the identity matrix, and the objective function is bounded from below. Suppose that an optimal vertex $(\overline{x}, 0)$ is found for (3.5) in which no y_i is basic variable. Then the corresponding columns of the matrix A are linear independent and form a feasible basis of (LP). If one of y_i is a basic variable, say, for the sake of simple writing, y_1 is the unique basic variable together with $m - 1$ basic variables $x_1, \cdots, x_{m-1}$, then by applying the simplex method one may arrive at the point that either the basic variable y_1 is replaced by a new basic variable x_j with $m \leq j \leq n$, or it is impossible to substitute y_1 by those x_j. The latter case can happen only when the coefficients $y_{1j}, j = 2m, \cdots, m+n$ of the matrix $B^{-1}N$ are all zero. This shows that the rank of the matrix A is $m - 1$, and so the constraints of (LP) are redundant, a contradiction.

The product form of the inverse

By looking at the basic matrices B_k and B_{k+1} we notice that they differ from each other only in one column, that is they are adjacent. This enables us to compute the inverse of B_{k+1} from the inverse of B_k. In fact, denote by D the elementary $m \times m$-matrix, called the *matrix for change of basis*, which is the identity matrix except for the ℓth column equal to the vector

$$\left(-\frac{\bar{a}_{1s}}{\bar{a}_{\ell s}}, \cdots, \frac{1}{\bar{a}_{\ell s}}, \cdots, -\frac{\bar{a}_{ms}}{\bar{a}_{\ell s}} \right)^T .$$

Namely,

$$D = \begin{bmatrix} 1 \cdots & -\bar{a}_{1s}/\bar{a}_{\ell s} & \cdots 0 \\ 0 \cdots & 1/\bar{a}_{\ell s} & \cdots 0 \\ 0 \cdots & -\bar{a}_{ms}/\bar{a}_{\ell s} & \cdots 1 \end{bmatrix}$$

Proposition 3.3.3 *With the matrix D above, one has*

$$B_{k+1}^{-1} = D B_k^{-1}.$$

In particular, if the first basis is the identity matrix, then

$$B_k^{-1} = D_k \cdots D_1$$

where D_i are change matrices.

Proof Let $\beta_1, \cdots, \beta_m$ be the columns of the matrix B_k. Then the columns of B_{k+1} are the same, with its ℓth column substituted by the column a_s of the matrix A. By multiplying B_{k+1} by D we obtain a matrix whose columns are exactly as B_k except for the ℓth one given by

$$\frac{1}{\bar{a}_{\ell s}} (B_k(-\bar{a}_s) + a_s) + \beta_\ell .$$

By definition, $\bar{a}_s = B_k^{-1} a_s$, we deduce that $B_k(-\bar{a}_s) + a_s = 0$ and the ℓth column of the product $B_{k+1} D$ is equal to β_ℓ. Consequently, $B_{k+1} D = B_k$ and the requested formula follows. $\qquad\square$

It would be noticed that each elementary matrix D is uniquely determined by the number ℓ and its ℓth column. Thus, it is sufficient to store $m + 1$ numbers $(\ell, -\bar{a}_{1s}/\bar{a}_{\ell s}, \cdots, 1/\bar{a}_{\ell s}, \cdots, -\bar{a}_{ms}/\bar{a}_{\ell s})$ in order to fully restitute D.

The simplex tableau

In order to solve the problem (LP):

$$\text{maximize} \ \langle c, x \rangle$$
$$\text{subject to} \ \ Ax = b \text{ and } x \geqq 0$$

we assume that b is a positive vector and the matrix A is written in the form $(B \ N)$ where B is a feasible basis. To simplify the writing, the cost vector c is set in row form. The simplex tableau is of the form, denoted T,

$c^T = (c_B^T \ c_N^T)$	0
$A = (B \ \ N)$	b

By pre-multiplying the tableau T by the extended inverse of B,

1	$-c_B^T B^{-1}$
0	B^{-1}

we obtain the tableau T^* as follows

0	$\bar{c}_N^T = c_N^T - c_B^T B^{-1} N$	$-c_B^T B^{-1} b$
I	$\bar{N} = \quad B^{-1} N$	$B^{-1} b$

The tableau T^* contains all information necessary for the simplex algorithm.

- The associated basic solution is found in the right bottom box: $x = \begin{pmatrix} x_B \\ x_N \end{pmatrix}$ with $x_B = B^{-1} b$ and $x_N = 0$.
- The value of the objective function at this basic solution is equal to $\langle c, x \rangle = (c_B)^T B^{-1} b$, the opposite of the value given in the upper right corner.
- The reduced cost $\bar{c}_N$ is given in the upper middle box. If all components of this vector are negative, then the current basic vertex is optimal.
- If some of the components of the reduced cost vector are positive, choose an index, say s, with $\bar{c}_s$ largest. The variable x_s will enter the basis. A variable x_ℓ with the index ℓ satisfying

$$\frac{\bar{b}_\ell}{\bar{a}_{\ell s}} = \min\left\{\frac{\bar{b}_i}{\bar{a}_{is}} : \bar{a}_{is} > 0\right\}$$

will leave the basis.

The simplex tableau of the next iteration is obtained from T^* by pre-multiplying it by the matrix

$$S = \begin{bmatrix} 1 & 0 & \cdots & -\bar{c}_s/\bar{a}_{\ell s} & \cdots & 0 \\ 0 & 1 & \cdots & -\bar{a}_{1s}/\bar{a}_{\ell s} & \cdots & 0 \\ & & \cdots & & & \\ \vdots & \vdots & & 1/\bar{a}_{\ell s} & & \vdots \\ & & & & \cdots & \\ 0 & 0 & \cdots & -\bar{a}_{ms}/\bar{a}_{\ell s} & \cdots & 1 \end{bmatrix}$$

We find again the matrix of change D in the right low box. The pivot of the simplex tableau is the element $\bar{a}_{\ell s}$.

Note that after identifying the pivot $\bar{a}_{\ell s}$, the simplex tableau of the next iteration is obtained from the current one by multiplying the pivot row by the inverse of the pivot and by adding the pivot row multiplied by a number to other rows so that the pivot column becomes a vector whose ℓth component is equal to one and the other components are zero. This is exactly what the pre-multiplication of T^* by S does.

Example 3.3.4 We consider the problem

$$\begin{aligned}
\text{maximize} \quad & x_1 + 2x_2 \\
\text{subject to} \quad & -3x_1 + 2x_2 \leq 2 \\
& -x_1 + 2x_2 \leq 4 \\
& x_1 + x_2 \leq 5 \\
& x_1, \ x_2 \geq 0.
\end{aligned}$$

It is equivalent to the following problem in standard form

$$\begin{aligned}
\text{maximize} \quad & x_1 + 2x_2 \\
\text{subject to} \quad & -3x_1 + 2x_2 + x_3 && = 2 \\
& -x_1 + 2x_2 && + x_4 && = 4 \\
& x_1 + x_2 && + x_5 = 5 \\
& x_1, \cdots, x_5 \geq 0.
\end{aligned}$$

The initial simplex tableau is given as

1	2	0	0	0	0
-3	2	1	0	0	2
-1	2	0	1	0	4
1	1	0	0	1	5

Choose an evident basis $B = I$ the identity matrix corresponding to the basic variables x_3, x_4 and x_5. Since the basic part of the cost vector c_B is null, the reduced cost vector of this basis is

$$\bar{c}_N = c_N - [B^{-1}N]^T c_B = c_N = \begin{pmatrix} 1 \\ 2 \end{pmatrix}.$$

In view of Theorem 3.1.4 this basis is not optimal. To move to a better solution we make a change of basis by introducing the non-basic variable x_2 corresponding to the biggest reduced cost $\bar{c}_2 = 2$ into the basic variables. We have

$$\bar{a}_2 = B^{-1}a_2 = a_2 = \begin{pmatrix} 2 \\ 2 \\ 1 \end{pmatrix}$$

$$\hat{t} = \min\left\{ \frac{\bar{b}_i}{\bar{a}_{i2}} : \bar{a}_{i2} > 0 \right\} = \min\left\{ \frac{2}{2}, \frac{4}{2}, \frac{5}{1} \right\} = 1$$

We see that $\hat{t}$ is reached at $i = 1$ which corresponds to the basic variable x_3. Thus, x_3 leaves the basis and x_2 enters it. The pivot is the element $\bar{a}_{12} = 2$ with the pivotal row $\ell = 1$ and the pivotal column $s = 2$. The matrix of change D and S are given by

$$D_1 = \begin{pmatrix} 1/2 & 0 & 0 \\ -2/2 & 1 & 0 \\ -1/2 & 0 & 1 \end{pmatrix} \text{ and } S_1 = \begin{array}{|c|ccc|} \hline 1 & -1 & 0 & 0 \\ \hline 0 & 1/2 & 0 & 0 \\ 0 & -1 & 1 & 0 \\ 0 & -1/2 & 0 & 1 \\ \hline \end{array}$$

The new tableau T_2 is obtained by the product $S_1 T_1$ and is displayed below

$$\begin{array}{|rrrrr|r|}
\hline
4 & 0 & -1 & 0 & 0 & -2 \\
\hline
-3/2 & 1 & 1/2 & 0 & 0 & 1 \\
2 & 0 & -1 & 1 & 0 & 2 \\
5/2 & 0 & -1/2 & 0 & 1 & 4 \\
\hline
\end{array}$$

The current basic variables are x_2, x_4 and x_5. We read from the simplex tableau that the reduced cost vector $\bar{c}_N = \begin{pmatrix} 4 \\ -1 \end{pmatrix}$ and by Theorem 3.1.4 the current basis is not optimal. The unique non-basic variable with the positive reduced cost is x_1 (the reduced cost $\bar{c}_1 = 4$). We have

$$\bar{a}_1 = B^{-1}a_1 = \begin{pmatrix} -3/2 \\ 2 \\ 5/2 \end{pmatrix}$$

$$\hat{t} = \min\left\{ \frac{\bar{b}_i}{\bar{a}_{i1}} : \bar{a}_{i1} > 0 \right\} = \min\left\{ \frac{2}{2}, \frac{4}{5/2} \right\} = 1.$$

The value $\hat{t}$ is reached at $i = 2$ which corresponds to the basic variable x_4. Thus, x_4 leaves the basis and x_1 enters it. The pivot is the element $\bar{a}_{21} = 2$ with the pivotal row $\ell = 2$ and the pivotal column $s = 1$. The matrix S is given by

$$S_2 = \begin{array}{|cccc|} \hline 1 & 0 & -4/2 & 0 \\ 0 & 1 & 3/4 & 0 \\ 0 & 0 & 1/2 & 0 \\ 0 & 0 & -5/4 & 1 \\ \hline \end{array}$$

The new tableau T_3 is obtained by the product $S_2 T_2$ and is displayed below

$$\begin{array}{|cccccc|} \hline 0 & 0 & 1 & -2 & 0 & -6 \\ 0 & 1 & -1/4 & 3/4 & 0 & 5/2 \\ 1 & 0 & -1/2 & 1/2 & 0 & 1 \\ 0 & 0 & 3/4 & -5/4 & 1 & 3/2 \\ \hline \end{array}$$

The current basic variables are x_1, x_2 and x_5. The reduced cost vector is $\bar{c}_N = \begin{pmatrix} 1 \\ -2 \end{pmatrix}$ and again by Theorem 3.1.4 the current basis is not optimal. The unique non-basic variable with the positive reduced cost is x_3 (the reduced cost $\bar{c}_3 = 1$). We have

$$\bar{a}_3 = B^{-1}a_3 = \begin{pmatrix} -1/4 \\ -1/2 \\ 3/4 \end{pmatrix}$$

$$\hat{t} = \min\left\{ \frac{\bar{b}_i}{\bar{a}_{i3}} : \bar{a}_{i3} > 0 \right\} = \frac{3/2}{3/4} = 2.$$

The value $\hat{t}$ is reached at $i = 3$ which corresponds to the basic variable x_5. Thus, x_5 leaves the basis and x_3 enters it. The pivot is the element $\bar{a}_{33} = 3/4$ with the pivotal row: $\ell = 3$, and the pivotal column: $s = 3$. The matrix S is given by

$$S_3 = \begin{array}{|cccc|} \hline 1 & 0 & 0 & -4/3 \\ 0 & 1 & 0 & 1/3 \\ 0 & 0 & 1 & 2/3 \\ 0 & 0 & 0 & 4/3 \\ \hline \end{array}$$

The new tableau T_4 is obtained by the product $S_3 T_3$ and is displayed below

$$
\begin{array}{|ccccc|c|}
\hline
0 & 0 & 0 & -1/3 & -4/3 & -8 \\
0 & 1 & 0 & 1/3 & 1/3 & 3 \\
1 & 0 & 0 & -1/3 & 2/3 & 2 \\
0 & 0 & 1 & -5/3 & 4/3 & 2 \\
\hline
\end{array}
$$

The current basic variables are x_1, x_2 and x_3. The reduced cost vector $\overline{c}_N = \begin{pmatrix} -1/3 \\ -4/3 \end{pmatrix}$
is negative, hence the current basis is optimal. We obtain immediately $x_1 = 2$, $x_2 = 3$ and $x_3 = 2$. The optimal value is 8 (the opposite of the number at the upper right corner of the tableau). At this iteration the algorithm terminates.

The Two-phase method

When a feasible basis for starting the simplex algorithm is not apparent, the auxiliary problem (3.5) with artificial variables represents Phase I of the simplex method. Once a basic feasible solution is found from Phase I, one applies Phase II to find an optimal solution. In Phase II the artificial variables and its objective function play no role and so they are omitted.

Example 3.3.5 We consider the problem

$$
\begin{aligned}
\text{maximize} \quad & x_1 + x_2 + x_3 \\
\text{subject to} \quad & 2x_1 + x_2 + 2x_3 = 4 \\
& 3x_1 + 3x_2 + x_3 = 3 \\
& x_1, x_2, \ x_3 \geq 0.
\end{aligned}
$$

Since a feasible basic solution is not evident, we introduce two artificial variables x_4 and x_5. Phase I of the simplex algorithm consists of solving the problem

$$
\begin{aligned}
\text{maximize} \quad & - x_4 - x_5 \\
\text{subject to} \quad & 2x_1 + x_2 + 2x_3 + x_4 = 4 \\
& 3x_1 + 3x_2 + x_3 + x_5 = 3 \\
& x_1, \cdots, x_5 \geq 0.
\end{aligned}
$$

The initial simplex tableau in Phase I is given as

$$
\begin{array}{|ccccc|c|}
\hline
0 & 0 & 0 & -1 & -1 & 0 \\
2 & 1 & 2 & 1 & 0 & 4 \\
3 & 3 & 1 & 0 & 1 & 3 \\
\hline
\end{array}
$$

Choose the evident basis $B = I$ the identity matrix corresponding to the basic variables x_4 and x_5. Since the non-basic component of the cost vector c is null, the reduced cost vector of this basis is

$$\bar{c}_N = -[B^{-1}N]^T c_B = \begin{pmatrix} 5 \\ 4 \\ 3 \end{pmatrix}.$$

In view of Theorem 3.1.4 this basis is not optimal. Since B is the identity matrix, the new tableau differs from the former one only by the first row equal to $(5, 4, 3, 0, 0, 7)$ and is displayed below

5	4	3	0	0	7
2	1	2	1	0	4
3	3	1	0	1	3

In fact the new tableau is obtained from the initial one by updating the first row so that components corresponding to the basic variables are zero. The first component of this row has the biggest positive value equal to 5, we choose the pivot $\bar{a}_{21} = 3$ with the pivotal row $\ell = 2$ and pivotal column $s = 1$ and obtain the next tableau

0	−1	4/3	0	−5/3	2
0	−1	4/3	1	−2/3	2
1	1	1/3	0	1/3	1

It is clear that a suitable pivot is $\bar{a}_{13} = 4/3$ with pivot row $\ell = 1$ and pivot column $s = 3$. The next tableau is given as

0	0	0	−1	−1	0
0	−3/4	1	3/4	−1/2	3/2
1	5/4	0	−1/4	1/2	1/2

Phase I of the algorithm terminated, we obtain a basic feasible solution $x_1 = 1/2$ and $x_3 = 3/2$. Now we proceed to Phase II by starting from the feasible basis $B = \begin{pmatrix} 2 & 2 \\ 3 & 1 \end{pmatrix}$ corresponding to the basic variables x_1 and x_3. The initial tableau of this phase is given below

1	1	1	0
0	−3/4	1	3/2
1	5/4	0	1/2

By subtracting the sum of the second and the third rows from the first row to make zero all components of the first row which correspond to the basic variables, we obtain a new tableau

0	1/2	0	−2
0	−3/4	1	3/2
1	5/4	0	1/2

It is clear that this basic feasible solution is not optimal. The pivot element is $\bar{a}_{22} = 5/4$. Making this column equal to $(0, 0, 1)^T$ by row operations we deduce the next tableau

−2/5	0	0	−11/5
3/5	0	1	9/5
4/5	1	0	2/5

At this iteration the reduced cost vector is negative, and so the algorithm terminates. The solution $x_1 = 0$, $x_2 = 2/5$ and $x_3 = 9/5$ is optimal.

Degeneracy

As we experienced in Example 3.1.6 when a basis is degenerate, there is a possibility that a new basis is also degenerate and the value of the objective function does not change and that even if a basis changes, the associated solution remains unchanged. A sequence of such degenerate feasible solutions may make no increase in the value of the objective function and causes cycling. The simplex algorithm then never terminates. To avoid cycling, there are some techniques to modify the algorithm. We cite below two that are frequently used. The first one is Bland's rule to tighten pivot choice. It consists of selecting the pivotal columns a_s with the smallest subscript among those with strictly positive entries and selecting the pivotal row corresponding to the basic variable with lowest subscript among rows i having the same minimum ratio $\bar{b}_i/\bar{a}_{is}$ equal to $\hat{x}_s$. With this rule the simplex method cannot cycle and hence is finite.

The second technique consists of perturbing the constraints, replacing $Ax = b$ by $Ax = b + A(\varepsilon, \varepsilon^2, \cdots, \varepsilon^n)^T$ with a small $\varepsilon > 0$. For a certain range of ε, a degenerate basis B will be non-degenerate for the perturbed system and leads to a new feasible basic solution. This way the simplex method avoids cycling and terminates in a finite number of iterations too.

It is worthwhile noticing that rigorous techniques to overcome cycling require additional operations at each iteration and become costly when solving very large sized problems. On the other hand experience indicates that cycling in the simplex algorithm is very rare. Therefore, most of commercial codes apply the method without paying any attention to it.

The primal-dual method

The primal-dual method is based on duality relations between the primal and the dual problems. Namely, if a feasible solution $\overline{y}$ of the dual problem is known, and if we succeed in finding a feasible solution $\overline{x}$ of the primal problem so that the complementary slackness condition $\langle A^T \overline{y} - c, \overline{x} \rangle = 0$ holds, then $\overline{x}$ is an optimal solution of (LP) and $\overline{y}$ is an optimal solution of (LD). In order to find a feasible solution of (LP) that satisfies the complementary slackness we evoke the auxiliary problem (3.5) with an additional constraint $x_i = 0$ for all i outside of the active index set $I(\overline{y})$.

Thus, given a feasible solution of (LD) we consider the restricted problem, denoted $(P_{\overline{y}})$

$$\text{maximize} \quad \left\langle d, \begin{pmatrix} x \\ y \end{pmatrix} \right\rangle$$
$$\text{subject to} \quad Ax + y = b$$
$$x \geq 0, \, y \geq 0$$
$$x_i = 0, \, i \notin I(\overline{y}),$$

where $d = (0, \cdots, 0, -1, \cdots, -1)^T$, and its dual, denoted $(D_{\overline{y}})$

$$\text{minimize} \quad \langle b, z \rangle$$
$$\text{subject to} \quad \langle a^i, z \rangle \geq 0, \, i \in I(\overline{y})$$
$$z \geq (-1, ..., -1)^T.$$

Here is a relationship between solutions of the primal and dual problems and their restricted problems.

Theorem 3.3.6 *Let (x^0, y^0) be an optimal solution of the restricted problem associated with a feasible solution $\overline{y}$ of the dual problem (LD) and B a non-degenerate basis associated with this optimal solution. The following statements hold.*

(i)   *If $y^0 = 0$, then x^0 is an optimal solution of (LP) and $\overline{y}$ is an optimal solution of (LD).*

(ii) *If $y^0 \neq 0$ and if the feasible solution $z^0 = (B^{-1})^T d_B$ of the dual problem $(D_{\overline{y}})$ satisfies*

$$\langle a^i, z^0 \rangle \geq 0 \text{ for all } i \notin I(\overline{y}), \tag{3.6}$$

then the primal problem (LP) is infeasible.

(iii)   *If $y^0 \neq 0$ and (3.6) does not hold, then the vector $\hat{y} = \bar{y} + tz^0$ with*

$$t = \min \left\{ \frac{c_i - \langle a^i, \bar{y} \rangle}{\langle a^i, z^0 \rangle} : \langle a^i, z^0 \rangle < 0 \right\}$$

is a feasible solution of (LD) satisfying $\langle b, \hat{y} \rangle < \langle b, \bar{y} \rangle$.

Proof For the first statement we observe that x^0 is a feasible solution of (LP) under $y^0 = 0$. Moreover, for active indices $i \in I(\bar{y})$ we have

$$\langle a^i, \bar{y} \rangle - c_i = 0$$

while for inactive indices $j \notin I(\bar{y})$ the components x_j^0 are zero. Consequently,

$$\langle A^T \bar{y} - c, x^0 \rangle = \sum_{i=1}^n \left(\langle a^i, \bar{y} \rangle - c_i \right) x_i^0 = 0.$$

By Theorem 3.2.3, x^0 is an optimal solution of (LP) and $\bar{y}$ is an optimal solution of (LD).

For (ii) we set

$$y_t = \bar{y} + tz^0 \text{ for } t \geq 0.$$

We make two observations. First, for every positive number t, the vector y_t is a feasible solution of (LD). In fact, by hypothesis, $A^T z^0 \geq 0$, and hence

$$A^T y_t = A^T \bar{y} + t A^T z^0 \geq A^T \bar{y} \geq c.$$

Second, the optimal values of the primal restricted problem $(P_{\bar{y}})$ and its dual $(D_{\bar{y}})$ being equal, we deduce from Theorem 3.1.4 and Corollary 3.2.5 that

$$\langle b, z^0 \rangle = -y_1^0 - \cdots - y_m^0 < 0.$$

It follows that

$$\lim_{t \to \infty} \langle b, y_t \rangle = -\infty.$$

Thus, the dual problem (LD) is unbounded below. In view of Theorem 3.2.3 the primal problem (LP) is infeasible.

To prove (iii), we find conditions on t such that y_t is a feasible solution of (LD). For those indices i with $\langle a^i, z^0 \rangle \geq 0$, the ith constraints are evidently satisfied when t is positive because

$$\langle a^i, y_t \rangle = \langle a^i, \bar{y} \rangle + t \langle a^i, z^0 \rangle \geq \langle a^i, \bar{y} \rangle \geq c_i.$$

For those i with $\langle a^i, z^0 \rangle < 0$, the ith constraints are evidently satisfied if

$$t\langle a^i, z^0 \rangle \geq c_i - \langle a^i, \overline{y} \rangle.$$

It follows readily that the value t given in (iii) is the biggest that makes $\hat{y}$ feasible for the dual problem (LD). Finally, we have

$$\langle b, \hat{y} \rangle = \langle b, \overline{y} \rangle + t\langle b, z^0 \rangle < \langle b, \overline{y} \rangle$$

because t is strictly positive and $\langle b, z^0 \rangle$ is strictly negative. $\square$

We shall proceed now to describe the primal-dual algorithm, assuming a feasible solution $\overline{y}$ of the dual problem (LD) at our disposition.

Step 1: Solve the restricted problem (P$_{\overline{y}}$) associated with $\overline{y}$.
If the optimal value is zero, then stop. The vector x^0, where $\begin{pmatrix} x^0 \\ y^0 \end{pmatrix}$ with $y^0 = 0$ is an optimal solution of (P$_{\overline{y}}$), is optimal for (LP) and $\overline{y}$ is optimal for (LD). Otherwise, go the the next step.
Step 2. Compute $z^0 = (B^{-1})^T d_B$ where B is an optimal non-degenerate basis associated with the optimal solution $\begin{pmatrix} x^0 \\ y^0 \end{pmatrix}$ with $y^0 \neq 0$ and d_B is the basic component of the vector d determining the objective function of the restricted problem.
If $\langle a^i, z^0 \rangle \geq 0$ for all $i \notin I(\overline{y})$, then stop. The primal problem (LP) is infeasible. Otherwise, go to the next step.
Step 3. Compute

$$t = \min \left\{ \frac{c_i - \langle a^i, \overline{y} \rangle}{\langle a^i, z^0 \rangle} : \langle a^i, z^0 \rangle < 0 \right\}$$

and return to Step 1, replacing $\overline{y}$ by $\overline{y} + tz^0$.

A few useful comments on the implementation of the algorithm are in order. First, to solve the restricted primal problem an evident basis corresponding to the basic variables $y_1, \cdots, y_m$ can be started with. Remember that $b \geq 0$ is assumed. Second, if an index k realizes the minimum in the formula of t in Step 3, then it is also active index at the new dual feasible solution $\overline{y} + tz^0$ because

$$\langle a^k, \overline{y} + tz^0 \rangle = \langle a^k, \overline{y} \rangle + \frac{c_k - \langle a^k, \overline{y} \rangle}{\langle a^k, z^0 \rangle} \langle a^k, z^0 \rangle = c_k.$$

Moreover, if a component x_j^0 of the optimal solution $\begin{pmatrix} x^0 \\ y^0 \end{pmatrix}$ obtained in Step 1 is strictly positive, then the complementary slackness $\langle a^j, z^0 \rangle = 0$ implies

$$\langle a^j, \overline{y} + tz^0 \rangle = \langle a^j, \overline{y} \rangle = c_j.$$

Hence j is an active index too.

Third, it follows from the preceding comment that when an index i is not active at the new feasible solution $\bar{y} + tz^0$, that is $\langle a^i, \bar{y} + tz^0 \rangle > c_i$, then the corresponding ith component of x^0 is zero. By this, the optimal solution $\begin{pmatrix} x^0 \\ y^0 \end{pmatrix}$ is feasible for the new restricted problem $(P_{\bar{y}+tz^0})$.

Finally, we may claim that under non-degeneracy the algorithm terminates after a finite number of iterations. Indeed, at each iteration we arrive at either the infeasibility of (LP), or a new feasible basic solution of the dual problem (LD) whose value strictly decreases. Since the number of bases of the dual problem is finite, the algorithm is finite too.

Example 3.3.7 We consider the problem

$$
\begin{aligned}
\text{maximize} \quad & -x_1 - x_2 - x_3 \\
\text{subject to} \quad & x_1 + 2x_2 + x_3 = 2 \\
& 2x_1 + 3x_2 + x_3 = 3 \\
& x_1, x_2, x_3 \geq 0,
\end{aligned}
$$

and its dual problem

$$
\begin{aligned}
\text{minimize} \quad & 2y_1 + 3y_2 \\
\text{subject to} \quad & y_1 + 2y_2 \geq -1 \\
& 2y_1 + 3y_2 \geq -1 \\
& y_1 + y_2 \geq -1.
\end{aligned}
$$

An evident feasible solution of the dual problem is $\bar{y} = \begin{pmatrix} 1 \\ 0 \end{pmatrix}$ whose active index set $I(\bar{y})$ is empty. The restricted primal problem associated with this solution is the following

$$
\begin{aligned}
\text{maximize} \quad & -z_1 - z_2 \\
\text{subject to} \quad & x_1 + 2x_2 + x_3 + z_1 \quad\quad = 2 \\
& 2x_1 + 3x_2 + x_3 \quad\quad + z_2 = 3 \\
& x_1 = x_2 = x_3 = 0 \\
& z_1, z_2 \geq 0.
\end{aligned}
$$

Since the latter problem has only one feasible solution, its optimal value is strictly negative. The solution $x^0 = (0, 0, 0)^T$ is not feasible for (LP). The optimal basis corresponding to the basic variables z_1 and z_2 is the identity matrix $B = \begin{pmatrix} 1 & 0 \\ 0 & 1 \end{pmatrix}$ and the basic component of the objective vector is $d_B = (-1, -1)^T$. By Corollary 3.2.5, the vector $u^0 = (B^{-1})^T d_B = d_B$ is a feasible solution of the dual problem of the restricted primal problem. We compute the value of t given by the formula in the Step 3 of the primal-dual algorithm, which is exactly the maximal t such that

$$\begin{pmatrix} 1 & 2 \\ 2 & 3 \\ 1 & 1 \end{pmatrix} \left[\begin{pmatrix} 1 \\ 0 \end{pmatrix} + t \begin{pmatrix} -1 \\ -1 \end{pmatrix} \right] \geq \begin{pmatrix} -1 \\ -1 \\ -1 \end{pmatrix},$$

which yields $t = 3/5$. The new feasible solution of the dual problem (LD) for the next iteration is $\bar{y} + tu^0 = \begin{pmatrix} 2/5 \\ -3/5 \end{pmatrix}$. Its active index set is $I(\bar{y} + tu^0) = \{2\}$. The restricted primal problem associated with it is now

$$
\begin{array}{lll}
\text{maximize} & -z_1 & -z_2 \\
\text{subject to} & x_1 + 2x_2 + x_3 + z_1 & = 2 \\
& 2x_1 + 3x_2 + x_3 \quad\quad + z_2 = 3 \\
& \quad x_1 = x_3 = 0 \\
& \quad x_2, z_1, z_2 \geq 0.
\end{array}
$$

Solving this problem by the simplex method we have the initial tableau without the variables x_1 and x_3 which are zero

0	−1	−1	0
2	1	0	2
3	0	1	3

Using the basis corresponding to the basic variables z_1 and z_2 the tableau becomes

5	0	0	5
2	1	0	2
3	0	1	3

The pivot $\bar{a}_{21} = 3$ leads to the final tableau

0	0	−5/3	0
0	1	−2/3	0
1	0	1/3	1

which indicates that the solution $x_1 = 0, x_2 = 1, x_3 = 0, z_1 = 0, z_2 = 0$ is optimal with the optimal value equal to zero. According to the algorithm the solution $x_1 = 0, x_2 = 1, x_2 = 0$ is feasible for (LP), and hence it is an optimal solution of it. The dual solution $y_1 = 2/5, y_2 = -3/5$ is an optimal solution of the dual problem (LD).

Part II
Theory

Chapter 4
Pareto Optimality

In a multi-dimensional Euclidean space there are several ways to classify elements of a given set of vectors. The componentwise order relation introduced in the very beginning of the second chapter seems to be the most appropriate for this classification purpose and leads to the concept of Pareto optimality or efficiency, a cornerstone of multiobjective optimization that we are going to study in the present chapter.

4.1 Pareto Maximal Points

In the space $\mathbb{R}^k$ with $k > 1$ the componentwise order $x \geqq y$ signifies that each component of x is bigger than or equal to the corresponding component of y. Equivalently, $x \geqq y$ if and only if the difference vector $x - y$ has non-negative components only. This order is not complete in the sense that not every couple of vectors is comparable, and hence the usual notion of maximum or minimum does not apply. We recall also that $x > y$ means that all components of the vector $x - y$ are strictly positive, and $x \geq y$ signifies $x \geqq y$ and $x \neq y$. The following definition lays the basis for our study of multiobjective optimization problems.

Definition 4.1.1 Let Q be a nonempty set in $\mathbb{R}^k$. A point $y \in Q$ is said to be a *(Pareto) maximal point* of the set Q if there is no point $y' \in Q$ such that $y' \geqq y$ and $y' \neq y$. And it is said to be a *(Pareto) weakly maximal point* if there is no $y' \in Q$ such that $y' > y$.

The sets of maximal points and weakly maximal points of Q are respectively denoted $\mathrm{Max}(Q)$ and $\mathrm{WMax}(Q)$ (Figs. 4.1 and 4.2). They are traditionally called the *efficient* and *weakly efficient sets* or the *non-dominated* and *weakly non-dominated sets* of Q. The set of minimal points $\mathrm{Min}(Q)$ and weakly minimal points $\mathrm{WMin}(Q)$ are defined in a similar manner. When no confusion likely occurs between maximal and minimal elements, the set $\mathrm{Min}(Q)$ and $\mathrm{WMin}(Q)$ are called the *efficient* and *weakly efficient sets* of Q too. The terminology of efficiency is advantageous in certain circumstances

© Springer International Publishing Switzerland 2016

D.T. Luc, *Multiobjective Linear Programming*,

DOI 10.1007/978-3-319-21091-9_4

Fig. 4.1 Max and Min

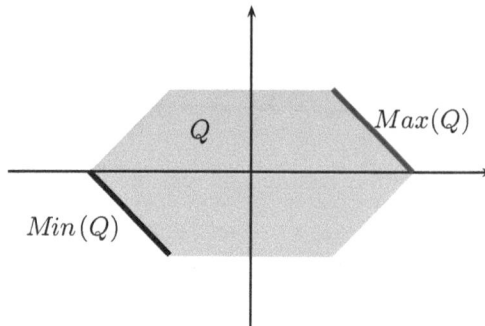

Fig. 4.2 WMax and WMin

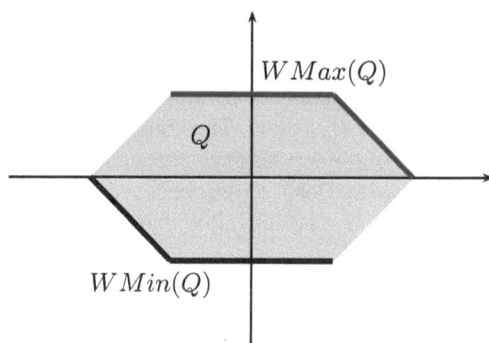

in which we deal simultaneously with maximal points of a set as introduced above and maximal elements of a family of subsets which are defined to be maximal with respect to inclusion. Thus, given a convex polyhedron, a face of it is efficient if it consists of maximal points only. When we refer to a maximal efficient face, it is understood that that face is efficient and maximal by inclusion which means that no efficient face of the polyhedron contains it as a proper subset. In some situations one is interested in an *ideal maximal point* (called also a *utopia* point), which is defined to be a point $y \in Q$ that satisfies

$$y \geqq y' \text{ for all } y' \in Q.$$

Such a point is generally unattainable, and if it exists it is unique and denoted by IMax(Q) (Fig. 4.3).

Geometrically, a point y of Q is an efficient (maximal) point if the intersection of the set Q with the positive orthant shifted at y consists of y only, that is,

$$Q \cap (y + \mathbb{R}_+^k) = \{y\}$$

and it is weakly maximal if the intersection of Q with the interior of the positive orthant shifted at y is empty, that is,

Fig. 4.3 IMax

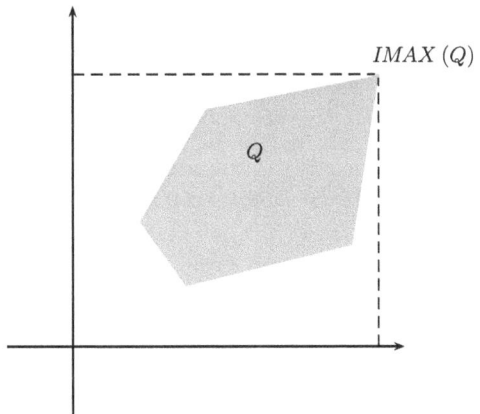

$$Q \cap (y + \text{int}(\mathbb{R}_+^k)) = \emptyset.$$

Of course, maximal points are weakly maximal, and the converse is not true in general. Here are some examples in $\mathbb{R}^2$.

Example 4.1.2 Let Q be the triangle of vertices $a = \begin{pmatrix} 0 \\ 0 \end{pmatrix}$, $b = \begin{pmatrix} 1 \\ 0 \end{pmatrix}$ and $c = \begin{pmatrix} 0 \\ 1 \end{pmatrix}$ in $\mathbb{R}^2$. Then $\text{Max}(Q) = \text{WMax}(Q) = [b, c]$, $\text{Min}(Q) = \{a\}$ and $\text{WMin}(Q) = [a, b] \cup [a, c]$.

Example 4.1.3 Let Q be the polytope in the space $\mathbb{R}^3$, determined by two inequalities

$$y_2 + y_3 \geqq 0$$
$$y_3 \geqq 0.$$

Then $\text{Max}(Q) = \text{WMax}(Q) = \emptyset$, $\text{Min}(Q) = \emptyset$ and $\text{WMin}(Q) = Q \setminus \text{int}(Q)$.

Existence of pareto maximal points

As we have already seen in Example 4.1.3, a polyhedron may have no weakly maximal points. This happens when some components of elements of the set are unbounded above. Positive functionals provide an easy test for such situations.

Theorem 4.1.4 *Let Q be a nonempty set and let λ be a nonzero vector in $\mathbb{R}^k$. Assume that $y \in Q$ is a maximizer of the functional $\langle \lambda, . \rangle$ on Q. Then*

(i) *y is a weakly maximal point of Q if λ is a positive vector;*
(ii) *y is a maximal point of Q if either λ is a strictly positive vector, or λ is a positive vector and y is the unique maximizer.*

In particular, if Q is a nonempty compact set, then it has a maximal point.

Proof Assume λ is a nonzero positive vector. If y were not weakly maximal, then there would exist another vector y' in Q such that the vector $y' - y$ is strictly positive. This would yield $\langle \lambda, y' \rangle > \langle \lambda, y \rangle$, a contradiction.

Now, if λ is strictly positive, then for any $y' \geq y$ and $y' \neq y$, one has $\langle \lambda, y' \rangle > \langle \lambda, y \rangle$ as well. Hence y is a Pareto maximal point of Q.

When λ is positive (not necessarily strictly positive) and not zero, the above inequality is not strict. Actually, we have equality because y is a maximizer. But, in that case y' is also a maximizer of the functional $\langle \lambda, . \rangle$ on Q, which contradicts the hypothesis.

When Q is compact, any strictly positive vector λ produces a maximizer on Q, hence a Pareto maximal point too.   □

Maximizers of the functional $\langle \lambda, . \rangle$ with λ positive, but not strictly positive, may produce no maximal points as seen in the following example.

Example 4.1.5 Consider the set Q in $\mathbb{R}^3$ consisting of the vectors $x = (x_1, x_2, x_3)^T$ with $x_3 \leq 0$. Choose $\lambda = (0, 0, 1)^T$. Then every element x of Q with $x_3 = 0$ is a maximizer of the functional $\langle \lambda, . \rangle$ on Q, hence it is weakly maximal, but not maximal, for the set Q has no maximal element.

Given a reference point a in the space, the set of all elements of a set Q that are bigger than the point a forms a dominant subset, called a *section* of Q at a. The lemma below shows that maximal elements of a section are also maximal elements of the given set.

Lemma 4.1.6 *Let Q be a nonempty set in $\mathbb{R}^k$. Then for every point a in $\mathbb{R}^k$ one has*

$$\text{Max}\left(Q \cap (a + \mathbb{R}^k_+)\right) \subseteq \text{Max}(Q)$$
$$\text{WMax}\left(Q \cap (a + \mathbb{R}^k_+)\right) \subseteq \text{WMax}(Q).$$

Proof Let y be a Pareto maximal point of the section $Q \cap (a + \mathbb{R}^k_+)$. If y were not maximal, then one would find some y' in Q such that $y' \geq y$ and $y' \neq y$. It would follow that y' belongs to the section $Q \cap (a + \mathbb{R}^k_+)$ and yield a contradiction. The second inclusion is proven by the same argument.   □

For convex polyhedra existence of maximal points is characterized by position of asymptotic directions with respect to the positive orthant of the space.

Theorem 4.1.7 *Let Q be a convex polyhedron in $\mathbb{R}^k$. The following assertions hold.*

(i) *Q has maximal points if and only if*

$$Q_\infty \cap \mathbb{R}^k_+ = \{0\}.$$

(ii) *Q has weakly maximal points if and only if*

$$Q_\infty \cap \text{int}(\mathbb{R}^k_+) = \emptyset.$$

In particular, every polytope has a maximal vertex.

Proof Let y be a maximal point of Q and let v be any nonzero asymptotic direction of Q. Since $y + v$ belongs to Q and $Q \cap (y + \mathbb{R}^k_+) = \{y\}$, we deduce that v does not belong to $\mathbb{R}^k_+$. Conversely, assume Q has no nonzero asymptotic direction. Then for a fixed vector y in Q the section $Q \cap (y + \mathbb{R}^k_+)$ is bounded; otherwise any nonzero asymptotic direction of that closed convex intersection, which exists due to Corollary 2.3.16, should be a positive asymptotic vector of Q. In view of Theorem 4.1.4 the compact section $Q \cap (y + \mathbb{R}^k_+)$ possesses a maximal point, hence, in view of Lemma 4.1.6, so does Q.

For the second assertion, the same argument as above shows that when Q has a weakly maximal point, no asymptotic direction of it is strictly positive. For the converse part, by the hypothesis we know that Q_∞ and $\mathbb{R}^k_+$ are two convex polyhedra without relative interior points in common. Hence, in view of Theorem 2.3.10 there is a nonzero vector $\lambda \in \mathbb{R}^k$ separating them, that is

$$\langle \lambda, v \rangle \leqq \langle \lambda, d \rangle \text{ for all } v \in Q_\infty \text{ and } d \in \mathbb{R}^k_+.$$

In particular, for $v = 0$ and for d being usual coordinate unit vectors, we deduce from the above relation that λ is positive. Moreover, the linear function $\langle \lambda, . \rangle$ is then non-positive on every asymptotic direction of Q. We apply Theorem 3.1.1 to obtain a maximum of $\langle \lambda, . \rangle$ on Q. In view of Theorem 4.1.4 that maximum is a weakly maximal point of Q.

Finally, if Q is a polytope, then its asymptotic cone is trivial. Hence, by the first assertion, it has maximal points. To prove that it has a maximal vertex, choose any strictly positive vector $\lambda \in \mathbb{R}^k$ and consider the linear problem of maximizing $\langle \lambda, . \rangle$ over Q. In view of Theorem 3.1.3 the optimal solution set contains a vertex, which, by Theorem 4.1.4, is also a maximal vertex of Q. □

In Example 4.1.5 a positive functional $\langle \lambda, . \rangle$ was given on a polyhedron having no maximizer that is maximal. This, however, is impossible when the polyhedron has maximal elements.

Corollary 4.1.8 *Assume that Q is a convex polyhedron and λ is a nonzero positive vector in $\mathbb{R}^k$. If Q has a maximal point and the linear functional $\langle \lambda, . \rangle$ has maximizers on Q, then among its maximizers there is a maximal point of Q.*

Proof Let us denote by Q_0 the nonempty intersection of Q with the hyperplane $\{y \in \mathbb{R}^k : \langle \lambda, y \rangle = d\}$ where d is the maximum of $\langle \lambda, . \rangle$ on Q. It is a convex polyhedron. Since Q has maximal elements, in view of Theorem 4.1.7 one has $Q_\infty \cap \mathbb{R}^k_+ = \{0\}$, which implies that $(Q_0)_\infty \cap \mathbb{R}^k_+ = \{0\}$ too. By the same theorem, Q_0 has a maximal element, say y_0. We show that this y_0 is also a maximal element of Q. Indeed, if not, one could find some $y \in Q$ such that $y \geq y_0$ and $y \neq y_0$. Since λ is positive, we deduce that $\langle \lambda, y \rangle \geq \langle \lambda, y_0 \rangle = d$. Moreover, as y does not belong to Q_0, this inequality must be strict which is a contradiction. $\square$

We say a set Q in the space $\mathbb{R}^k$ has the *domination property* if its elements are dominated by maximal elements, that is, for every $y \in Q$ there is some maximal element a of Q such that $a \geq y$. The weak domination property refers to domination by weakly maximal elements.

Corollary 4.1.9 *A convex polyhedron has the domination property (respectively weak domination property) if and only if it has maximal elements (respectively weakly maximal elements).*

Proof The "only if" part is clear. Assume a convex polyhedron Q has maximal elements. In view of Theorem 4.1.7, the asymptotic cone of Q has no nonzero vector in common with the positive orthant $\mathbb{R}^k_+$. Hence so does the section of Q at a given point $a \in Q$. Again by Theorem 4.1.7 that section has maximal points that dominate a and by Lemma 4.1.6 they are maximal points of Q. Hence Q has the domination property. The weak domination property is proven by the same argument. $\square$

We learned in Sect. 2.3 how to compute the normal cone at a given point of a polyhedron. It turns out that by looking at the normal directions it is possible to say whether a given point is maximal or not.

Theorem 4.1.10 *Let Q be a convex polyhedron in $\mathbb{R}^k$. The following assertions hold.*

(i) $y \in Q$ is a maximal point if and only if the normal cone $N_Q(y)$ to Q at y contains a strictly positive vector.

(ii) $y \in Q$ is a weakly maximal point if and only if the normal cone $N_Q(y)$ to Q at y contains a nonzero positive vector.

Proof Let y be a point in Q. If the normal cone to Q at y contains a strictly positive vector, say λ, then by the definition of normal vectors, the functional $\langle \lambda, . \rangle$ attains its maximum on Q at y. In view of Theorem 4.1.4, y is a maximal point of Q. The proof of the "only if" part of (i) is based on Farkas' theorem. We assume that y is a maximal point of Q and suppose to the contrary that the normal cone to Q at that point has no vector in common with the interior of the positive orthant $\mathbb{R}^k_+$. We may assume that Q is given by a system of inequalities

$$\langle a^i, z \rangle \leq b_i, \ i = 1, \cdots, m. \tag{4.1}$$

The active index set at y is denoted $I(y)$. By Theorem 2.3.24, the normal cone to Q at y is the positive hull of the vectors $a^i, i \in I(y)$. Its empty intersection with $\text{int}(\mathbb{R}^k_+)$ means that the following system has no solution

$$A_{I(y)}\lambda \geq e$$
$$\lambda \geq 0,$$

where $A_{I(y)}$ denotes the matrix whose columns are a^i, $i \in I(y)$ and e is the vector whose components are all equal to one. By introducing artificial variables z, the above system is equivalent to the system

$$[A_{I(y)} \; (-I)] \begin{pmatrix} \lambda \\ z \end{pmatrix} = e$$

$$\lambda \geqq 0$$

$$z \geqq 0.$$

Apply Farkas' theorem (Theorem 2.2.3) to obtain a nonzero positive vector v such that

$$\langle a^i, v \rangle \leqq 0 \text{ for all } i \in I(y).$$

The inequalities (4.1) corresponding to the inactive indices at y being strict, we may find a strictly positive number t such that

$$\langle a^i, y + tv \rangle \leqq b_i \text{ for all } i = 1, \cdots, m.$$

In other words, the point $y + tv$ belongs to Q. Moreover, $y + tv \geqq y$ and $y + tv \neq y$ which contradicts the hypothesis. This proves (i).

As to the second assertion, the "if" part is clear, again, Theorem 4.1.4 is in use. For the converse part, we proceed the same way as in (i). The fact that the intersection of $N_Q(y)$ with the positive orthant $\mathbb{R}^k_+$ consists of the zero vector only, means that the system

$$A_{I(y)} \lambda \geqq 0$$

$$\lambda \geqq 0$$

has no nonzero solution. Applying Corollary 2.2.5 we deduce the existence of a strictly positive vector v such that

$$\langle a^i, v \rangle \leqq 0 \text{ for all } i \in I(y).$$

Then, as before, the vector $y + tv$ with $t > 0$ sufficiently small, belongs to Q and $y + tv > y$, which is a contradiction. $\square$

Example 4.1.11 Consider a convex polyhedron Q in $\mathbb{R}^3$ determined by the system

$$\begin{pmatrix} 1 & 1 & 1 \\ 0 & 1 & 1 \\ 1 & 0 & 1 \\ 0 & 0 & -1 \\ 0 & 0 & 1 \end{pmatrix} \begin{pmatrix} x_1 \\ x_2 \\ x_3 \end{pmatrix} \leqq \begin{pmatrix} 1 \\ 1 \\ 1 \\ 0 \\ 1 \end{pmatrix}.$$

We analyze the point $y = (1/3, 1/3, 1/3)^T \in Q$. Its active index set is $I(y) = \{1\}$. By Theorem 2.3.24 the normal cone to Q at that point is generated by the vector $(1, 1, 1)^T$. According to Theorem 4.1.10 the point y is a maximal point of Q. Now, take another point of Q, say $z = (-1, 0, 1)^T$. Its active index set consists of two indices 2 and 5. The normal cone to Q at z is generated by two directions $(0, 1, 1)^T$ and $(0, 0, 1)^T$. It is clear that this normal cone contains no strictly positive vector, hence the point z is not a maximal point of Q because $z^T \leq (0, 0, 1)^T$. It is a weakly maximal point, however, because normal directions at z are positive. Finally, we choose a point $w = (0, 0, 0)^T$ in Q. Its active index set is $I(w) = \{4\}$. The normal cone to Q at w is the cone generated by the direction $(0, 0, -1)^T$. This cone contains no positive vector, hence the point w is not weakly maximal. This can also be seen from the fact that w is strictly dominated by y.

Scalarizing vectors

In remaining of this section we shall use the terminology of efficient points instead of (Pareto) maximal points in order to avoid possible confusion with the concept of maximal element of a family of sets by inclusion. Given a family $\{A_i : i \in I\}$ of sets, we say that A_{i_0} is maximal (respectively minimal) if there is no element A_i of the family such that $A_i \neq A_{i_0}$ and $A_{i_0} \subset A_i$ (respectively $A_{i_0} \supset A_i$). Another formulation of Theorem 4.1.10 is viewed by maximizing linear functionals on the set Q.

Corollary 4.1.12 *Let Q be a convex polyhedron in $\mathbb{R}^k$. Then the following statements hold.*

(i) *$y \in Q$ is an efficient point if and only if there is a strictly positive vector $\lambda \in \mathbb{R}^k$ such that y maximizes the functional $\langle \lambda, . \rangle$ on Q.*

(ii) *$y \in Q$ is a weakly efficient point if and only if there is a nonzero positive vector $\lambda \in \mathbb{R}^k$ such that y maximizes the functional $\langle \lambda, . \rangle$ on Q.*

Proof This is immediate from the definition of normal cones and from Theorem 4.1.10.   □

The vector λ mentioned in this corollary is called a *scalarizing vector* (or *weakly scalarizing vector* in (ii)) of the set Q. We remark that not every strictly positive vector is a scalarizing vector of Q like not every strictly positive functional attains its maximum on Q. Moreover, an efficient point of Q may maximize a number of scalarizing vectors that are linearly independent, and vice versa, a scalarizing vector may determine several maximizers on Q. For a given polyhedron Q that has efficient elements, the question of how to choose a vector λ so that the functional associated with it furnishes a maximizer is not evident. Analytical choice of positive directions such as the one discussed in Example 4.1.11 is conceivable and will be given in details later. Random generating methods or uniform divisions of the standard simplex do not work in many instances. In fact, look at a simple problem of finding efficient points of the convex polyhedral set given by the inequality

$$x_1 + \sqrt{2}x_2 \leqq 1$$

in the two-dimensional space $\mathbb{R}^2$. Except for one direction, every positive vector λ leads to a linear problem of maximizing $\langle \lambda, x \rangle$ over that polyhedron with unbounded objective. Hence, using positive vectors λ_i of a uniform partition

$$\lambda_i = \frac{i}{p} \begin{pmatrix} 1 \\ 0 \end{pmatrix} + \frac{p-i}{p} \begin{pmatrix} 0 \\ 1 \end{pmatrix}$$

of the simplex $[(1, 0)^T, (0, 1)^T]$ of the space $\mathbb{R}^2$ for whatever the positive integer p be, will never generate efficient points of the set.

Any nonzero positive vector of the space $\mathbb{R}^k$ is a positive multiple of a vector from the standard simplex Δ. This combined with Corollary 4.1.12 yields the following equalities

$$\text{Max}(Q) = \bigcup_{\lambda \in \text{ri}\Delta} \text{argmax}_Q \langle \lambda, . \rangle$$

$$\text{WMax}(Q) = \bigcup_{\lambda \in \Delta} \text{argmax}_Q \langle \lambda, . \rangle$$

where $\text{argmax}_Q \langle \lambda, . \rangle$ is the set of all maximizers of the functional $\langle \lambda, . \rangle$ on Q. Given a point $y \in Q$ denote

$$\Delta_y = \left\{ \lambda \in \Delta : y \in \text{argmax}_Q \langle \lambda, . \rangle \right\}$$
$$\Delta_Q = \bigcup_{y \in Q} \Delta_y.$$

The set Δ_Q is called the *weakly scalarizing set* of Q and Δ_y is the *weakly scalarizing set* of Q at y (Fig. 4.4). By Corollary 4.1.12 the set Δ_y is nonempty if and only if the point y is a weakly efficient element of Q. Hence when Q has weakly efficient points, the set Δ_Q can be expressed as

$$\Delta_Q = \bigcup_{y \in \text{WMax}(Q)} \Delta_y, \qquad (4.2)$$

in which every set Δ_y is nonempty. By definition a vector $\lambda \in \Delta$ belongs to Δ_y if and only if

$$\langle \lambda, y' - y \rangle \leqq 0 \text{ for all } y' \in Q.$$

The latter inequality signifies that λ is a normal vector to Q at y, and so (4.2) becomes

$$\Delta_Q = \bigcup_{y \in \text{WMax}(Q)} N_Q(y) \cap \Delta.$$

Fig. 4.4 Scalarizing set at y

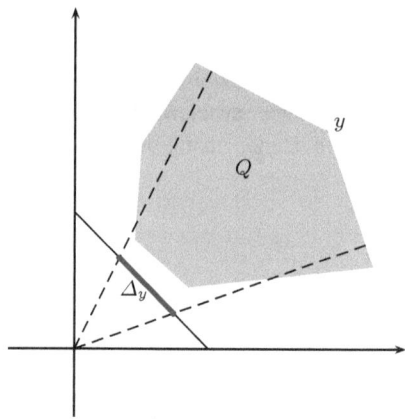

Let $\mathcal{F} = \{F_1, \cdots, F_q\}$ be the collection of all faces of Q and let $N(F_i)$ be the normal cone to F_i, which, by definition, is the normal cone to Q at a relative interior point of F_i. Since each element of Q is a relative interior point of some face, the decomposition (4.2) produces the following decomposition of Δ_Q:

$$\Delta_Q = \bigcup_{i \in I} \Delta_i, \tag{4.3}$$

where $\Delta_i = N(F_i) \cap \Delta$ and I is the set of those indices i from $\{1, \cdots, q\}$ such that the faces F_i are weakly efficient. We note that when a face is not weakly efficient, the normal cone to it does not meet the simplex Δ. Remember that a face of Q is *weakly efficient* if all elements of it are weakly efficient elements of Q, or equivalently if a relative interior point of it is a weakly efficient element. A face that is not weakly efficient may contain weakly efficient elements on its proper faces.

We say a face of Q is a *maximal weakly efficient face* if it is weakly efficient and no weakly efficient face of Q contains it as a proper subset. It is clear that when a convex polyhedron has weakly efficient elements, it does have maximal weakly efficient faces. Below we present some properties of the decompositions (4.2) and (4.3) of the weakly scalarizing set.

Lemma 4.1.13 *If P and Q are convex polyhedra with $P \cap Q \neq \emptyset$, then there are faces $P' \subseteq P$ and $Q' \subseteq Q$ such that $P \cap Q = P' \cap Q'$ and $\mathrm{ri}(P') \cap \mathrm{ri}(Q') \neq \emptyset$. Moreover, if the interior of Q is nonempty and contains some elements of P, then $\mathrm{ri}(P) \cap \mathrm{int}(Q) \neq \emptyset$ and $\mathrm{ri}(P \cap Q) = \mathrm{ri}(P) \cap \mathrm{int}(Q)$.*

Proof Let x be a relative interior point of the intersection $P \cap Q$. Let $P' \subseteq P$ and $Q' \subseteq Q$ be faces that contain x in their relative interiors. These faces meet the requirements of the lemma. Indeed, it suffices to show that every point y from $P \cap Q$ belongs to $P' \cap Q'$. Since x is a relative interior point of $P \cap Q$, the segment $[x - \varepsilon(x - y), x + \varepsilon(x - y)]$ belongs to that intersection when $\varepsilon > 0$ is sufficiently small. Moreover, as P' is a face, this segment must lie in P' which implies that y lies in P'. The same argument shows that y lies in Q', proving the first part of the lemma.

For the second part it suffices to observe that P is the closure of its relative interior. Hence it has relative interior points inside the interior of Q. The last equality of the conclusion is then immediate. $\qquad\square$

Theorem 4.1.14 *The weakly scalarizing set Δ_Q is a polytope. Moreover, if Δ_Q is nonempty, the elements of the decomposition (4.2) and (4.3) are polytopes and satisfy the following conditions:*

(i) *If $\Delta_y = \Delta_z$ for some weakly efficient elements y and z, then there is $i \in I$ such that $y, z \in F_i$ and $\Delta_y = \Delta_z = \Delta_i$.*

(ii) *If F_i is a maximal weakly efficient face of Q, then Δ_i is a minimal element of the decomposition (4.3). Conversely, if the polytope Δ_i is minimal among the polytopes of the decomposition (4.3), then there is a maximal weakly efficient face F_j such that $\Delta_j = \Delta_i$.*

(iii) *For all $i, j \in I$ with $i \neq j$, one has either $\Delta_i = \Delta_j$ or $\mathrm{ri}(\Delta_i) \cap \mathrm{ri}(\Delta_j) = \emptyset$.*

(iv) *Let F_i and F_j be two weakly efficient adjacent vertices (zero-dimensional faces) of Q. Then the edge joining them is weakly efficient if and only if $\Delta_i \cap \Delta_j \neq \emptyset$.*

Proof Since Δ_y is empty when y is not a weakly efficient point of Q, we may express Δ_Q as

$$\Delta_Q = \bigcup_{y \in Q} \Delta_y = \bigcup_{y \in Q} (N_Q(y) \cap \Delta) = N_Q \cap \Delta$$

which proves that Δ_Q is a bounded polyhedron because the normal cone N_Q is a polyhedral cone. Likewise, the sets $\Delta_y = N_Q(y) \cap \Delta$ and $\Delta_i = N(F_i) \cap \Delta$ are convex polytopes.

To establish (i) we apply Lemma 4.1.13 to the intersections $\Delta_y = N_Q(y) \cap \Delta$ and $\Delta_z = N_Q(z) \cap \Delta$. There exist faces $N \subseteq N_Q(y)$, $M \subseteq N_Q(z)$ and $\Delta^y, \Delta^z \subseteq \Delta$ such that

$$N_Q(y) \cap \Delta = N \cap \Delta^y, \quad \mathrm{ri}(N) \cap \mathrm{ri}(\Delta^y) \neq \emptyset$$
$$N_Q(z) \cap \Delta = M \cap \Delta^z, \quad \mathrm{ri}(M) \cap \mathrm{ri}(\Delta^z) \neq \emptyset.$$

Choose any vector ξ from the relative interior of Δ_y. Then it is also a relative interior vector of the faces N, M, Δ^y and Δ^z. This implies that $N = M$ and $\Delta^y = \Delta^z$. Using Theorem 2.3.26 we find a face F_i of Q such that $N(F_i) = N$. Then F_i contains y and z and satisfies

$$\Delta_i = N(F_i) \cap \Delta = N \cap \Delta = \Delta_y = \Delta_z.$$

For (ii) assume F_i is a maximal weakly efficient face. Assume that Δ_j is a subset of Δ_i for some $j \in I$. We choose any vector ξ from Δ_j and consider the face F'

consisting of all maximizers of $\langle \xi, . \rangle$ on Q. Then F' is a weakly efficient face and contains F_j and F_i. As F_i is maximal, we must have $F' = F_i$. Thus, $F_j \subseteq F_i$ and

$$\Delta_i = N(F_i) \cap \Delta \subseteq N(F_j) \cap \Delta = \Delta_j.$$

Conversely, let Δ_i be a minimal element among the polytopes Δ_j, $j \in I$. If F_i is maximal weakly efficient face, we are done. If it is not, we find a maximal weakly efficient face F_j containing F_i. Then $\Delta_j = N(F_j) \cap \Delta \subseteq N(F_i) \cap \Delta = \Delta_i$ and $\Delta_j = \Delta_i$ by hypothesis.

We proceed to (iii). Assume that the relative interior of Δ_i and the relative interior of Δ_j have a vector ξ in common. In view of Lemma 4.1.13 one can find four faces: N^i of $N(F_i)$, N^j of $N(F_j)$, Δ^i and Δ^j of Δ such that

$$N(F_i) \cap \Delta = N^i \cap \Delta^i, \text{ ri}(N^i) \cap \text{ri}(\Delta^i) \neq \emptyset$$
$$N(F_j) \cap \Delta = N^j \cap \Delta^j, \text{ ri}(N^j) \cap \text{ri}(\Delta^j) \neq \emptyset.$$

According to Theorem 2.3.26 there are faces F_ℓ and F_m of Q which respectively contain F_i and F_j with $N(F_\ell) = N^i$ and $N(F_m) = N^j$. Then ξ is a relative interior vector of the faces $N(F_\ell)$, $N(F_m)$, Δ^i and Δ^j. We deduce that the face Δ^i coincides with Δ^j, and F_ℓ coincides with F_m. Consequently, $\Delta_i = \Delta_j$.

To prove the last property we assume F_i and F_j are adjacent vertices (zero-dimensional faces) of Q. Let a one-dimensional face F_l be the edge joining them. According to Corollary 2.3.28 we have $N(F_l) = N(F_i) \cap N(F_j)$. Then $\Delta_l = \Delta_i \cap \Delta_j$ which shows that F_l is weakly efficient if and only if the latter intersection is nonempty.   □

Note that two different faces of Q may have the same weakly scalarizing set. For instance the singleton $\{(0, 0, 1)^T\}$ is the weakly scalarizing set for all weakly efficient faces of the polyhedron $\mathbb{R}^2_+ \times \{0\}$ in $\mathbb{R}^3$.

In order to treat efficient elements of Q we need to work with the relative interior of Δ. Corresponding notations will be set as follows

$$\Delta^r_Q = \Delta_Q \cap \text{ri}(\Delta)$$
$$\Delta^r_y = \Delta_y \cap \text{ri}(\Delta)$$
$$\Delta^r_i = \Delta_i \cap \text{ri}(\Delta).$$

The set Δ^r_Q is called the *scalarizing set* of Q. It is clear that $y \in Q$ is efficient if and only if Δ^r_y is nonempty, and it is weakly efficient, but not efficient if and only if Δ_y lies on the border of Δ. The decompositions of the weakly scalarizing set induce the following decompositions of the scalarizing set

$$\Delta^r_Q = \bigcup_{y \in \text{Max}(Q)} \Delta^r_y \tag{4.4}$$

and

$$\Delta_Q^r = \bigcup_{i \in I_0} \Delta_i^r \tag{4.5}$$

where I_0 consists of those indices i from $\{1, \cdots, q\}$ for which F_i are efficient.

Theorem 4.1.15 *Assume that the scalarizing set Δ_Q^r is nonempty. Then*

$$\Delta_Q = \mathrm{cl}(\Delta_Q^r).$$

Moreover the decompositions (4.4) and (4.5) of Δ_Q^r satisfy the following properties:

(i) *If $\Delta_y^r = \Delta_z^r$ for some efficient elements y and z, then there is $s \in I_0$ such that $y, z \in \mathrm{ri}(F_s)$ and $\Delta_y^r = \Delta_z^r = \Delta_s^r$.*
(ii) *For $i \in I_0$ the face F_i is a maximal efficient face if and only if Δ_i^r is a minimal element of the decomposition (4.5).*
(iii) *For all $i, j \in I_0$ with $i \neq j$, one has $\mathrm{ri}(\Delta_i^r) \cap \mathrm{ri}(\Delta_j^r) = \emptyset$.*
(iv) *Let F_i and F_j be two efficient adjacent vertices (zero-dimensional efficient faces) of Q. Then the edge joining them is efficient if and only if $\Delta_i^r \cap \Delta_j^r \neq \emptyset$.*

Proof Since the set Δ_Q^r is nonempty, the set Δ_Q does not lie on the border of Δ. Being a closed convex set, Δ_Q is the closure of its relative interior. Hence the relative interior of Δ_Q and the relative interior of Δ have at least one point in common and we deduce

$$\begin{aligned}
\Delta_Q = \Delta_Q \cap \Delta &= \mathrm{cl}\big(\mathrm{ri}(\Delta_Q \cap \Delta)\big) \\
&= \mathrm{cl}\big(\mathrm{ri}\Delta_Q \cap \mathrm{ri}\Delta\big) \subseteq \mathrm{cl}\big(\Delta_Q \cap \mathrm{ri}\Delta\big) \\
&\subseteq \mathrm{cl}\big(\Delta_Q^r\big).
\end{aligned}$$

The converse inclusion being evident, we obtain equality $\Delta_Q = \mathrm{cl}(\Delta_Q^r)$.

To prove (i) we apply the second part of Lemma 4.1.13 to have

$$\mathrm{ri}\big[\mathrm{cone}(\Delta_y^r)\big] = \mathrm{ri}\big[N_Q(y) \cap \mathbb{R}_+^k\big] = \mathrm{ri}\big[N_Q(y)\big] \cap \mathrm{int}(\mathbb{R}_+^k)$$
$$\mathrm{ri}\big[\mathrm{cone}(\Delta_z^r)\big] = \mathrm{ri}\big[N_Q(z) \cap \mathbb{R}_+^k\big] = \mathrm{ri}\big[N_Q(z)\big] \cap \mathrm{int}(\mathbb{R}_+^k).$$

If y and z were relative interior points of two different faces, in view of Theorem 2.3.26 we would have $\mathrm{ri}[N_Q(y)] \cap \mathrm{ri}[N_Q(z)] = \emptyset$ that contradicts the hypothesis. Hence they are relative interior points of the same face, say F_s. By definition $N(F_s) = N_Q(y)$ and we deduce $\Delta_s^r = \Delta_y^r$.

For (ii) assume F_i is a maximal efficient face. If for some $j \in I_0$ one has $\Delta_j^r \subseteq \Delta_i^r$, then by Lemma 4.1.13 there is some strictly positive vector that lies in the relative interior of the normal cone $N(F_j)$ and in the normal cone $N(F_i)$. We deduce that either $F_i = F_j$, or F_i is a proper face of F_j. The last case is impossible because F_j is also an efficient face and F_i is maximal. Conversely, if F_i is not maximal, then there is a face F_j that is efficient and contains F_i as a proper face. We have $\Delta_j \subseteq \Delta_i$. This inclusion is strict because the relative interiors of $N(F_i)$ and $N(F_j)$ do not meet each other. Thus, Δ_i is not minimal.

We proceed to (iii). If $ri(\Delta_i^r) \cap ri(\Delta_j^r) \neq \emptyset$, in view of Theorem 4.1.14 one has $\Delta_i = \Delta_j$, and hence $\Delta_i^r = \Delta_j^r$. By (i), there is some face that contains relative interior points of F_i and F_j in its relative interior. This implies $F_i = F_j$ a contradiction.

For the last property we know that the normal cone to the edge joining the vertices F_i and F_j satisfies $N([F_i, F_j]) = N(F_i) \cap N(F_j)$. Hence the edge $[F_i, F_j]$ is efficient if and only if the normal cone to it meets the set $ri(\Delta)$, or equivalently $\Delta_i^r \cap \Delta_j^r$ is nonempty. $\square$

A practical way to compute the weakly scalarizing set is to solve a system of linear equalities when the polyhedron Q is given by a system of linear inequalities.

Corollary 4.1.16 *Assume the polyhedron Q in $\mathbb{R}^k$ is determined by the system*

$$\langle a^i, y \rangle \leq b_i, i = 1, \cdots, m.$$

Then for every $y \in Q$, the set Δ_y consists of all solutions z to the following system

$$z_1 + \cdots + z_k = 1$$
$$\sum_{i \in I(y)} \alpha_i a^i = z$$
$$z_i \geq 0, i = 1, \cdots, k, \alpha_i \geq 0, i \in I(y).$$

In particular the weakly scalarizing set Δ_Q is the solution set to the above system with $I = \{1, \cdots, m\}$.

Proof According to Theorem 2.3.24 the normal cone to Q at y is the positive hull of the vectors $a^i, i \in I(y)$. Hence the set Δ_y is the intersection of the positive hull of these vectors and the simplex Δ, which is exactly the solution set to the system described in the corollary. For the second part of the corollary it suffices to observe that the normal cone of Q is the polar cone of the asymptotic cone of Q (Theorem 2.3.26) which, in view of Theorem 2.3.19, is the positive hull of the vectors $a^i, i = 1, \cdots, m$. $\square$

Example 4.1.17 Consider the polyhedron defined by

$$y_1 - y_2 - y_3 \leq 1$$
$$2y_1 \quad\quad + y_3 \leq 0.$$

By Corollary 4.1.16 the weakly scalarizing set Δ_Q is the solution set to the system

$$z_1 + z_2 + z_3 = 1$$
$$\alpha_1 + 2\alpha_2 = z_1$$
$$-\alpha_1 = z_2$$
$$-\alpha_1 + \alpha_2 = z_3$$
$$z_1, z_2, z_3, \alpha_1, \alpha_2 \geqq 0.$$

This produces a unique solution z with $z_1 = 2/3$, $z_2 = 0$ and $z_3 = 1/3$. Then Δ_Q consists of this solution only. The scalarizing set Δ_Q^r is empty, which shows that Q has no efficient point. Its weakly efficient set is determined by the problem

$$\text{maximize} \quad \tfrac{2}{3} y_1 + \tfrac{1}{3} y_3$$
$$\text{subject to} \quad y \in Q.$$

It follows from the second inequality determining Q that the maximum value of the objective function is zero and attained on the face given by the equations $2y_1 + y_3 = 0$ and $y_1 - y_2 - y_3 \leqq 1$.

In the next example we show how to compute the scalarizing set when the polyhedron is given by a system of equalities (see also Exercise 4.4.13 at the end of this chapter).

Example 4.1.18 Let Q be a polyhedron in $\mathbb{R}^3$ determined by the sytem

$$y_1 + y_2 + y_3 = 1$$
$$y_1 - y_2 \qquad = 0$$
$$y_3 \geqq 0.$$

We consider the solution $\overline{y} = (1/2, 1/2, 0)^T$ and want to compute the scalarizing set at this solution if it exists. As the proof of Theorem 4.1.14 indicates, a vector $\lambda \in \Delta$ in $\mathbb{R}^3$ is a weakly scalarizing vector of Q at $\overline{y}$ if and only if it is normal to Q at that point. Since the last component of $\overline{y}$ is zero, a vector λ is normal to Q at $\overline{y}$ if and only if there are real numbers α, β and a positive number γ such that

$$\lambda = \alpha \begin{pmatrix} 1 \\ 1 \\ 1 \end{pmatrix} + \beta \begin{pmatrix} 1 \\ -1 \\ 0 \end{pmatrix} - \gamma \begin{pmatrix} 0 \\ 0 \\ 1 \end{pmatrix}.$$

We deduce $\lambda \in \Delta_{\overline{y}}$ if and only if

$$\alpha + \beta \geq 0$$
$$\alpha - \beta \geq 0$$

$$\alpha - \gamma \geqq 0$$
$$(\alpha + \beta) + (\alpha - \beta) + (\alpha - \gamma) = 1$$

and hence $\Delta_{\bar{y}}$ consists of vectors λ whose components satisfy $0 \leqq \lambda_3 \leqq 1/3$, $\lambda_1 + \lambda_2 = 1 - \lambda_3$ and $\lambda_1, \lambda_2 \geqq 0$.

To obtain the scalarizing vectors, it suffices to choose λ as above with an additional requirement that $\lambda_i > 0$ for $i = 1, 2, 3$.

Structure of the set of efficient points

Given a convex polyhedron Q in the space $\mathbb{R}^k$, the set of its efficient elements is not simple. For instance, it is generally not convex, and an edge of it is not necessarily efficient even if its two extreme end-points are efficient vertices. Despite of this, a number of nice properties of this set can be scrutinized.

Corollary 4.1.19 *Let Q be a convex polyhedron in $\mathbb{R}^k$. The following statements hold.*

(i) *If a relative interior point of a face of Q is efficient or weakly efficient, then so is every point of that face.*

(ii) *If Q has vertices, it has an efficient vertex (respectively a weakly efficient vertex) provided that it has efficient (respectively weakly efficient) elements.*

Proof Since the normal cone to Q at every point of a face contains the normal cone at a relative interior point, the first statement follows directly from Theorem 4.1.10.

For the second statement let y be an efficient point of the polyhedron Q. By Theorem 4.1.10 one can find a strictly positive vector λ such that y is a maximizer of the linear functional $\langle \lambda, . \rangle$ on Q. The face which contains y in its relative interior maximizes the above functional. According to Corollary 2.3.14 there is a vertex of Q inside that face and in view of Theorem 4.1.10 this vertex is an efficient point of Q. The case of weakly efficient points is proven by the same argument. $\square$

A subset P of $\mathbb{R}^k$ is called *arcwise connected* if for any pair of points y and z in P, there are a finite number of points $y^0, \cdots, y^\ell$ in P such that $y^0 = y$, $y^\ell = z$ and the segments $[y^i, y^{i+1}]$, $i = 0, \cdots, \ell - 1$ lie all in P.

Theorem 4.1.20 *The sets of all efficient points and weakly efficient points of a convex polyhedron consist of faces of the polyhedron and are closed and arcwise connected.*

Proof By analogy, it suffices to prove the theorem for the efficient set. According to Corollary 4.1.19, if a point $\bar{y}$ in Q is efficient, then the whole face containing y in its relative interior is a face of efficient elements. Hence, $\mathrm{Max}(Q)$ consists of faces of Q if it is nonempty. Moreover, as faces are closed, their union is a closed set.

Now we prove the connectedness of this set by assuming that Q has efficient elements. Let y and z be any pair of efficient points of Q. We may assume without loss of generality that y is a relative interior point of a face Q_y and z is a relative interior point of a face Q_z. Consider the decomposition (4.5) of the scalarizing set

Δ_Q^r. For a face F of Q, the scalarizing set $N(F) \cap \mathrm{ri}(\Delta)$ is denoted by $\Delta_{i(F)}^r$. Let λ_y be a relative interior point of the set $\Delta_{i(Q_y)}^r$ and λ_z a relative interior point of $\Delta_{i(Q_z)}^r$. Then the segment joining λ_y and λ_z lies in Δ_Q^r. The decomposition of the latter set induces a decomposition of the segment $[\lambda_y, \lambda_z]$ by $[\lambda_i, \lambda_{i+1}], i = 0, \cdots, \ell - 1$ where $\lambda_0 = \lambda_y, \lambda_\ell = \lambda_z$. Let $Q_1, \cdots, Q_\ell$ be faces of Q such that

$$[\lambda_j, \lambda_{j+1}] \subseteq \Delta_{i(Q_{j+1})}^r \ j = 0, \cdots, \ell - 1.$$

For every j, we choose a relative interior point y^j of the face Q_j. Then λ_j belongs to the normal cones to Q at y^j and y^{j+1}. Consequently, the points y^j and y^{j+1} lie in the face $\mathrm{argmax}_Q \langle \lambda_j, . \rangle$ and so does the segment joining them. As $\lambda_j \in \Delta_Q^r$, by Theorem 4.1.10 the segment $[y^j, y^{j+1}]$ consists of efficient points of Q. Moreover, as the vector λ_0 belongs to the normal cones to Q at y and at y_1, we conclude that the segment $[y, y^1]$ is composed of efficient points of Q. Similarly we have that $[y^{\ell-1}, z]$ lies in the set $\mathrm{Max}(Q)$. Thus, the union $[y, y^1] \cup [y^1, y^2] \cup \cdots [y^{\ell-1}, z]$ forms a path of efficient elements joining y and z. This completes the proof. □

We know that every efficient point of a convex polyhedron is contained in a maximal efficient face. Hence the set of efficient points is the union of maximal efficient faces. Dimension of a maximal efficient face may vary from zero to $k - 1$.

Corollary 4.1.21 *Let Q be a convex polyhedron in $\mathbb{R}^k$. The following statements hold.*

(i) *Q has a zero-dimensional maximal efficient face if and only if its efficient set is a singleton.*

(ii) *Every $(k - 1)$-dimensional efficient face of Q, if any exists, is maximal. In particular in the two dimensional space $\mathbb{R}^2$ every efficient edge of Q is maximal if the efficient set of Q consists of more than two elements.*

(iii) *An efficient face F of Q is maximal if and only if the restriction of the decomposition of Δ_Q^r on Δ_F consists of one element only.*

Proof The first statement follows from the arcwise connectedness of the efficient set of Q. In $\mathbb{R}^k$ a proper face of Q is of dimension at most $k - 1$. Moreover, a k-dimensional polyhedron cannot be efficient, for its interior points are not maximal. Hence, if the dimension of an efficient face is equal to $k - 1$, it is maximal.

The last statement follows immediately from Theorem 4.1.15. □

Example 4.1.22 Let Q be a polyhedron in $\mathbb{R}^3$ defined by the system

$$\begin{aligned} x_1 \quad\ \ + x_3 &\leq 1 \\ x_2 + x_3 &\leq 1 \\ x_1, x_2, x_3 &\geq 0. \end{aligned}$$

Since Q is bounded, it is evident that the weakly scalarizing set Δ_Q is the whole standard simplex Δ and the scalarizing set is the relative interior of Δ. Denote

$$q^1 = \begin{pmatrix} 1 \\ 0 \\ 0 \end{pmatrix}, \; q^2 = \begin{pmatrix} 0 \\ 1 \\ 0 \end{pmatrix}, \; q^3 = \begin{pmatrix} 0 \\ 0 \\ 1 \end{pmatrix}, \; q^4 = \begin{pmatrix} 1/2 \\ 0 \\ 1/2 \end{pmatrix}, \; q^5 = \begin{pmatrix} 0 \\ 1/2 \\ 1/2 \end{pmatrix}.$$

Applying Corollary 4.1.16 we obtain the following decomposition of Δ_Q^r:

(i) $\mathrm{ri}[q^4, q^5]$ is the scalarizing set of the face determined by the equalities $x_1 + x_3 = 1$ and $x_2 + x_3 = 1$;

(ii) $\mathrm{ri}(\mathrm{co}([q^3, q^4, q^5]))$ is the scalarizing set of the face determined by the equalities $x_1 + x_3 = 1$, $x_2 + x_3 = 1$ and $x_1 = x_2 = 0$;

(iii) $\mathrm{ri}(\mathrm{co}([q^1, q^2, q^4, q^5]))$ is the scalarizing set of the face determined by the equalities $x_1 + x_3 = 1$, $x_2 + x_3 = 1$ and $x_3 = 0$.

In view of Corollary 4.1.21 the one dimensional face (edge) determined by $x_1 + x_3 = 1$ and $x_2 + x_3 = 1$ is a maximal efficient face.

4.2 Multiobjective Linear Problems

The central multiobjective linear programming problem which we propose to study throughout is denoted (MOLP) and written in the form :

$$\text{Maximize} \quad Cx$$
$$\text{subject to } x \in X,$$

where X is a nonempty convex polyhedron in $\mathbb{R}^n$ and C is a real $k \times n$-matrix. This problem means finding a *Pareto efficient (Pareto maximal) solution* $\bar{x} \in X$ such that $C\bar{x} \in \mathrm{Max}(C(X))$. In other words, a feasible solution $\bar{x}$ solves (MOLP) if there is no feasible solution $x \in X$ such that

$$C\bar{x} \leq Cx \text{ and } C\bar{x} \neq Cx.$$

The efficient solution set of (MOLP) is denoted $S(MOLP)$. When x is an efficient solution, the vector Cx is called an *efficient or maximal value* of the problem. In a similar manner one defines the set of *weakly efficient solutions* WS(MOLP) to be the set of all feasible solutions whose image by C belong to the weakly efficient set WMax($C(X)$). It is clear that an efficient solution is a weakly efficient solution, but not vice versa as we have already discussed in the preceding section. When the feasible set X is given by the system

$$Ax = b$$
$$x \geq 0,$$

where A is a real $m \times n$-matrix and b is a real m-vector, we say that (MOLP) is given in *standard form*, and it is given in *canonical form* if X is determined by the system

$$Ax \leqq b.$$

The matrix C is also considered as a linear operator from $\mathbb{R}^n$ to $\mathbb{R}^k$, and so its kernel consists of vectors x with $Cx = 0$.

Theorem 4.2.1 *Assume that the problem (MOLP) has feasible solutions. Then the following assertions hold.*

(i) *(MOLP) admits efficient solutions if and only if*

$$C(X_\infty) \cap \mathbb{R}_+^k = \{0\}.$$

(ii) *(MOLP) admits weakly efficient solutions if and only if*

$$C(X_\infty) \cap \text{int}(\mathbb{R}_+^k) = \emptyset.$$

In particular, if all asymptotic rays of X belong to the kernel of C, then (MOLP) has an efficient solution.

Proof By definition, (MOLP) has an efficient solution if and only if the set $C(X)$ has an efficient point, which, in virtue of Theorem 4.1.7, is equivalent with

$$[C(X)]_\infty \cap \mathbb{R}_+^k = \{0\}.$$

Now the first assertion is deduced from this equivalence and from the fact that the asymptotic cone of $C(X)$ coincides with the cone $C(X_\infty)$ (Corollary 2.3.17). The second assertion is proven by a similar argument.   □

Example 4.2.2 Assume that the feasible set X of the problem (MOLP) is given by the system

$$\begin{aligned} x_1 + x_2 - x_3 &= 5 \\ x_1 - x_2 &= 4 \\ x_1, x_2, x_3 &\geqq 0. \end{aligned}$$

It is nonempty and parametrically presented as

$$X = \left\{ \begin{pmatrix} t+4 \\ t \\ 2t-1 \end{pmatrix} : t \geqq \frac{1}{2} \right\}.$$

Its asymptotic cone is given by

$$X_\infty = \left\{ \begin{pmatrix} t \\ t \\ 2t \end{pmatrix} : t \geqq 0 \right\}.$$

Consider an objective function C with values in $\mathbb{R}^2$ given by the matrix

$$C = \begin{pmatrix} 1 & 0 & 1 \\ -2 & -4 & 0 \end{pmatrix}.$$

Then the image of X_∞ under C is the set

$$C(X_\infty) = \left\{ \begin{pmatrix} 3t \\ -6t \end{pmatrix} : t \geqq 0 \right\},$$

that has only the zero vector in common with the positive orthant. In view of Theorem 4.2.1 the problem has maximal solutions.

Now we choose another objective function C' given by

$$C' = \begin{pmatrix} -1 & 1 & 0 \\ 0 & 0 & 1 \end{pmatrix}.$$

Then the image of X_∞ under C' is the set

$$C'(X_\infty) = \left\{ \begin{pmatrix} 0 \\ 2t \end{pmatrix} : t \geqq 0 \right\},$$

that has no common point with the interior of the positive orthant. Hence the problem admits weakly efficient solutions. It has no efficient solution because the intersection of $C'(X_\infty)$ with the positive orthant does contain positive vectors.

Definition 4.2.3 The objective function of the problem (MOLP) is said to be bounded (respectively weakly bounded) from above if there is no vector $v \in X_\infty$ such that

$$Cv \geq 0 \text{ (respectively } Cv > 0).$$

We shall simply say that (MOLP) is bounded if its objective function is bounded from above. Of course, a bounded problem is weakly bounded and not every weakly bounded problem is bounded. A sufficient condition for a problem to be bounded is given by the inequality

$$Cx \leqq a \text{ for every } x \in X,$$

where a is some vector from $\mathbb{R}^k$. This condition is also necessary when $k = 1$, but not so when $k > 1$.

Example 4.2.4 Consider the bi-objective problem

$$\text{Maximize} \quad \begin{pmatrix} -3 & 1 & 1 \\ 0 & 1 & 0 \end{pmatrix} \begin{pmatrix} x_1 \\ x_2 \\ x_3 \end{pmatrix}$$

$$\text{subject to} \quad \begin{pmatrix} 1 & -1 & 0 \\ 0 & 0 & 1 \end{pmatrix} \begin{pmatrix} x_1 \\ x_2 \\ x_3 \end{pmatrix} = \begin{pmatrix} 0 \\ 1 \end{pmatrix}$$

$$x_1, x_2, x_3 \geqq 0.$$

The feasible set and its asymptotic cone are given respectively by

$$X = \left\{ \begin{pmatrix} t \\ t \\ 1 \end{pmatrix} \in \mathbb{R}^3 : t \geqq 0 \right\}$$

and

$$X_\infty = \left\{ \begin{pmatrix} t \\ t \\ 0 \end{pmatrix} \in \mathbb{R}^3 : t \geqq 0 \right\}$$

Then for every asymptotic direction $v = (t, t, 0)^T \in X_\infty$ one has

$$Cv = \begin{pmatrix} -2t \\ t \end{pmatrix} \not\geqq 0.$$

By definition the objective function is bounded. Nevertheless the value set of the problem consists of vectors

$$C(X) = \left\{ \begin{pmatrix} -2t + 1 \\ t \end{pmatrix} : t \geqq 0 \right\}$$

for which no vector $a \in \mathbb{R}^2$ satisfies $Cx \leqq a$ for all $x \in X$.

Corollary 4.2.5 *The problem (MOLP) has efficient solutions (respectively weakly efficient solutions) if and only if its objective function is bounded (respectively weakly bounded).*

Proof This is immediate from Theorem 4.2.1. ☐

The following theorem provides a criterion for efficiency in terms of normal directions.

Theorem 4.2.6 *Let $\bar{x}$ be a feasible solution of (MOLP). Then*

(i) *$\bar{x}$ is an efficient solution if and only if the normal cone $N_X(\bar{x})$ to X at $\bar{x}$ contains some vector $C^T\lambda$ with λ a strictly positive vector of $\mathbb{R}^k$;*
(ii) *$\bar{x}$ is a weakly efficient point if and only if the normal cone $N_X(\bar{x})$ to X at $\bar{x}$ contains some vector $C^T\lambda$ with λ a nonzero positive vector of $\mathbb{R}^k$.*

Proof If the vector $C^T\lambda$ with λ strictly positive, is normal to X at $\bar{x}$, then

$$\langle C^T\lambda, x - \bar{x}\rangle \le 0 \text{ for every } x \in X$$

which means that

$$\langle \lambda, Cx\rangle \le \langle \lambda, C\bar{x}\rangle \text{ for every } x \in X.$$

By Theorem 4.1.4 the vector $C\bar{x}$ is an efficient point of the set $C(X)$. By definition, $\bar{x}$ is an efficient solution of (MOLP).

Conversely, if $C\bar{x}$ is an efficient point of $C(X)$, then by Theorem 4.1.10, the normal cone to $C(X)$ at $C\bar{x}$ contains a strictly positive vector, denoted by λ. We deduce that

$$\langle C^T\lambda, x - \bar{x}\rangle \le 0 \text{ for all } x \in X.$$

This shows that the vector $C^T\lambda$ is normal to X at $\bar{x}$. The second assertion is proven similarly.   □

Example 4.2.7 We reconsider the bi-objective problem given in Example 4.2.2

$$\text{Maximize } \begin{pmatrix} 1 & 0 & 1 \\ -2 & -4 & 0 \end{pmatrix}\begin{pmatrix} x_1 \\ x_2 \\ x_3 \end{pmatrix}$$

$$\text{subject to } \begin{array}{rcl} x_1 + x_2 - x_3 &=& 5 \\ x_1 - x_2 &=& 4 \\ x_1, x_2, x_3 &\ge& 0. \end{array}$$

Choose a feasible solution $\bar{x} = (9/2, 1/2, 0)^T$ corresponding to $t = 1/2$. The normal cone to the feasible set at $\bar{x}$ is the positive hull of the hyperplane of basis $\{(1, 1, -1)^T, (1, -1, 0)^T\}$ (the row vectors of the constraint matrix) and the vector $(0, 0, -1)^T$ (the constraint $x_3 \ge 0$ is active at this point). In other words, this normal cone is the half-space determined by the inequality

$$x_1 + x_2 + 2x_3 \le 0. \tag{4.6}$$

The image of the positive orthant of the value space $\mathbb{R}^2$ under C^T is the positive hull of the vectors

$$v_1 = \begin{pmatrix} 1 & -2 \\ 0 & -4 \\ 1 & 0 \end{pmatrix} \begin{pmatrix} 1 \\ 0 \end{pmatrix} = \begin{pmatrix} 1 \\ 0 \\ 1 \end{pmatrix} \text{ and } v_2 = \begin{pmatrix} 1 & -2 \\ 0 & -4 \\ 1 & 0 \end{pmatrix} \begin{pmatrix} 0 \\ 1 \end{pmatrix} = \begin{pmatrix} -2 \\ -4 \\ 0 \end{pmatrix}.$$

Using inequality (4.6) we deduce that the vector v_2 lies in the interior of the normal cone to the feasible set at $\bar{x}$. Hence that normal cone does contain a vector $C^T \lambda$ with some strictly positive vector λ. By Theorem 4.2.6 the solution $\bar{x}$ is efficient. It is routine to check that the solution $\bar{x}$ is a vertex of the feasible set.

If we pick another feasible solution, say $\bar{z} = (5, 1, 1)^T$, then the normal cone to the feasible set at $\bar{z}$ is the hyperplane determine by equation

$$x_1 + x_2 + 2x_3 = 0.$$

Direct calculation shows that the vectors v_1 and v_2 lie in different sides of the normal cone at $\bar{z}$. Hence there does exist a strictly positive vector λ in $\mathbb{R}^2$ such that $C^T \lambda$ is contained in that cone. Consequently, the solution $\bar{z}$ is efficient too.

4.3 Scalarization

We associate with a nonzero k-vector λ a scalar linear problem, denoted (LP$_\lambda$)

$$\text{maximize } \langle \lambda, Cx \rangle$$
$$\text{subject to } x \in X.$$

This problem is referred to as a *scalarized problem* of (MOLP) and λ is called a *scalarizing vector*. Now we shall see how useful scalarized problems are in solving multiobjective problems.

Theorem 4.3.1 *The following statements hold.*

(i) *A feasible solution $\bar{x}$ of (MOLP) is efficient if and only if there is a strictly positive k-vector λ such that $\bar{x}$ is an optimal solution of the scalarized problem (LP$_\lambda$).*

(ii) *A feasible solution $\bar{x}$ of (MOLP) is weakly efficient if and only if there is a nonzero positive k-vector λ such that $\bar{x}$ is an optimal solution of the scalarized problem (LP$_\lambda$).*

Proof If $\bar{x}$ is an efficient solution of (MOLP), then, in view of Theorem 4.2.6, there is a strictly positive vector λ such that $C^T \lambda$ is a normal vector to X at $\bar{x}$. This implies that $\bar{x}$ maximizes the linear functional $\langle \lambda, C(.) \rangle$ on X, that is, $\bar{x}$ is an optimal solution of (LP$_\lambda$).

Conversely, if $\bar{x}$ solves the problem (LP$_\lambda$) with λ strictly positive, then $C^T \lambda$ is a normal cone to X at $\bar{x}$. Again, in view of Theorem 4.2.6, the point $\bar{x}$ is an efficient solution of (MOLP). The proof of the second statement follows the same line. □

We notice that Theorem 4.3.1 remains valid if the scalarizing vector λ is taken from the standard simplex, that is $\lambda_1 + \cdots + \lambda_k = 1$. Then another formulation of the theorem is given by equalities

$$S(MOLP) = \bigcup_{\lambda \in ri\Delta} S(LP_\lambda) \tag{4.7}$$

$$WS(MOLP) = \bigcup_{\lambda \in \Delta} S(LP_\lambda) \tag{4.8}$$

where $S(LP_\lambda)$ denotes the optimal solution set of (LP_λ). It was already mentioned that a weakly efficient solution is not necessarily an efficient solution. Consequently a positive, but not strictly positive vector λ may produce weakly efficient solutions which are not efficient. Here is an exception.

Corollary 4.3.2 *Assume for a positive vector λ, the set consisting of the values Cx with x being optimal solution of (LP_λ) is a singleton, in particular when (LP_λ) has a unique solution. Then every optimal solution of (LP_λ) is an efficient solution of (MOLP).*

Proof Let x be an optimal solution of (LP_λ) and let y be a feasible solution of (MOLP) such that $Cy \geq Cx$. Since λ is positive, one has

$$\langle \lambda, Cx \rangle \leqq \langle \lambda, Cy \rangle.$$

Actually we have equality because x solves (LP_λ). Hence y solves (LP_λ) too. By hypothesis $Cx = Cy$ which shows that x is an efficient solution of (MOLP). □

Equalities (4.7) and (4.8) show that efficient and weakly efficient solutions of (MOLP) can be generated by solving a family of scalar problems. It turns out that a finite number of such problems are sufficient to generate the whole efficient and weakly efficient solution sets of (MOLP).

Corollary 4.3.3 *There exists a finite number of strictly positive vectors (respectively positive vectors) $\lambda^i, i = 1, \cdots, p$ such that*

$$S(MOLP) = \bigcup_{i=1}^{p} S(LP_{\lambda^i})$$

$$(\text{respectively } WS(MOLP) = \bigcup_{i=1}^{p} S(LP_{\lambda^i}))$$

Proof It follows from Theorem 3.1.3 that if an efficient solution is a relative interior of a face of the feasible polyhedron and an optimal solution of (LP_λ) for some strictly positive vector λ, then the whole face is optimal for (LP_λ). Since the number of faces is finite, a finite number of such vectors λ is sufficient to generate all efficient solutions of (MOLP). The case of weakly efficient solutions is treated in the same way.   □

Corollary 4.3.4 *Assume that (MOLP) has an efficient solution and (LP$_\lambda$), where λ is a nonzero positive vector, has an optimal solution. Then there is an efficient solution of (MOLP) among the optimal solutions of (LP$_\lambda$).*

Proof Apply Theorem 4.3.1 and the method of Corollary 4.1.8.   □

Corollary 4.3.5 *Assume that the scalarized problems*

$$\text{maximize } \langle c^i, x \rangle$$
$$\text{subject to } x \in X$$

where $c^i, i = 1, \cdots, k$ are the columns of the matrix C^T, are solvable. Then (MOLP) has an efficient solution.

Proof The linear problems mentioned in the corollary correspond to the scalarized problems (LP$_\lambda$) with $\lambda = (0, \cdots, 1, \cdots, 0)^T$ where the one is on the ith place, $i = 1, \cdots, k$. These problems provide weakly efficient solutions of (MOLP). The linear problem whose objective is the sum $\langle c^1, x \rangle + \cdots + \langle c^k, x \rangle$ is solvable too. It is the scalarized problem with $\lambda = (1, \cdots, 1)^T$, and hence by Theorem 4.3.1, (MOLP) has efficient solutions.   □

Decomposition of the scalarizing set

Given a feasible solution x of (MOLP) we denote the set of all vectors $\lambda \in \Delta$ such that x solves (LP$_\lambda$) by $\Lambda(x)$, and the union of all these $\Lambda(x)$ over $x \in X$ by $\Lambda(X)$. We denote also

$$\Lambda^r(x) = \Lambda(x) \cap \mathrm{ri}(\Delta)$$
$$\Lambda^r(X) = \Lambda(X) \cap \mathrm{ri}(\Delta).$$

The sets $\Lambda^r(X)$ and $\Lambda(X)$ are respectively called the *scalarizing and weakly scalarizing sets* of (MOLP). The decomposition results for efficient elements (Theorems 4.1.14 and 4.1.15) are easily adapted to decompose the weakly scalarizing and scalarizing sets of the problem (MOLP). We deduce a useful corollary below for computing purposes.

Corollary 4.3.6 *The following assertions hold for (MOLP).*

  (i) *A feasible solution $x \in X$ is efficient (respectively weakly efficient) if and only if $\Lambda^r(x)$ (respectively $\Lambda(x)$) is nonempty.*
  (ii) *If X has vertices, then the set $\Lambda^r(X)$ (respectively $\Lambda(X)$) is the union of the sets $\Lambda^r(x^i)$ (respectively $\Lambda(x^i)$) where x^i runs over the set of all efficient (respectively weakly efficient) vertices of (MOLP).*

(iii) *If X is given by system*

$$\langle a^i, x \rangle \leq b_i, i = 1, \cdots, m$$

and x is a feasible solution, then the set $\Lambda(x)$ consists of all solutions λ to the following system

$$\lambda_1 + \cdots + \lambda_k = 1$$
$$\sum_{i \in I(x)} \alpha_i a^i = \lambda_1 c^1 + \cdots + \lambda_k c^k$$
$$\lambda_i \geq 0, i = 1, \cdots, k, \alpha_i \geq 0, i \in I(x).$$

In particular the weakly scalarizing set $\Lambda(X)$ is the solution set to the above system with $I = \{1, \cdots, m\}$.

Proof The first assertion is clear from Theorem 4.3.1. For the second assertion we observe that when X has vertices, every face of X has vertices too (Corollary 2.3.6). Hence the normal cone of X is the union of the normal cones to X at its vertices. Moreover, by writing the objective function $\langle \lambda, C(.) \rangle$ of (LP_λ) in the form $\langle C^T \lambda, . \rangle$, we deduce that

$$\Lambda(x) = \{\lambda \in \mathbb{R}^k : C^T \lambda \in N_X(x) \cap C^T(\Delta)\}. \tag{4.9}$$

Consequently,

$$\Lambda(X) = \bigcup_{x \in X} \Lambda(x)$$
$$= \bigcup \{\lambda : C^T \lambda \in N_X(x) \cap C^T(\Delta), x \in X\}$$
$$= \bigcup \{\lambda : C^T \lambda \in N_X(x) \cap C^T(\Delta), x \text{ is a vertex of } X\}$$
$$= \bigcup \{\Lambda(x) : x \text{ is a weakly efficient vertex of } X\}.$$

The proof for efficient solutions is similar. The last assertion is derived from (4.9) and Corollary 4.1.16. $\square$

Example 4.3.7 We reconsider the bi-objective problem

$$\text{Maximize} \quad \begin{pmatrix} 1 & 1 \\ 2 & -1 \end{pmatrix} \begin{pmatrix} x_1 \\ x_2 \end{pmatrix}$$

$$\text{subject to} \quad \begin{aligned} x_1 + x_2 &\leq 1 \\ 3x_1 + 2x_2 &\leq 2. \end{aligned}$$

We wish to find the weakly scalarizing set of this problem. According to the preceding corollary, it consists of positive vectors λ from the standard simplex of $\mathbb{R}^2$, solutions to the following system:

$$\lambda_1 + \lambda_2 = 1$$

$$\alpha_1 \begin{pmatrix} 1 \\ 1 \end{pmatrix} + \alpha_2 \begin{pmatrix} 3 \\ 2 \end{pmatrix} = \lambda_1 \begin{pmatrix} 1 \\ 1 \end{pmatrix} + \lambda_2 \begin{pmatrix} 2 \\ -1 \end{pmatrix}$$

$$\alpha_1, \alpha_2 \geqq 0, \lambda_1, \lambda_2 \geqq 0.$$

Solving this system we obtain $\lambda = \begin{pmatrix} t \\ 1 - t \end{pmatrix}$ with $7/8 \leqq t \leqq 1$. For $t = 1$, the scalarized problem associated with λ is of the form

$$\begin{aligned} \text{maximize} \quad & x_1 + x_2 \\ \text{subject to} \quad & x_1 + x_2 \leqq 1 \\ & 3x_1 + 2x_2 \leqq 2. \end{aligned}$$

It can be seen that x solves this problem if and only if $x_1 + x_2 = 1$ and $x_1 \leqq 0$. These solutions form the set of weakly efficient solutions of the multiobjective problem.

For $t = 7/8$, the scalarized problem associated with $\lambda = (7/8, 1/8)^T$ is of the form

$$\begin{aligned} \text{maximize} \quad & \frac{9}{8}x_1 + \frac{3}{4}x_2 \\ \text{subject to} \quad & x_1 + x_2 \leqq 1 \\ & 3x_1 + 2x_2 \leqq 2. \end{aligned}$$

Its optimal solutions are given by $3x_1 + 2x_2 = 2$ and $x_1 \geqq 0$. Since λ is strictly positive, these solutions are efficient solutions of the multiobjective problem. If we choose $\lambda = (1/2, 1/2)^T$ outside of the scalarizing set, then the associated scalarized problem has no optimal solution.

Structure of the efficient solution set

We knew in Chap. 3 that the optimal solution set of a scalar linear problem is a face of the feasible set. This property, unfortunately, is no longer true when the problem is multiobjective. However, a few interesting properties of the efficient set of a polyhedron we established in the first section are still valid for the efficient solution set and exposed in the next theorem.

Theorem 4.3.8 *The efficient solutions of the problem (MOLP) have the following properties.*

(i) *If a relative interior point of a face of X is an efficient or weakly efficient solution, then so is every point of that face.*

(ii) *If X has a vertex and (MOLP) has an efficient (weakly efficient) solution, then it has an efficient (weakly efficient) vertex solution.*

(iii) *The efficient and weakly efficient solution sets of (MOLP) consist of faces of the feasible polyhedron and are closed and arcwise connected.*

Proof Since the normal cone to X at every point of a face contains the normal cone at a relative interior point, the first property follows directly from Theorem 4.2.6.

Further, under the hypothesis of (ii) there is a strictly positive vector $\lambda \in \mathbb{R}^k$ such that the scalarized problem (LP$_\lambda$) is solvable. The argument in proving (ii) of Corollary 4.1.19 is applicable to obtain an optimal vertex of (LP$_\lambda$) which is also an efficient vertex solution of (MOLP).

The proof of the last property is much similar to the one of Theorem 4.1.15. We first notice that in view of (i) the efficient and weakly efficient solution sets are composed of faces of the feasible set X, and as the number of faces of X is finite, they are closed. We now prove the arcwise connectedness of the weakly efficient solution set, the argument going through for efficient solutions too. Let x and y be two weakly efficient solutions, relative interior points of efficient faces X_x and X_y of X. Let λ_x and λ_y be relative interior vectors of the weakly scalarizing sets $\Lambda(X_x)$ and $\Lambda(X_y)$. The decomposition of the weakly scalarizing set $\Lambda(X)$ induces a decomposition of the segment joining λ_x and λ_y by

$$[\lambda_x, \lambda_y] = [\lambda_1, \lambda_2] \cup [\lambda_2, \lambda_3] \cup \cdots \cup [\lambda_{\ell-1}, \lambda_\ell]$$

with $\lambda_1 = \lambda_x$, $\lambda_\ell = \lambda_y$ and $[\lambda_i, \lambda_{i+1}] \subseteq \Lambda(X_i)$ for some face X_i of X, $i = 1, ..., \ell - 1$. Since λ_i belongs simultaneously to $\Lambda(X_i)$ and $\Lambda(X_{i+1})$, there is some common point $x^i \in X_i \cap X_{i+1}$, $i = 1, ..., \ell - 1$. It is clear that $[x, x_1] \cup [x_1, x_2] \cup \cdots \cup [x_{\ell-1}, y]$ is an arcwise path joining x and y and each member segment $[x_i, x_{i+1}]$ is efficient because being in the face X_i, $i = 1, ..., \ell$ with $x_\ell = y$.   □

4.4 Exercises

4.4.1 *Find maximal elements of the sets determined by the following systems*

(a)
$$\begin{cases} 2x + \ y \le 15 \\ x + 3y \le 20 \\ x, \ \ y \ge 0. \end{cases}$$

(b)
$$\begin{cases} x + 4y \le 12 \\ -2x + \ y \le 0 \\ x, \ \ y \ge 0. \end{cases}$$

(c)
$$\begin{cases} x + 2y \qquad \le 20 \\ 7x \qquad + \ z \le 6 \\ 3y + 4z \ \le 30 \\ x, \ y, \ z \ge 0. \end{cases}$$

(d)
$$\begin{cases} x + 2y + 3z \le 70 \\ x + \ y + \ z \le 50 \\ - \ y + \ z \le 0 \\ x, \ y, \ \ z \ge 0. \end{cases}$$

4.4.2 *Find maximal and weakly maximal elements of the following sets*

$$Q_1 = \left\{ \begin{pmatrix} x_1 \\ x_2 \\ x_3 \end{pmatrix} \in \mathbb{R}^3 : x_1 \geq 0,\, x_2 \geq 0,\, x_3 \geq 0,\, x_2^2 + x_3^2 \leq 1 \right\}$$

$$Q_2 = co(A, B) \text{ with } A = \left\{ \begin{pmatrix} 1 \\ 0 \\ s \end{pmatrix} \in \mathbb{R}^3 : 0 \leq s \leq 1 \right\}$$

$$\text{and } B = \left\{ \begin{pmatrix} 0 \\ x_2 \\ x_3 \end{pmatrix} \in \mathbb{R}^3 : x_2 \geq 0,\, x_3 \geq 0,\, x_2^2 + x_3^2 \leq 1 \right\}.$$

4.4.3 *We say a real function g on $\mathbb{R}^k$ is increasing if $x, y \in \mathbb{R}^k$ and $x \geq y$ imply $g(x) > g(y)$, and it is weakly increasing if $x > y$ implies $g(x) > g(y)$. Prove that g is increasing (respectively weakly increasing) if and only if for every nonempty subset Q of $\mathbb{R}^k$, every maximizer of g on Q is an efficient (respectively weakly maximal) element of Q.*

4.4.4 *Let Q be a closed set in $\mathbb{R}^k$. Prove the following statements.*

(i) *The set $WMax(Q)$ is closed.*
(ii) *The set $Max(Q)$ is closed provided that $k = 2$ and $Q - \mathbb{R}^2_+$ is convex.*
(iii) *$Max(-Q) = -Min(Q)$ and $Max(\alpha Q) = \alpha\, Max(Q)$ for every $\alpha > 0$.*

4.4.5 *Let P and Q be two convex polyhedra in $\mathbb{R}^k$.*

(i) *Prove that $Max(P + Q) \subseteq Max(P) + Max(Q)$.*
(ii) *Find conditions under which equality holds in (i).*

4.4.6 *Prove that the set of maximal elements of a convex polytope is included in the convex hull of the maximal vertices. Is the converse true?*

4.4.7 *An element x of a set P in $\mathbb{R}^k$ is said to be dominated if there is some $x' \in P$ such that $x' \geq x$. Prove that the set of dominated elements of a convex polyhedral set is convex and if a face contains a dominated element, its relative interior points are dominated too.*

4.4.8 A diet problem. *A multiobjective version of the diet problem in hospital consists of finding a combination of foods for a patient to minimize simultaneously the cost of the menu and the number of calories under certain nutritional requirements prescribed by a treating physician. Assume a menu is composed of three main types of foods: meat with potatoes, fish with rice and vegetables. The nutrition facts, calories in foods and price per servings are given below*

	Fats	Carbohydrates	Vitamin	Calories	Prices/serving
Meat + potatoes	0.2	0.2	0.06	400	1.5
Fish + rice	0.1	0.2	0.08	300	1.5
Vegetables	0	0.05	0.8	50	0.8

Using three variables: x= number of servings of meat, y= number of servings of fish and z= number of servings of vegetables, formulate a bi-objective linear problem whose objective functions are the cost and the number of calories of the menu while maintaining the physician's prescription of at least one unit and at most one and half unit for each nutritional substance. Discuss the menus that minimize the cost and the number of calories separately.

4.4.9 An investment problem. *An investor disposes a budget of 20,000 USD and wishes to invest into three product projects with amounts x, y and z respectively. The total profit is given by*

$$P(x, y, z) = 20x + 10y + 100z$$

and the total sale is given by

$$S(x, y, z) = 10x + 2y + z.$$

Find x, y and z to maximize the total profit and total sale.

4.4.10 Bilevel linear programming problem. *A typical bilevel programming problem consists of two problems: the upper level problem of the form*

$$\text{maximize} \quad \langle c, x \rangle + \langle d, y \rangle$$
$$\text{subject to} \quad A_1 x \leqq b_1$$
$$x \geqq 0$$

and the lower level problem for which y is an optimal solution:

$$\text{maximize} \quad \langle p, z \rangle$$
$$\text{subject to} \quad A_2 x + A_3 z \leqq b_2$$
$$z \geqq 0.$$

Here c, p, d, b_1 and b_2 are vectors of dimension n_1, n_2, n_2, m_1 and m_2 respectively; A_1, A_2 and A_3 are matrices of dimension $m_1 \times n_1$, $m_2 \times n_1$ and $m_2 \times n_2$ correspondingly.

Consider the following multiobjective problem

$$\text{Maximize} \quad \begin{pmatrix} x \\ -\langle e, x \rangle \\ \langle p, y \rangle \end{pmatrix}$$
$$\text{subject to} \quad A_1 x \leqq b_1$$
$$A_2 x + A_3 y \leqq b_2$$
$$x \geq 0, y \geqq 0$$

where e is the vector whose components are all equal to one. Prove that $(\overline{x}, \overline{y})$ *is an efficient solution of this latter problem if and only if it is a feasible solution of the upper level problem described above.*

4.4.11 *Apply Theorem 4.1.15 to find a decomposition of the scalarizing set for the polyhedron defined by the system*

$$2x_1 + x_2 + 2x_3 \leqq 5$$
$$x_1 + 2x_2 + 2x_3 \leqq 5$$
$$x_1, x_2, x_3 \geqq 0.$$

4.4.12 *Find the weakly scalarizing set of a polyhedron in* $\mathbb{R}^k$ *determined by the system*

$$\langle a^i, y \rangle = b_i, \ i = 1, \cdots, m$$
$$y \geqq 0,$$

and apply it to find the weakly scalarizing set of a multiobjective problem given in standard form.

4.4.13 Scalarizing set at a vertex solution. *Consider the problem (MOLP) in standard form*

$$\text{Maximize} \quad Cx$$
$$\text{subject to } Ax = b$$
$$x \geqq 0,$$

where C is a $k \times n$*-matrix, A is an* $m \times n$*-matrix and b is an m-vector. Assume* $\overline{x}$ *is a feasible solution associated with a non-degenerate basis B. The non-basic part of A is denoted N, the basic and non-basic parts of C are denoted* C_B *and* C_N *respectively. Prove the following statements.*

(a) A vector λ *belongs to* $\Lambda(\overline{x})$ *if and only if it belongs to* Δ *and solves the following system*

$$[C_N^T - (B^{-1}N)^T C_B^T]\lambda \leqq 0.$$

(b) If the vector on the left hand side of the system in (a) is strictly negative for some λ*, then* $\overline{x}$ *is a unique solution of the scalarized problem*

$$\text{maximize} \quad \langle \lambda, Cx \rangle$$
$$\text{subject to} \quad Ax = b$$
$$x \geqq 0.$$

In particular, if in addition λ *is positive, then* $\overline{x}$ *is an efficient solution of (MOLP).*

4.4.14 Pascoletti-Serafini's method. *Let $\lambda \in \mathbb{R}^k$ be a strictly positive vector and $Cx \geqq 0$ for every feasible solution $x \in X$ of (MOLP). Show that if $(\overline{x}, \overline{\alpha})$ is an optimal solution of the problem*

$$\begin{aligned} \text{maximize} \quad & \alpha \\ \text{subject to} \quad & \langle c^j, x \rangle \geqq \alpha, \ j = 1, ..., k \\ & x \in X, \end{aligned}$$

then $\overline{x}$ is a weakly efficient solution of (MOLP).

4.4.15 Weighted constraint method. *Prove that a feasible solution $\overline{x} \in X$ is a weakly efficient solution of (MOLP) if and only if there is some strictly positive vector $\lambda \in \mathbb{R}^k$ such that $\overline{x}$ solves*

$$\begin{aligned} \text{maximize} \quad & \lambda_\ell \langle c^\ell, x \rangle \\ \text{subject to} \quad & \lambda_j \langle c^j, x \rangle \geqq \lambda_\ell \langle c^\ell, x \rangle, \ j = 1, \cdots, k, j \neq \ell \\ & x \in X \end{aligned}$$

for $\ell = 1, \cdots, k$.

4.4.16 Constraint method. *Choose $\ell \in \{1, \cdots, k\}$, $L_j \in \mathbb{R}$, $j = 1, \cdots, k, j \neq \ell$, and solve the scalar problem (P_ℓ):*

$$\begin{aligned} \text{maximize} \quad & \langle c^\ell, x \rangle \\ \text{subject to} \quad & \langle c^j, x \rangle \geq L_j, j = 1, \cdots, k, j \neq \ell \\ & Ax = b, x \geq 0. \end{aligned}$$

Note that if L_j are big, then (P_ℓ) may have no feasible solution. A constraint $\langle c^j, x \rangle \geq L_j$ is called binding if equality $\langle c^j, x \rangle = L_j$ is satisfied at every optimal solution of (P_ℓ). Prove that

(a) *every optimal solution of (P_ℓ) is a weakly efficient solution of (MOLP);*
(b) *if an optimal solution of (P_ℓ) is unique or all constraints of (P_ℓ) are binding, then it is an efficient solution of (MOLP);*
(c) *a feasible solution x^0 of (MOLP) is efficient if and only if it is optimal for all (P_ℓ), $\ell = 1, ..., k$ and*

$$L_\ell = (\langle c^1, x^0 \rangle, \cdots, \langle c^{\ell-1}, x^0 \rangle, \langle c^{\ell+1}, x^0 \rangle, \cdots, \langle c^k, x^0 \rangle).$$

4.4.17 *Let d be a k-vector such that $Cx \geqq d$ for some feasible solution x of (MOLP). Consider the problem (P)*

$$\text{maximize} \qquad \langle e, y \rangle$$
$$\text{subject to} \qquad Cx = d + y$$
$$Ax = b, x \geqq 0, y \geqq 0,$$

where e is the vector of ones in $\mathbb{R}^k$. Show that

(a) a feasible solution x^0 of (MOLP) is efficient if and only if the optimal value of (P) with $d = Cx^0$ is equal to zero;

(b) (MOLP) has efficient solutions if and only if the optimal value of (P) is finite.

4.4.18 Let $\bar{x}$ be a feasible solution of the problem

$$\text{Maximize} \quad Cx$$
$$\text{subject to} \ \ Ax \leqq b.$$

Show that the following statements are equivalent.

(i) $\bar{x}$ is a weak Pareto maximal solution.

(ii) The system
$$\begin{cases} Ax \leqq b \\ Cx > C\bar{x} \end{cases}$$

is inconsistent.

(iii) For every $t > 0$, the system
$$\begin{cases} Ax \leqq b - A\bar{x} \\ Cx \geqq te \end{cases}$$

is inconsistent, where e is the vector of ones.

(iv) For every $t > 0$, the system
$$\begin{cases} C^T \lambda - A^T \mu = 0 \\ \langle A\bar{x} - b, \mu \rangle + t \langle e, \lambda \rangle = 1 \\ \lambda, \mu \geqq 0 \end{cases}$$

is consistent.

4.4.19 Consider the multiobjective problem described in the preceding exercise. Assume that the cone $\mathrm{pos}\{c^1, \cdots, c^k\}$ contains the origin in its relative interior. Prove that if the interior of the feasible set is nonempty, then every feasible solution of (MOLP) is an efficient solution.

4.4.20 Let X denote the feasible set of the problem (MOLP) given in Exercise 4.4.18. Consider the following function

$$h(x) = \max_{x' \in X} \min_{\lambda \in \Delta} \langle \lambda, Cx' - Cx \rangle.$$

Prove that $\overline{x}$ is a weakly maximal solution of (MOLP) if and only if $h(\overline{x}) = 0$.

4.4.21 Geoffrion's proper efficient solutions. *Let X be a nonempty set in $\mathbb{R}^n$ and let f be a vector-valued function from $\mathbb{R}^n$ to $\mathbb{R}^k$. Consider the following multiobjective problem*

$$\text{Maximize} \quad f(x)$$
$$\text{subject to } x \in X.$$

A feasible solution $\overline{x}$ of this problem is said to be a proper efficient solution if there exists a constant $\alpha > 0$ such that for every $i \in \{1, \cdots, k\}$ and $x \in X$ satisfying $f_i(x) > f_i(\overline{x})$ there exists some $j \in \{1, \cdots, k\}$ for which $f_j(x) < f_j(\overline{x})$ and

$$\frac{f_i(x) - f_i(\overline{x})}{f_j(\overline{x}) - f_j(x)} \leq \alpha.$$

(i) *Justify that every proper efficient solution is efficient. Give an example of efficient solutions that are not proper.*
(ii) *Prove that when f is linear and X is a polyhedral set, every efficient solution is proper.*

4.4.22 Maximality with respect to a convex cone. *Let C be a convex cone in $\mathbb{R}^k$ with $C \cap (-C) = \{0\}$ (one says C is pointed). For $y, z \in \mathbb{R}^k$ define $y \geq_C z$ by $y - z \in C$. A point z of a set A is called C-maximal if there is no $y \in A$ such that $y \geq_C z$ and $y \neq z$. Prove the following properties:*

(i) *A point $z \in A$ is C-maximal if and only if $(A - a) \cap \mathbb{R}^k = \{0\}$;*
(ii) *If $\mathbb{R}^k_+ \subseteq C$, then every C-maximal point is Pareto maximal, and if $\mathbb{R}^k_+ \supseteq C$, then every Pareto maximal point is C-maximal;*
(iii) *If A is a polyhedral set, then there is a polyhedral cone C satisfying $\mathbb{R}^k_+ \subseteq \text{int}(C) \cup \{0\}$ such that a point of A is C-maximal if and only if it is Pareto maximal. Find such a cone for the sets in Exercise 4.4.1 (a) and (b).*

4.4.23 Lexicographical order. *The lexicographical order $\geq_{lex}$ in $\mathbb{R}^k$ is defined as $y \geq_{lex} z$ for $y, z \in \mathbb{R}^k$ if and only if either $y = z$ or there is some $j \in \{1, \cdots, k\}$ such that $y_i = z_i$ for $i < j$ and $y_j > z_j$. A point z of a nonempty set A in $\mathbb{R}^k$ is called lex-maximal if there is no $y \in A$ such that $y \geq_{lex} z$ and $y \neq z$.*

(i) *Show that the lexicographical order is total in the sense that for every $y, z \in \mathbb{R}^k$ one has either $y \geq_{lex} z$ or $z \geq_{lex} y$.*
(ii) *Find a convex cone C such that $y \geq_{lex} z$ if and only if $y - z \in C$.*
(iii) *Prove that every lex-maximal element of a set is Pareto maximal.*

Do the same for the colexicographical order: $y \geq_{colex} z$ if and only if either $y = z$ or there is some $j \in \{1, \cdots, k\}$ such that $y_i = z_i$ for $i > j$ and $y_j > z_j$.

Chapter 5
Duality

In linear programming associated with every maximization problem is a dual mini-mization problem whose feasible solutions yield upper bounds on the values of the former problem. This property is known as weak duality between the primal and dual problems. When the optimal values of the two problems are finite, they are equal, and so strong duality holds. The interpretation of weak duality is not unique in a multi-dimensional Euclidean space, therefore, there are different dual constructions for a given primal problem. This chapter is concerned with three most important duality developments. The first duality is based on dual sets in the value space, the second duality is obtained from the Lagrangian function associated with a given multiobjec-tive linear problem and the last one uses polar cones and normal cones of the value set.

5.1 Dual Sets and Dual Problems

Given a nonempty convex polyhedral set W in $\mathbb{R}^k$, we consider a nonempty subset W^d of $\mathbb{R}^k$ and three relations connecting them

$$y \geq w \text{ for all } w \in W, y \in W^d \tag{5.1}$$

$$y \nleq w \text{ for all } w \in W, y \in W^d \tag{5.2}$$

$$y \nless w \text{ for all } w \in W, y \in W^d. \tag{5.3}$$

Under the above relations the set W^d is called respectively an *ideal dual set*, a *strong dual set* and a *weak dual set* of W. A dual set of W is referred to any of these three types. It is clear that (5.1) is equivalent to each of the following inclusions

$$W^d \subseteq \bigcap_{w \in W} (w + \mathbb{R}^k_+),$$

$$W \subseteq \bigcap_{y \in W^d} (y - \mathbb{R}^k_+).$$

© Springer International Publishing Switzerland 2016
D.T. Luc, *Multiobjective Linear Programming*,
DOI 10.1007/978-3-319-21091-9_5

Similarly, equivalent inclusions for (5.2) and (5.3) are respectively given by

$$W^d \cap \left(W - \mathbb{R}_+^k \setminus \{0\}\right) = \emptyset,$$
$$W \cap \left(W^d + \mathbb{R}_+^k \setminus \{0\}\right) = \emptyset,$$

and

$$W^d \cap \left(W - \mathrm{int}(\mathbb{R}_+^k)\right) = \emptyset,$$
$$W \cap \left(W^d + \mathrm{int}(\mathbb{R}_+^k)\right) = \emptyset.$$

The next theorem expresses a weak duality relation between W and its dual sets.

Theorem 5.1.1 *Let W be a convex polyhedral set and W^d a subset in $\mathbb{R}^k$. Then the following statements hold.*

  (i) $W \cap W^d = \mathrm{IMax}(W) \cap \mathrm{IMin}(W^d)$ *if W^d is an ideal dual set of W.*
 (ii) $W \cap W^d = \mathrm{Max}(W) \cap \mathrm{Min}(W^d)$ *if W^d is a strong dual set of W.*
(iii) $W \cap W^d = \mathrm{WMax}(W) \cap \mathrm{WMin}(W^d)$ *if W^d is a weak dual set of W.*

Proof It is evident that in each statement the set on the right hand side is included in the set on the left hand side. The converse inclusions follow from the relations (5.1)–(5.3). $\qquad\square$

We notice that given a nonempty polyhedral set W it is not necessary that a dual set exists (nonempty). On the other hand when W has an ideal element (respectively maximal and weakly maximal elements), ideal dual set (respectively strong dual and weak dual sets) does exist and it is not unique in general. Moreover, the union of dual sets of any type is again a dual set of the same type. Therefore, we may speak about the biggest dual sets for a given polyhedral set W. To this end let us define the following dual sets associated with W (Figs. 5.1, 5.2 and 5.3),

$$W^* = \sup(W) + \mathbb{R}_+^k$$
$$W^{**} = \left(\mathbb{R}^k \setminus \left(W - \mathbb{R}_+^k\right)\right) \cup \mathrm{Max}(W)$$
$$W^{***} = \mathbb{R}^k \setminus \left(W - \mathrm{int}(\mathbb{R}_+^k)\right).$$

In this definition $\sup(W)$ stands for the vector whose components are the suprema of the projection of W on the coordinate axes. The value $\sup(W)$ is infinite if at least one of its components is not finite, in which case the set W^* is empty by convention. When the value $\sup(W)$ is finite and belongs to the set W, it is the ideal maximal point of W, that is

$$\sup(W) = \mathrm{IMax}(W).$$

It is clear that the sets W^*, W^{**} and W^{***} satisfy respectively (5.1)–(5.3). Moreover, they are the biggest among dual sets of their types. Indeed, let W^d be a nonempty ideal dual set of W. Then $\sup(W)$ is finite. Let $y \in W^d$. By definition $y \geq w$ for all

Fig. 5.1 The first dual set

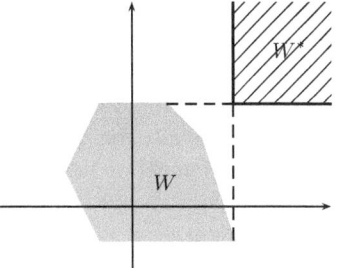

Fig. 5.2 The second dual set

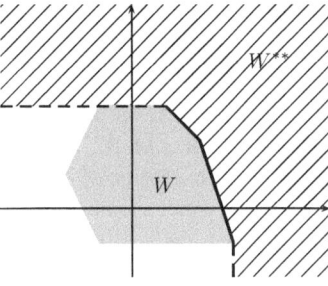

Fig. 5.3 The third dual set

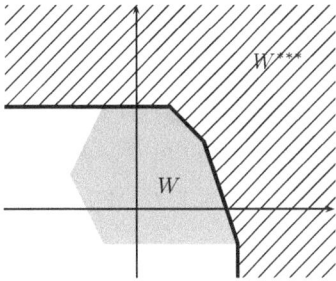

$w \in W$ which implies $y \geqq \sup(W)$. Hence $y \in W^*$ and W^* contains W^d. Now, if W^d is a nonempty strong dual set of W, then every $y \in W^d$ belongs either to W or not. If y belongs to W, then it is a maximal element of W. If y does not belong to W, then it does not belong to $W - \mathbb{R}^k_+$ either, because otherwise there would exist some $w \in W$ such that $w \geq y$ contradicting the hypothesis. Thus, y belongs to W^{**} and W^{**} contains W^d. The proof for the set W^{***} uses the same argument.

The following result presents a relationship between efficient elements of W^*, W^{**}, W^{***} and W.

Theorem 5.1.2 *Let W be a convex polyhedral set in $\mathbb{R}^k$. Then we have*

$$\sup(W) = \mathrm{IMin}(W^*)$$
$$\mathrm{Max}(W) = \mathrm{Min}(W^{**})$$
$$\mathrm{WMax}(W - \mathbb{R}^k_+) = \mathrm{WMin}(W^{***}).$$

Proof The first equality is clear. For the second equality, notice that if Max(W) is empty, then the set W^{**} is open, hence it cannot have minimal points. Let a be a maximal point of W, and let z be any element of W^{**}. If z belongs to Max(W), then $z \leq a$ is false. If z does not belong to Max$W$, then $z \notin W - \mathbb{R}_+^k$, which implies that $z \leq a$ is false too. Hence a is a minimal element of W^{**}. Conversely, if a is a minimal element of W^{**}, then it cannot belong to $\mathbb{R}^k \setminus (W - \mathbb{R}_+^k)$ because the latter set is open, hence it belongs to Max(W). For the last equality, let a be a weakly maximal element of $W - \mathbb{R}_+^k$, say $a = z - v$ for some $z \in W$ and $v \in \mathbb{R}_+^k$. Then a clearly belongs to W^{***}. If it were not a weakly minimal element of W^{***}, there would exist some a' from W^{***} such that $a - a' > 0$. Consequently,

$$a' = a - (a - a') = z - (v + (a - a')) \in W - \mathrm{int}(\mathbb{R}_+^k)$$

which contradicts the fact that a' belongs to W^{***}. Conversely, let a be a weakly minimal element of W^{***}. Then a belongs to $W - \mathbb{R}_+^k$, because otherwise it would belong to the interior of W^{***} and could not be weakly minimal. If it were not a weakly maximal element of $W - \mathbb{R}_+^k$, then there would exist some $z \in W$ and $v \in \mathbb{R}_+^k$ such that $z - v > a$. We deduce

$$a = (z - v) + (a - (z - v)) = z - (v + (z - v - a)) \in z - \mathrm{int}(\mathbb{R}_+^k) \subseteq W - \mathrm{int}(\mathbb{R}_+^k)$$

which contradicts the fact that a belongs to W^{***}. $\qquad\square$

The ideal dual set W^* is a convex polyhedral set, while the dual sets W^{**} and W^{***} are not. But all of them can be generated by closed half-spaces.

Theorem 5.1.3 *Let W be a convex polyhedral set in $\mathbb{R}^k$. Then we have*

$$W^* = \bigcap_{\lambda \in \Delta} \left\{ y \in \mathbb{R}^k : \langle \lambda, y \rangle \geq \sup_{w \in W} \langle \lambda, w \rangle \right\}$$

$$W^{**} = \bigcup_{\lambda \in \mathrm{ri}\Delta} \left\{ y \in \mathbb{R}^k : \langle \lambda, y \rangle \geq \sup_{w \in W} \langle \lambda, w \rangle \right\} \text{ provided that } \mathrm{Max}(W) \neq \emptyset$$

$$W^{***} = \bigcup_{\lambda \in \Delta} \left\{ y \in \mathbb{R}^k : \langle \lambda, y \rangle \geq \sup_{w \in W} \langle \lambda, w \rangle \right\}.$$

Proof If the set W^* is empty, then sup(W) is infinite. There is at least one index i such that

$$\sup_{x \in W} \langle e^i, w \rangle = \infty, \tag{5.4}$$

where e^i is the ith coordinate unit vector. This shows that the set in the right hand side of the first equality of the theorem is empty. Conversely, if the latter set is empty, then there is some index i such that (5.4) holds true because Δ is the convex hull of the coordinate unit vectors $e^1, \cdots, e^k$, which implies that sup(W) is infinite. Assume W^* is nonempty. Then the ith coordinate of the vector $a := \sup(W)$ is the finite maximum of (5.4). We have

$$\langle e^i, a \rangle \geq \langle e^i, w \rangle \text{ for all } w \in W, i = 1, \cdots, k.$$

Hence a belongs to the set on the right hand side of the first equality of the theorem, and so does any element of W^*. Conversely, assume that y satisfies

$$\langle \lambda, y \rangle \geq \langle \lambda, w \rangle \text{ for all } w \in W, \lambda \in \Delta.$$

Then by setting $\lambda = e^i, i = 1, \cdots, k$ we deduce that $\sup(W)$ is finite and $y \geq \sup(W)$. Consequently, y belongs to W^* and the first equality of the theorem follows.

To prove the second equality, assume that y satisfies

$$\langle \lambda, y \rangle \geq \langle \lambda, w \rangle \text{ for all } w \in W \text{ and some } \lambda \in \mathrm{ri}(\Delta). \tag{5.5}$$

If y is a maximal element of W, we are done. If not, it does not belong to $W - \mathbb{R}^k_+$, because otherwise one should express $y = y' - u$ with $y' \in W$ and $u \geq 0$. Then $\langle \lambda, y \rangle = \langle \lambda, y' \rangle - \langle \lambda, u \rangle$. If $u \neq 0$, the value $\langle \lambda, u \rangle$ being strictly positive, the latter equality contradicts (5.5). Thus, y belongs to W^{**}. Conversely, if y is an element of W^{**}, then either it is a maximal element of W, or y does not belong to the convex polyhedral set $W - \mathbb{R}^k_+$. In the first case, in view of Theorem 4.3.1 there is some strictly positive vector $\lambda \in \Delta$ such that

$$\langle \lambda, y \rangle \geq \langle \lambda, w \rangle \text{ for all } w \in W$$

which shows that y belongs to the set on the right hand side of the second equality of the theorem. In the second case we claim that for $\varepsilon > 0$ sufficiently small,

$$(W - \mathbb{R}^k_+) \cap (y + \mathbb{R}^k_+(\varepsilon)) = \emptyset, \tag{5.6}$$

where

$$\mathbb{R}^k_+(\varepsilon) = \mathrm{pos}\{e^i - \varepsilon e : i = 1, \cdots, k\}$$

with $e = (1, \cdots, 1)^T$. Indeed, suppose on the contrary that for every $s > 0$ we find some $w^s \in W$, u^s and $v^s \in \mathbb{R}^k_+$ such that

$$y = w^s - (u^s + v^s) + \frac{1}{s}\Big(\sum_{i=1}^k v^s_i\Big)e. \tag{5.7}$$

If the sequence $\{\sum_{i=1}^k v^s_i / s\}_{s \geq 1}$ converges to zero, then $y = \lim_{s \to \infty}(w^s - (u^s + v^s))$ and belongs to $W - \mathbb{R}^k_+$. This contradicts the hypothesis. If that consequence does not converge to zero, we may assume that the sequence $\{\sum_{i=1}^k (u^s_i + v^s_i)\}_s$ converges to ∞. Dividing (5.7) by the positive number $\sum_{i=1}^{k}(u_i^s + v_i^s)$ and passing to the limit as $s \to \infty$, we deduce

$$\lim_{s\to\infty} \frac{w^s}{\sum_{i=1}^{k}(u_i^s + v_i^s)} = \lim_{s\to\infty} \frac{u^s + v^s}{\sum_{i=1}^{k}(u_i^s + v_i^s)}.$$

In view of Theorem 2.3.11 the limit on the left hand side is an asymptotic direction of W and the limit on the right hand side is a nonzero positive vector. This contradicts the hypothesis that W has maximal elements (see Theorem 4.1.7) and hence (5.6) is true. We now apply the separation theorem (Theorem 2.3.10) to obtain a nonzero vector $\lambda \in \mathbb{R}^k$ such that

$$\langle \lambda, w \rangle \leq \langle \lambda, y + v \rangle \text{ for all } w \in W, v \in \mathbb{R}_+^k(\varepsilon).$$

In particular this implies $\lambda > 0$ that may be assumed to be in $\mathrm{ri}(\Delta)$ and

$$\langle \lambda, y \rangle \geq \sup_{w\in W} \langle \lambda, w \rangle$$

as requested. The third equality of the theorem is proven by a similar argument. □

We observe that the second equality of the above theorem is not true when the set W has no maximal point (see Example 5.2.4).

General scheme of matrix dual construction

Consider the multiobjective linear problem (MOLP)

$$\begin{array}{ll} \text{Maximize} & Cx \\ \text{subject to} & Ax = b \\ & x \geqq 0, \end{array}$$

where C is a real $k \times n$-matrix, A is a real $m \times n$-matrix and b is a real m-vector. A dual problem, denoted by (VD), of (MOLP) is a minimization problem of the form

$$\begin{array}{ll} \text{Minimize} & Yb \\ \text{subject to} & Y \in \mathcal{Y}(A, C), \end{array}$$

where $\mathcal{Y}(A, C)$ consists of $k \times m$-matrices satisfying certain inequalities involving the matrices A and C. Weak duality between (MOLP) and (VD) expresses the fact that the value Cx of a feasible solution x of the primal problem (MOLP) cannot be bigger than the value Yb of the dual problem. There are several ways to interpret the weak duality requirement. Most relevant ones, based on (5.1)–(5.3), are given below:

$$Cx \leqq Yb, \qquad (5.8)$$
$$Cx \not\geqq Yb, \qquad (5.9)$$
$$Cx \not> Yb. \qquad (5.10)$$

It is clear that (5.8) implies (5.9), which in its turn implies (5.10), but the converse is not true except for the case when $k = 1$. We shall exploit these duality inequalities to construct different dual problems for (MOLP). Notice further that if the primal maximization problem is given in canonical form

$$\text{Maximize} \quad Cx$$
$$\text{subject to} \quad Ax \leqq b$$
$$x \geqq 0,$$

then by introducing a slack variable v of dimension m we may rewrite the primal problem in standard form

$$\text{Maximize} \quad \overline{C}x$$
$$\text{subject to} \quad \overline{A}\begin{pmatrix} x \\ v \end{pmatrix} = b$$
$$\begin{pmatrix} x \\ v \end{pmatrix} \geqq 0,$$

where $\overline{A} = (A I)$ with I being the identity $m \times m$-matrix and $\overline{C} = (C O)$ with O being the null $m \times m$-matrix. Its dual is then

$$\text{Minimize} \quad Yb$$
$$\text{subject to} \quad Y \in \mathcal{Y}(\overline{A}, \overline{C}),$$

where again $\mathcal{Y}(\overline{A}, \overline{C})$ is a set of $(k + m) \times m$-matrices involving $\overline{A}$ and $\overline{C}$ to be specified. To this purpose let us define three dual feasible sets in the space of $k \times m$-matrices:

$$\mathcal{Y}_1 = \{Y : YA \geqq C\}$$
$$\mathcal{Y}_2 = \{Y : \lambda^T YA \geqq \lambda^T C \text{ for some } \lambda \in \mathbb{R}^k, \lambda > 0\}$$
$$\mathcal{Y}_3 = \{Y : \lambda^T YA \geqq \lambda^T C \text{ for some } \lambda \in \mathbb{R}^k, \lambda \geq 0\}.$$

It is clear that $\mathcal{Y}_1 \subseteq \mathcal{Y}_2 \subseteq \mathcal{Y}_3$, and the inclusions are strict except for the case $k = 1$. The dual problems (VD) associated with these dual feasible sets will be respectively denoted by (VD1), (VD2) and (VD3) and displayed below.
The first dual (VD1)

$$\text{Minimize} \quad Yb$$
$$\text{subject to} \quad YA \geq C.$$

The second dual (VD2)

$$\text{Minimize} \qquad\qquad Yb$$
$$\text{subject to} \quad \lambda^T Y A \geq \lambda^T C \text{ for some } \lambda \in \mathbb{R}^k, \lambda > 0.$$

The third dual (VD3)

$$\text{Minimize} \qquad\qquad Yb$$
$$\text{subject to} \quad \lambda^T Y A \geq \lambda^T C \text{ for some } \lambda \in \mathbb{R}^k, \lambda \geq 0.$$

We denote the *value set* of (MOLP) by Q, that is

$$Q = \{Cx : Ax = b, x \geq 0\}$$

and the value sets of (VD1)–(VD3) respectively by Q^{D1}, Q^{D2} and Q^{D3}. The *maximal solution set* of (MOLP) consists of those feasible vectors x for which Cx is a maximal element of Q.

Lemma 5.1.4 *Assume that (MOLP) has feasible solutions. If the set Q^{D1} (respectively Q^{D2} and Q^{D3}) is nonempty, it is an ideal dual set (respectively a strong dual set and a weak dual set) of Q.*

Proof Let y be an arbitrary element of Q, say $y = Cx$ for some feasible solution x of (MOLP). If z is an element of Q^{D1}, say $z = Yb$ for some Y satisfying $YA \geq C$, then

$$Yb = YAx \geq Cx = y$$

because x is a positive vector and $Ax = b$. Consequently, $z \geq y$, which shows that Q^{D1} is an ideal dual set of Q. Furthermore, if z is an element of Q^{D2}, then there are some Y and $\lambda \in \text{ri}(\Delta)$ such that $z = Yb$ and

$$\lambda^T Y A \geq \lambda^T C.$$

Multiplying both sides of the latter inequality by the positive feasible vector x we deduce

$$\lambda^T z = \lambda^T Yb = \lambda^T Y Ax \geq \lambda^T Cx = \lambda^T y.$$

This implies that $y \not\geq z$ and proves that Q^{D2} is a strong dual set of Q. The proof for Q^{D3} is similar. $\qquad\qquad\qquad\qquad\qquad\qquad\qquad\qquad\qquad\qquad\qquad$ $\square$

Here is a weak duality relation between the primal problem (MOLP) and its dual problems (VD1)–(VD3).

Theorem 5.1.5 (Weak duality) *For every pair of feasible solutions x and Y of the problems (MOLP) and (VD1) (respectively (VD2) and (VD3)) the weak duality relation (5.8) (respectively (5.9) and (5.10)) holds.*

Moreover, if equality holds in (5.8), then x and Y are ideal efficient solutions of the respective problems; and if equality holds in (5.9) (respectively in (5.10)), then Y is a minimal solution of (VD2) (respectively weakly minimal solution of (VD3)) and x is a maximal solution (respectively weakly maximal solution) of (MOLP).

Proof The weak duality relations (5.8)–(5.10) are obtained from the fact that Q^{D1}, Q^{D2} and Q^{D3} are dual sets of Q as stated in Lemma 5.1.4. The second part of the theorem is derived from Lemma 5.1.4 and Theorem 5.1.1. $\qquad\square$

5.2 Ideal Dual Problem

Consider the linear multiobjective problem (MOLP) described in the previous section. Our aim is to study the dual problem (VD1), called the ideal dual (or Corley's dual):

$$\text{Minimize} \quad Yb$$
$$\text{subject to} \quad YA \geq C.$$

As we have already seen in Lemma 5.1.4 the value set Q^{D1} of (VD1) is an ideal dual set of the value set Q of (MOLP). According to Theorem 5.1.5 for every feasible solutions x and Y, one has weak duality relation $Cx \leq Yb$, and if in addition equality holds, both solutions are ideal, which means that x is ideal for (MOLP) and Y is ideal for (VD1). This dual is a linear problem and shares a certain symmetry of the duality of linear programming. It provides the best upper bound for the primal problem as we shall see later. In what follows $c^1, \cdots, c^k$ denote the columns of the matrix C^T. The columns of an $m \times k$-matrix Y^T are denoted $y^1, \cdots, y^k$.

Lemma 5.2.1 *Assume that (MOLP) has feasible solutions. Then*

$$Q^* = Q^{D1} + \mathbb{R}^k_+.$$

Consequently, the ideal minimal points of Q^* and Q^{D1} coincide if they exist.

Proof Consider the scalar problem (P_i) for $i = 1, \cdots, k$,

$$\text{maximize} \quad \langle c^i, x \rangle$$
$$\text{subject to} \quad Ax = b$$
$$x \geq 0$$

and denote its optimal value by α_i. If this problem is unbounded, α_i takes the infinite value. If all α_i's are finite, the vector composed of these values, denoted α, is exactly sup(Q). Further, consider the dual problem (D_i) of (P_i):

$$\text{minimize} \quad \langle b, y \rangle$$
$$\text{subject to} \quad A^T y \geq c^i.$$

If α_i takes the infinite value, then (D_i) has no feasible solutions (see Theorem 3.2.3). Hence (VD1) has no feasible solutions either. This is because the ith row of a feasible matrix of (VD1) would be a feasible solution for (D_i). If all α_i's are finite, by the strong duality (Theorem 3.2.3) there are couples of optimal solutions $(x(i), y(i))$ such that $\langle c^i, x(i) \rangle = \langle b, y(i) \rangle$. Let $\bar{Y}$ be the matrix whose rows are the transposes of the vectors $y(1), \cdots, y(k)$. It is a feasible solution of (VD1) and $\bar{Y}b = (\alpha_1, \cdots, \alpha_k)^T$. Moreover, for a feasible solution Y, every column y^i of Y^T is a feasible solution of (D_i), hence by the weak duality, $\langle b, y^i \rangle \geq \alpha_i$. By this, equality stated in the lemma is true. The second part of the lemma is direct from the first part. □

In general the set Q^{D1} is smaller than Q^*. For instance when b is the zero vector, the set Q^{D1} reduces to the singleton $\{0\}$ while Q^* contains all positive vectors. Now we deduce a strong duality relation between (MOLP) and (VD1).

Theorem 5.2.2 *Let x be a feasible solution of (MOLP). Then it is an ideal maximal solution if and only if there is a feasible solution Y of (VD1) such that $Cx = Yb$, in which case Y is an ideal minimal solution of (VD1).*

Similarly, let Y be a feasible solution of (VD1). Then it is an ideal minimal solution of (VD1) if and only if sup(Q) $= Yb$.

Proof Let x be a feasible solution of (MOLP). If it is an ideal solution of (MOLP), then Cx is the supremum of Q. By Theorem 5.1.2, it is an ideal minimal element of Q^*. In view of Lemma 5.2.1 this supremum is the ideal minimal value of (VD1), and hence there is a feasible solution Y such that $Cx = Yb$. Now, if Y is a feasible solution of (VD1) such that $Yb = Cx$, by Theorem 5.1.2 both x and Y are ideal solutions.

We proceed to the second part of the theorem. Let $Y \in \mathcal{Y}_1$. By the weak duality, the value set Q of (MOLP) is bounded from above, and its supremum is finite. If Y is an ideal solution of (VD1), then in view of Theorem 5.1.2 and Lemma 5.2.1, the supremum of Q is equal to Yb. Conversely, if Yb coincides with sup(Q), then $Y'b \geq \text{sup}(Q)$ for all $Y' \in \mathcal{Y}_1$. By this, Y is an ideal minimal solution of (VD1). □

When we say that the objective function of (MOLP) is *unbounded* (from above) we means $Cv \geq 0$ for some asymptotic direction v of the feasible set. Similarly, the objective function of (VD1) is unbounded (from below) if there is an asymptotic direction Y^0 of the feasible set of (VD1) such that $Y^0 b \leq 0$.

Theorem 5.2.3 *For the couple of primal and dual problems (MOLP) and (VD1) the following assertions hold.*

(i) *If either of the problems (MOLP) and (VD1) has an unbounded objective, the other has no feasible solution.*

(ii) *The primal problem (MOLP) has maximal solutions and its maximal value set is bounded if and only if the dual problem (VD1) has ideal minimal solutions. Moreover there exist k maximal solutions $x_1^*, \cdots, x_k^*$ of (MOLP) such that the ideal minimal value of (VD1) has its components $\langle c^i, x_i^* \rangle, i = 1, \cdots, k$.*

Proof The first assertion is obtained from Theorem 5.1.5. We prove the second assertion. If (MOLP) has maximal solutions and its maximal value set is bounded, then the problems $(P_1), \cdots, (P_k)$ have finite optimal values $\alpha_1, \cdots, \alpha_k$. By Lemma 5.2.1 the dual problem (VD1) has minimal solutions and its minimal value set consists of the vector $(\alpha_1, \cdots, \alpha_k)^T$ only. It is clear that optimal solutions of each problem (P_i) are weakly maximal solutions for (MOLP). However, as (MOLP) has maximal solutions, among those optimal solutions there are maximal solutions of as well. Choosing any optimal solution x_i^* of (P_i) which is maximal for (MOLP) we obtain that the unique minimal value of (VD1) has components $\langle c^i, x_i^* \rangle, i = 1, \cdots, k$.

Conversely, if (VD1) has minimal solutions, then (MOLP) has a bounded objective on its feasible domain, and therefore it admits maximal solutions. We prove that the set of all maximal values of (MOLP) is bounded. Indeed, if not, there would exist a strictly positive vector $\lambda \in \mathbb{R}^k$ such that the face F of Q consisting of all maximizers of the linear function $\langle \lambda, . \rangle$ on Q is unbounded. Let z be any element and v a nonzero recession direction of F. Then $\langle \lambda, v \rangle = 0$ and the points $z + tv \in F$ with $t \geq 0$ are maximal values for (MOLP). Since λ is strictly positive, some component of v must be strictly positive. Then the corresponding component of $z + tv$ tends to infinity as t goes to ∞, which contradicts the fact that it is bounded from above by the corresponding component of the minimal value of (VD1). □

Let us remark that when x^* is a maximal solution to (MOLP), it is not necessary that a minimal solution Y^* to (VD1) exists for which $Cx^* = Y^*b$ and $Y^*(b - Ax^*) = 0$.

Example 5.2.4 Consider a multiobjective problem

$$\text{Maximize} \quad \begin{pmatrix} 1 & 0 \\ 0 & 1 \end{pmatrix} \begin{pmatrix} x_1 \\ x_2 \end{pmatrix}$$

$$\text{subject to} \quad \begin{pmatrix} 1 & 1 \\ 0 & 0 \end{pmatrix} \begin{pmatrix} x_1 \\ x_2 \end{pmatrix} = \begin{pmatrix} 1 \\ 0 \end{pmatrix}$$

$$x_1, x_2 \geqq 0.$$

Then the dual problem is written as

$$\text{Minimize} \quad \begin{pmatrix} y_1 & y_2 \\ y_3 & y_4 \end{pmatrix} \begin{pmatrix} 1 \\ 0 \end{pmatrix}$$

$$\text{subject to} \quad \begin{pmatrix} y_1 & y_2 \\ y_3 & y_4 \end{pmatrix} \begin{pmatrix} 1 & 1 \\ 0 & 0 \end{pmatrix} \geqq \begin{pmatrix} 1 & 0 \\ 0 & 1 \end{pmatrix}.$$

It is clear that the optimal solution set of the primal problem is the segment

$$\left[\begin{pmatrix} 1 \\ 0 \end{pmatrix}, \begin{pmatrix} 0 \\ 1 \end{pmatrix} \right]$$

and the optimal solution set of the dual problem consists of the matrices

$$\begin{pmatrix} 1 & y_2 \\ 1 & y_4 \end{pmatrix}, \quad y_2, y_4 \in \mathbb{R}.$$

In the outcome (value) space $\mathbb{R}^2$, the optimal value set of the primal problem coincides with the optimal solution set because the objective function is the identity map; while the optimal value set of the dual problem is the point $(1, 1)^T$. Weak duality inequality is satisfied, but equality is not.

Extended dual problem

We introduce an auxiliary real variable t and write (MOLP) in the following equivalent form, denoted (MOLP'):

$$\text{Maximize} \quad (C \ 0) \begin{pmatrix} x \\ t \end{pmatrix}$$

$$\text{subject to} \quad \begin{pmatrix} A & 0 \\ 0 & 1 \end{pmatrix} \begin{pmatrix} x \\ t \end{pmatrix} = \begin{pmatrix} b \\ 1 \end{pmatrix}$$

$$x \geqq 0, t \geqq 0.$$

The variable t takes a constant value and has no effect on the objective function. It is clear that a vector x is feasible for (MOLP) if and only if the vector $\begin{pmatrix} x \\ 1 \end{pmatrix}$ is feasible for (MOLP'); and it is an ideal maximal solution (respectively maximal solution and weakly maximal solution) of (MOLP) if and only if $\begin{pmatrix} x \\ 1 \end{pmatrix}$ is so for (MOLP'). Then the ideal dual of (MOLP') is written as

$$\text{Minimize} \quad (Y Y_{m+1}) \begin{pmatrix} b \\ 1 \end{pmatrix}$$

$$\text{subject to} \quad (Y Y_{m+1}) \begin{pmatrix} A & 0 \\ 0 & 1 \end{pmatrix} \geqq (C, \ 0),$$

which is the same as

$$\text{Minimize } Yb + Y_{m+1}$$
$$\text{subject to } YA \geqq C$$
$$Y_{m+1} \geqq 0.$$

The latter problem is called the ideal *extended dual* of (MOLP) and denoted (VD1′). The variable of (VD1) is a $k \times m$-matrix and the variable of (VD1′) is a $k \times (m+1)$-matrix. Problems (VD1) and (VD2) are equivalent in the sense that a matrix Y is a feasible solution of (VD1) if and only if the matrix $(Y\ Y_{m+1})$ with Y_{m+1} a positive vector is a feasible solution of (VD1′), and a feasible solution Y of (VD1) is ideal if and only if the matrix $(Y\ 0)$ is ideal for (VD1′). We notice, however, that the value sets of these problems do not coincide in general. Actually it follows from Lemma 5.2.1 that the value set of (VD1′) is exactly the ideal dual set Q^*.

5.3 Strong Dual Problem

In this section we study the dual problem (VD2), called the strong dual (or Isermann's dual):

$$\text{Minimize } Yb$$
$$\text{subject to } \lambda^T Y A \geq \lambda^T C \text{ for some } \lambda \in \mathbb{R}^k, \lambda > 0.$$

A weak duality relation between the primal problem (MOLP) and the strong dual problem has already presented in Theorem 5.1.5. Namely, if x and Y are feasible solutions, then $Cx \not\geq Yb$, and if in addition $Cx = Yb$, then x is a maximal solution of (MOLP) and Y is a minimal solution of (VD2). The dual problem (VD2) is not linear because the constraint set is not a convex polyhedron. Our main task is to establish a strong duality relation: the maximal value set of (MOLP) coincides with the minimal value set of (VD2). We begin with some lemmas. The value set of (VD2) is already denoted by Q^{D2}.

Lemma 5.3.1 *Assume that the value set Q of (MOLP) has maximal elements. Then for every $a \in [\mathbb{R}^k \setminus (Q - \mathbb{R}^k_+)] \cap [\mathbb{R}^k \setminus (\mathrm{Max}(Q) + \mathbb{R}^k_+)]$ there is a strictly positive vector $\lambda \in \mathbb{R}^k_+$ such that*

$$\langle \lambda, a \rangle = \max_{z \in Q} \langle \lambda, z \rangle.$$

Proof Let us decompose Q into two parts: $Q = Q_0 + Q_\infty$ according to Corollary 2.3.16, where Q_0 is a bounded convex polytope and Q_∞ is the asymptotic cone of Q. Then it can be verified that the closed convex hull of Q with a (denoted by Q_a) is a convex polyhedron which is the sum of the convex hull of Q_0 with a and the asymptotic cone Q_∞. We claim that a belongs to the maximal set of Q_a. Indeed, suppose on the contrary that there is some positive vector $v \neq 0$ such that $a + v \in Q_a$. Then there are some $t \in [0, 1]$, $z \in Q_0$ and $u \in Q_\infty$ such that

$$a + v = ta + (1 - t)z + u. \tag{5.11}$$

Since Q has maximal elements, in view of Theorem 4.2.1, $Q_\infty \cap \mathbb{R}^k_+ = \{0\}$. This implies that $t \neq 1$ and in view of (5.11),

$$a = z + \frac{1}{1-t}u - \frac{1}{1-t}v \in Q - \mathbb{R}_+^k.$$

The latter inclusion is a contradiction with the choice of a. We claim also that Max(Q) has an element that belongs to Max(Q_a). Indeed, observe first that a cannot be the unique maximal element of Q_a, because otherwise it should belong to Max$(Q)+\mathbb{R}_+^k$, which contradicts its choice. Furthermore, by Theorem 4.1.20 the set Max(Q_a) being a connected set consisting of faces of Q_a, there must be some maximal face F that is included in Max(Q) and containing both a and an element of Q at least. To complete the proof it remains to apply Theorem 4.2.6 to obtain a strictly positive vector λ from the normal cone to Q_a at a relative interior point of the maximal face F. This vector λ fulfils our request.   □

Lemma 5.3.2 *Assume that the value set Q of (MOLP) is nonempty, b is nonzero and that* Max(Q) *is nonempty. Then*

$$Q^{**} = Q^{D2} + \mathbb{R}_+^k. \tag{5.12}$$

Proof We observe that in view of Theorem 5.1.5, Yb belongs to Q^{**} whenever Y is a feasible solution of (VD2). Moreover, as $Q^{**} + \mathbb{R}_+^k$ is contained in Q^{**}, the set Q^{**} contains the set in the right hand side of (5.12). For the converse, let a be an arbitrary element of Q^{**}. We distinguish three possible cases

$$a \in \mathrm{Max}(Q) \tag{5.13}$$

$$a \in \mathrm{Max}(Q) + \mathbb{R}_+^k \tag{5.14}$$

$$a \in \left[\mathbb{R}^k \setminus (Q - \mathbb{R}_+^k)\right] \cap \left[\mathbb{R}^k \setminus \left(\mathrm{Max}(Q) + \mathbb{R}_+^k\right)\right]. \tag{5.15}$$

Our aim is to prove existence of $Y \in \mathcal{Y}_2$ such that $a = Yb$ when a is given in (5.13) and (5.15). The case (5.14) follows from (5.13) and needs no proof. Under (5.13), by scalarization (see Theorem 4.3.1) there is a strictly positive vector λ of $\mathbb{R}^k$ such that

$$\langle \lambda, a \rangle = \max_{z \in Q} \langle \lambda, z \rangle. \tag{5.16}$$

Under (5.15), in view of Lemma 5.3.1, there exists a strictly positive vector λ satisfying (5.16) too. With this λ in hand, we consider a linear system of $k \times m$ variables $y_{ij}, i = 1, \cdots, k; j = 1, \cdots, m$ (the entries of the matrix Y):

$$Yb = a$$
$$\lambda^T Y A \geqq \lambda^T C.$$

To prove the solvability of this system, let us expose it in a familiar form

$$
\begin{bmatrix}
b_1 & \cdots & b_m & \cdots & 0 & \cdots & 0 \\
\vdots & & \vdots & & \vdots & & \vdots \\
0 & \cdots & 0 & \cdots & b_1 & \cdots & b_m \\
-b_1 & \cdots & -b_m & \cdots & 0 & \cdots & 0 \\
\vdots & & \vdots & & \vdots & & \vdots \\
0 & \cdots & 0 & \cdots & -b_1 & \cdots & -b_m \\
\lambda_1 a_{11} & \cdots & \lambda_1 a_{m1} & \cdots & \lambda_k a_{11} & \cdots & \lambda_k a_{m1} \\
\vdots & & \vdots & & \vdots & & \vdots \\
\lambda_1 a_{1n} & \cdots & \lambda_1 a_{mn} & \cdots & \lambda_k a_{1n} & \cdots & \lambda_k a_{mn}
\end{bmatrix}
\begin{pmatrix} y_{11} \\ \vdots \\ y_{1m} \\ \vdots \\ y_{k1} \\ \vdots \\ y_{km} \end{pmatrix}
\geqq
\begin{pmatrix} a \\ -a \\ C^T \lambda \end{pmatrix}.
$$

Suppose to the contrary that this system has no solution. Direct application of Corollary 2.2.4 shows that there exists a positive vector (u, v, w) from $\mathbb{R}^k \times \mathbb{R}^k \times \mathbb{R}^n$ such that

$$\lambda_i A w = b(v_i - u_i), \, i = 1, \cdots, k \tag{5.17}$$

$$\langle \lambda, C w \rangle = 1 + \langle a, v - u \rangle. \tag{5.18}$$

Since all $\lambda_i's$ are strictly positive and b is nonzero, there is a real number t such that $t = (v_i - u_i)/\lambda_i$ for all $i = 1, \cdots, k$ and (5.17) and (5.18) become

$$A w = t b \tag{5.19}$$

$$\langle \lambda, C w \rangle = 1 + t \langle \lambda, a \rangle. \tag{5.20}$$

Indeed, if $t = 0$, then (5.19) and (5.20) prove that the vector w is a recession direction of the feasible set of (MOLP) on which the function $\langle \lambda^T C, . \rangle$ is strictly positive. This contradicts (5.16). If $t < 0$, choose a feasible solution x of (MOLP) that realizes the maximum in (5.16) and set

$$q := x + \frac{1}{|t|} w.$$

The relations (5.19) and (5.20) yield

$$A q = A x + \frac{1}{|t|} A w = b - b = 0$$

$$\langle \lambda, C q \rangle = \langle \lambda, C x \rangle + \frac{1}{|t|} \langle \lambda, C w \rangle$$

$$= \langle \lambda, C x \rangle + \frac{1}{|t|} - \langle \lambda, a \rangle$$

$$= \frac{1}{|t|} > 0.$$

Consequently, q is a recession direction of the feasible set of (MOLP) on which the function $\langle \lambda^T C, . \rangle$ is strictly positive. This again leads to a contradiction with (5.16).

Now we assume $t > 0$. It follows that the vector w/t is a feasible solution of (MOLP) and $C(w/t)$ belongs to Q. We deduce from (5.20) that

$$\langle \lambda, C(\tfrac{w}{t}) \rangle = \frac{1}{t} + \langle \lambda, a \rangle > \langle \lambda, a \rangle$$

which contradicts (5.16), and (5.12) follows.   □

Notice that equality in (5.12) is not true without the non-emptiness of Max(Q). This is illustrated by the following example.

Example 5.3.3 Consider the following problem

$$\text{Maximize } \begin{pmatrix} 1 & 1 \\ 0 & 0 \end{pmatrix} \begin{pmatrix} x_1 \\ x_2 \end{pmatrix}$$

$$\text{subject to } \begin{pmatrix} 1 & -1 \end{pmatrix} \begin{pmatrix} x_1 \\ x_2 \end{pmatrix} = 0$$

$$x_1, x_2 \geq 0.$$

In the value space $\mathbb{R}^2$, the value set Q consists of the vectors $(2x_1 \ 0)^T, x_1 \geq 0$. Hence it has no maximal element. We have $Q^{**} = \{(x_1, x_2)^T : x_2 > 0\}$, which is evidently nonempty. Let $Y = (y_1 \ y_2)^T$ satisfy

$$\lambda^T Y (1 \ -1) \geq \lambda^T \begin{pmatrix} 1 & 1 \\ 0 & 0 \end{pmatrix}$$

for some $\lambda > 0$, which is equivalent to the system

$$\lambda_1 y_1 + \lambda_2 y_2 \geq \lambda_1$$
$$-\lambda_1 y_1 - \lambda_2 y_2 \geq \lambda_1.$$

However the latter system is not solvable because both λ_1 and λ_2 are strictly positive. Thus, the set $\mathcal{Y}_2$ is empty, and equality in (5.12) is impossible.

We are now able to prove a strong duality relation between (MOLP) and (VD2).

Theorem 5.3.4 *Assume b is nonzero. Then a feasible solution x of (MOLP) is a maximal solution if and only if there is a feasible solution Y of (VD2) such that $Cx = Yb$, in which case Y is a minimal solution of (VD2).*

Similarly, let Y be a feasible solution of (VD2). Then it is a minimal solution of (VD2) if and only if there is a feasible solution x of (MOLP) such that $Cx = Yb$, in which case x is a maximal solution of (MOLP).

Proof For the first part of the theorem, let x be a feasible solution of (MOLP). If x is a maximal solution of (MOLP), then Cx is a maximal element of Q. By Theorem 5.1.2, Cx is a minimal element of Q^{**} and by Lemma 5.3.2 there is some $Y \in \mathcal{Y}_2$ such that $Cx = Yb$. The converse follows from Theorem 5.1.5.

We proceed to the second part of the theorem. Let $Y \in \mathcal{Y}_2$, say with a corresponding $\lambda > 0$. Then

$$\langle \lambda, Yb \rangle \geqq \langle \lambda, z \rangle$$

for all $z \in Q$. Hence $Q_\infty \cap \mathbb{R}^k_+ = \{0\}$. By Theorem 4.1.7 the set Q has maximal elements. In view of Theorem 5.1.2 and Lemma 5.3.2, Y is a minimal solution of (VD2) if and only if there is a maximal solution x of (MOLP) such that $Yb = Cx$. $\hspace{2cm}\square$

We recall that the objective function of (MOLP) is said to be bounded from above (respectively bounded from below) on the feasible set X if

$$C(X_\infty) \cap \mathbb{R}^k_+ = \{0\} \text{ (respectively } C(X_\infty) \cap (-\mathbb{R}^k_+) = \{0\}).$$

This definition is equivalent, as the reader is aware from Corollary 2.3.17, to the fact that the asymptotic one Q_∞ contains no positive vector. When $k = 1$ this is equivalent to the statement that $\sup(Q)$ is finite. When $k > 1$, the condition $\sup(Q) \leqq a$ for some $a \in \mathbb{R}^k$ means that the set Q is bounded from above and implies that the objective function C is bounded from above on X, but the converse is not true as already explained in Sect. 4.2.

Theorem 5.3.5 *Assume b is nonzero. For the pair of dual problems (MOLP) and (VD2) the following assertions hold.*

 (i) *Each problem has feasible solutions if and only if they have the same (nonempty) efficient values.*
 (ii) *Let (MOLP) have feasible solutions. Then (VD2) has no feasible solution if and only if the objective function of (MOLP) is unbounded from above on its feasible set. Similarly, let (VD2) have feasible solutions. Then (MOLP) has no feasible solution if and only if the objective function of (VD2) is unbounded from below on its feasible set.*
 (iii) *A feasible solution x of the primal problem (MOLP) is maximal if and only if there exists a minimal solution Y of the dual (VD2) such that $Cx = Yb$. Similarly, a feasible solution Y of the dual problem (VD2) is minimal if and only if there exists a maximal solution x of the primal problem (MOLP) such that $Cx = Yb$.*

Proof For (i) it suffices to show the "only if" part. Assume that both problems (MOLP) and (VD2) have feasible solutions. Then there is some matrix Y and a vector $\lambda > 0$ such that $\lambda^T Y A \geqq \lambda^T C$ which implies that the function $\langle \lambda, . \rangle$ is bounded from above on Q. Consequently Q has maximal elements. By Theorem 5.1.2 and Lemma 5.3.2 the problems (MOLP) and (VD2) have the same efficient value set. For (ii) assume that Q is nonempty. If it is unbounded from above, then it has no maximal element, and by (i), (VD2) has no feasible solutions. Conversely, if (VD2) has no feasible solution, then Q cannot be bounded from above, because otherwise it should have maximal elements which are also minimal values of (VD2) by Theorem 5.1.2

and Lemma 5.3.2. Finally, assume $\mathcal{Y}_2$ nonempty. If the objective function of (VD2) is unbounded from below, then by (i), Q is empty. For the converse, let Y be a feasible solution of (VD2). If Q is empty, then by Farkas' theorem there is some vector y such that $A^T y \geq 0$ and $\langle b, y \rangle < 0$. Let Y_0 be the $k \times m$ matrix ey^T with e being the vector of ones. Then for every $t \geq 0$, $Y + tY_0$ is feasible. For (VD2), by considering the value $(Y + tY_0)b$ as t tends to ∞, we conclude that the objective function of (VD2) is unbounded from below. The last assertion is a part of Theorem 5.3.4. □

Example 5.3.6 Consider the multiobjective problem

$$\text{Maximize} \quad \begin{pmatrix} 1 & 0 \\ 0 & 1 \end{pmatrix} \begin{pmatrix} x_1 \\ x_2 \end{pmatrix}$$

$$\text{subject to} \quad \begin{pmatrix} 2 & 1 \\ 1 & 2 \end{pmatrix} \begin{pmatrix} x_1 \\ x_2 \end{pmatrix} \leqq \begin{pmatrix} 6 \\ 6 \end{pmatrix}$$

$$x_1, x_2 \geqq 0.$$

For $\alpha > 0$ denote by $\mathcal{Y}_\alpha$ the set of all 2×2 matrices Y satisfying

$$(1 \ \alpha)\left(Y \begin{pmatrix} 2 & 1 \\ 1 & 2 \end{pmatrix} - \begin{pmatrix} 1 & 0 \\ 0 & 1 \end{pmatrix} \right) \geqq 0 \text{ and } Y \geqq 0.$$

The dual (VD2) is written in the form

$$\text{Minimize} \quad Y \begin{pmatrix} 6 \\ 6 \end{pmatrix}$$

$$\text{subject to} \quad Y \in \bigcup_{\alpha > 0} \mathcal{Y}_\alpha.$$

It is clear that a matrix Y with entries $y_1, \cdots, y_4$ belongs to $\mathcal{Y}_\alpha$ if and only if

$$2y_1 + y_2 + \alpha(2y_3 + y_4) \geqq 1 \qquad (5.21)$$
$$y_1 + 2y_2 + \alpha(y_3 + 2y_4) \geqq \alpha \qquad (5.22)$$
$$y_1, y_2, y_3, y_4 \geqq 0. \qquad (5.23)$$

We wish to find the value set of the dual (VD2) that we denote by Q^{D2}. It consists of the vectors $Yb = 6(y_1 + y_2, y_3 + y_4)^T$ with Y feasible, that is, Y satisfying (5.21)–(5.23) for some $\alpha > 0$. Observe that if inequalities (5.21) and (5.22) are strict for Y, then there is a feasible solution Y' such that $Y'b \leq Yb$. On the other hand, for a given feasible solution Y and a vector $w \geq Yb$, by increasing y_1 and y_3 only, we may obtain a new feasible solution Y' such that $Y'b = w$. In other words, the value set Q^{D2} is expressed by

$$Q^{D2} = \bigcup_{\alpha > 0} \left\{ 6 \begin{pmatrix} y_1 + y_2 \\ y_3 + y_4 \end{pmatrix} : \begin{array}{l} y_i \geqq 0, i = 1, \cdots, 4, \text{ satisfying (5.21)}, \\ \text{and (5.22) with at least one equality} \end{array} \right\} + \mathbb{R}_+^2.$$

We distinguish three possible cases of α: 1) $0 < \alpha \leq 1/2$; 2) $1/2 < \alpha \leq 2$; and 3) $\alpha > 2$. In the first case (5.22) is superfluous, while in the third case (5.22) is superfluous. In other words, the value set Q^{D2} is composed of the sum of $\mathbb{R}^2_+$ and three sets Q^1, Q^2, Q^3 and Q^4 below

$$Q^1 = \bigcup_{0 < \alpha \leq 1/2} \left\{ 6 \left(\frac{y_1 + y_2}{y_3 + y_4} \right) : Y \geqq 0 \text{ satisfying (5.21) as equality} \right\}$$

$$Q^2 = \bigcup_{1/2 < \alpha \leq 2} \left\{ 6 \left(\frac{y_1 + y_2}{y_3 + y_4} \right) : Y \text{ is feasible with (5.21) being equality} \right\}$$

$$Q^3 = \bigcup_{1/2 < \alpha \leq 2} \left\{ 6 \left(\frac{y_1 + y_2}{y_3 + y_4} \right) : Y \text{ is feasible with (5.22) being equality} \right\}$$

$$Q^4 = \bigcup_{\alpha > 2} \left\{ 6 \left(\frac{y_1 + y_2}{y_3 + y_4} \right) : Y \geqq 0 \text{ satisfying (5.22) as equality} \right\}.$$

Actually the set Q^1 is composed of the vectors $6(y_1 + y_2, y_3 + y_4)^T$ with Y solution of either a) the system (5.21) and (5.22) with the first equation being equality, or b) the system (5.21) and (5.22) with the second equation being equality. However, by setting $p = 2y_1 + y_2$ and $q = 2y_3 + y_4$ the case a) is written as

$$p + \alpha q = 1$$
$$p + \alpha q \geqq 2\alpha - 3(y_2 + \alpha y_4).$$

Since the entries y_2 and y_4 are positive and $\alpha \leq 1/2$, the second inequality above is superfluous. Furthermore, in the case b), if the inequality (5.21) is strict, the second one cannot be equality. This explains the description of Q^1 by equality $2y_1 + y_2 + \alpha(2y_3 + y_4) = 1$ only. A similar argument proves the expression of Q^4 when $\alpha > 2$. To compute the set Q^1, we put

$$u = y_1 + y_2$$
$$v = y_3 + y_4$$
$$t = y_1 + \alpha y_3.$$

Then Q^1 is composed of vectors $6(u, v)^T$ with u, v and t positive and satisfying

$$u + \alpha v = 1 - t.$$

We remember also that $u + \alpha v \geqq t$, therefore $t \leq 1/2$ and the minimum value of $1 - t$ is equal to $1/2$. Consequently,

$$Q^1 + \mathbb{R}^2_+ = \bigcup_{0 < \alpha \leq 1/2} \left\{ 6 \left(\frac{u}{v} \right) : u + \alpha v = \frac{1}{2}, u \geqq 0, v \geqq 0 \right\} + \mathbb{R}^2_+.$$

Similarly,

$$Q^4 + \mathbb{R}_+^2 = \bigcup_{\alpha>2} \left\{ 6 \binom{u}{v} : u + \alpha v = 2, u \geq 0, v \geq 0 \right\} + \mathbb{R}_+^2.$$

Moreover,

$$Q^2 \cup Q^3 + \mathbb{R}_+^2 \subseteq Q^1 \cup Q^4 + \mathbb{R}_+^2.$$

It follows that the minimal values of the dual (DV2) consists of the segments

$$\left[\binom{3}{0}, \binom{2}{2} \right] \cup \left[\binom{0}{3}, \binom{2}{2} \right],$$

which clearly coincides with the maximal values of the primal problem (MOLP).

Extended dual problem

It is evident that Lemma 5.3.2, Theorems 5.3.4 and 5.3.5 are not true without the assumption that b is nonzero. The extended dual problem allows us to remedy this gap. We consider the problem (MOLP'):

$$\text{Maximize } (C0) \binom{x}{t}$$

$$\text{subject to } \begin{pmatrix} A & 0 \\ 0 & 1 \end{pmatrix} \binom{x}{t} = \binom{b}{1}$$

$$x \geq 0, t \geq 0.$$

In this problem the vector on the right hand side of the equality constraint is always nonzero, and so we may apply strong duality to obtain the following dual problem, denoted (VD2'):

$$\text{Minimize } Yb + Y_{m+1}$$
$$\text{subject to } \lambda^T Y A \geq \lambda^T C$$
$$\lambda^T Y_{m+1} \geq 0 \text{ for some } \lambda \in \mathbb{R}^k, \lambda > 0.$$

It is evident that a $k \times m$-matrix Y is a feasible solution of (VD2) for some $\lambda > 0$ if and only if the $k \times (m + 1)$-matrix $(Y \ Y_{m+1})$ with $\lambda^T Y_{m+1} \geq 0$ is feasible for (VD2'). Less evident is the fact that the value set of (VD2') coincides with the dual set Q^{**}. Notice that when $b = 0$, the value set of (VD2) consists of the zero vector only, and when $b \neq 0$, by Lemma 5.3.7 below and Theorem 5.1.2 one has

$$Q^{D2} \subseteq Q^{D2'} \tag{5.24}$$
$$\text{Min}(Q^{D2'}) = \text{Min}(Q^{D2}) = \text{Max}(Q).$$

Lemma 5.3.7 *If (MOLP) admits maximal solutions, then the value set of the extended problem (VD2′) coincides with the dual set Q^{**}.*

Proof Since the vector in the right hand side of the equality constraint of (MOLP) is nonzero, we apply Lemma 5.3.2 to obtain equality

$$Q^{**} = Q^{D2'} + \mathbb{R}^k_+.$$

Observe that if a matrix $(Y\ Y_{m+1})$ is a feasible solution of (VD2′), then so is the matrix $(Y\ (Y_{m+1} + u))$ where u is any positive vector because λ being positive,

$$\lambda^T (Y_{m+1} + u) = \lambda^T Y_{m+1} + \lambda^T u \geq \lambda^T Y_{m+1} \geq 0.$$

This implies

$$Q^{D2'} + \mathbb{R}^k_+ \subseteq Q^{D2'}$$

and equality $Q^{**} = Q^{D2'}$ follows.   □

Corollary 5.3.8 *Assume that (MOLP) has feasible solutions and that b is nonzero. Then a feasible solution Y of (VD2) is minimal if and only if $(Y\ 0)$ is a minimal solution of (VD2′).*

Proof In view of Theorem 5.3.5, the primal problem (MOLP) has maximal solutions. It is clear that if $(Y\ 0)$ is a minimal solution of (VD2′), then by (5.24), Y is a minimal solution of (VD2). For the converse, suppose on the contrary that Y is a minimal solution of (VD2), but there is a feasible solution $(Y'\ y)$ of (VD2′) such that $Y'b + y \leq Yb$. By Lemmas 5.3.2 and 5.3.7 there are a feasible solution Z of (VD2) and a positive vector u such that $Y'b + y = Zb + u$. Then

$$Zb \leq Zb + u = Y'b + y \leq Yb$$

which contradicts the hypothesis on Y.   □

Under the condition that $b \neq 0$, minimal solutions of the extended problem (VD2′) do not necessarily have last column Y_{m+1} null, but it follows from Corollary 5.3.8 that for a minimal solution $(Y\ Y_{m+1})$ there does exist another minimal solution of the form $(Y'\ 0)$ such that $Y'b = Yb + Y_{m+1}$.

5.4 Weak Dual Problem

In this section we study the dual problem (VD3), called the weak dual problem:

Minimize Yb
subject to $\lambda^T Y A \geq \lambda^T C$ for some $\lambda \in \mathbb{R}^k$, $\lambda > 0$.

A weak duality relation between the primal problem (MOLP) and the weak dual problem has already been presented in Theorem 5.1.5. Namely, if x and Y are feasible solutions, then $Cx \not> Yb$, and if in addition $Cx = Yb$, then x is a weakly maximal solution of (MOLP) and Y is a weakly minimal solution of (VD3). Similar to the strong dual case, the dual problem (VD3) is not linear because the constraint set is not a convex polyhedron. Our aim is to establish a strong duality relation: the weakly maximal value set of (MOLP) is included in the weakly minimal value set of (VD3). We begin with some lemmas. The value set of (VD3) is already denoted by Q^{D3}.

Lemma 5.4.1 *Assume that the value set Q of (MOLP) is nonempty and that b is nonzero. Then*

$$Q^{***} = Q^{D3} + \mathbb{R}_+^k.$$

Proof If Y belongs to $\mathcal{Y}_3$, then Yb cannot belong to $Q - int(\mathbb{R}_+^k)$, because otherwise one would have

$$\lambda^T Cx > \lambda^T Yb = \lambda^T YAx$$

for some positive vector x and for every $\lambda \geq 0$ and arrive at an evident contradiction. Moreover, as $Q^{***} + \mathbb{R}_+^k$ is included in Q^{***}, the set $Q^{D3} + \mathbb{R}_+^k$ is contained in Q^{***}. To prove the converse inclusion we notice first that if Q has no weakly maximal element, then $Q - int(\mathbb{R}_+^k)$ is the whole space and Q^{***} is empty, hence equality holds trivially. If Q has weakly minimal elements, then by Theorem 5.1.2 $\text{WMax}(Q) = \text{WMin}(Q^{***})$. We claim that Q^{***} has the weak domination property, that is, every element of it is dominated by some weakly minimal element. In other words, $Q^{***} = \text{WMin}(Q^{***}) + \mathbb{R}_+^k$. Indeed, let z be any element of Q^{***}. Consider the maximization problem

$$\text{maximize } t$$
$$\text{subject to } z + te \in Q - \mathbb{R}_+^k,$$

where e is the vector whose components are all equal to 1. It is clear that the feasible set of this problem is nonempty, closed and bounded from above by 0. Hence it admits an optimal solution $t_0 \leq 0$. Direct verification shows that $z + t_0 e$ is a weakly maximal element of $Q - \mathbb{R}_+^k$, and by Theorem 5.1.2 it is also a weakly minimal element of Q^{***}. Thus, z belongs to the set $\text{WMin}(Q^{***}) + \mathbb{R}_+^k$. By this, to complete the proof it suffices to show that for every element a from $\text{WMin}(Q^{***})$ there is some $Y \in \mathcal{Y}_3$ such that $a = Yb$. To this end, as already mentioned, by Theorem 5.1.2, the vector a is a weakly maximal element of $Q - \mathbb{R}_+^k$. By scalarization there is a positive vector λ from $\mathbb{R}^k$ such that

$$\langle \lambda, a \rangle = \max_{z \in Q - \mathbb{R}_+^k} \langle \lambda, z \rangle.$$

By supposing the contrary as in the proof of (5.12), we arrive at relations (5.17) and (5.18). Notice that not all $\lambda_i's$ are strictly positive, but at least one of them is.

Consequently (5.19) and (5.20) are still available. The remaining part of the proof of Lemma 5.3.2 goes through. The proof is complete.   □

We are now able to prove a strong duality relation between (MOLP) and (VD3).

Theorem 5.4.2 *Assume b is nonzero. Then a feasible solution x of (MOLP) is a weakly maximal solution if and only if there is a feasible solution Y of (VD3) such that $Cx = Yb$, in which case Y is a weakly minimal solution of (VD3).*

Similarly, let Y be a feasible solution of (VD3). Then it is a weakly minimal solution of (VD3) if and only if there is a feasible solution x of (MOLP) such that $Cx \geq Yb$, in which case x is a weakly maximal solution of (MOLP).

Proof Let x be a weakly maximal solution of (MOLP), then Cx is a weakly maximal element of Q and of $Q - \mathbb{R}_+^k$ as well. By Theorem 5.1.2, Cx is a weakly minimal element of Q^{***}. According to Lemma 5.4.1, there is some weakly minimal solution Y from $\mathcal{Y}_3$ such that $Cx \geq Yb$. Actually, setting $a = Cx$ in the last part of the proof of Lemma 5.4.1, we obtain some $Y \in \mathcal{Y}_3$ such that $Cx = Yb$. Conversely, if there is some $Y \in \mathcal{Y}_3$ such that $Cx = Yb$, then by Theorem 5.1.5 the solution x is weakly maximal.

Let $Y \in \mathcal{Y}_3$. If Y is a weakly minimal solution of (VD3), then by Lemma 5.4.1, Yb is a weakly minimal element of Q^{***}. In view of Theorem 5.1.2, there is a weakly maximal solution of (MOLP) such that $Cx \geq Yb$. Conversely, assume that the latter inequality is true for some feasible solution x of (MOLP). If the solution Y were not weakly minimal to (VD3), there would exist some $Y' \in \mathcal{Y}_3$ such that $Yb > Y'b$. This implies $Cx > Y'b$ and contradicts the weak duality (Theorem 5.1.5).   □

It is interesting to notice that in the strong duality relation of the above theorem equality $Cx = Yb$ is not true in general. This can be seen by the example below.

Example 5.4.3 Consider a modified problem of Example 5.2.4

$$\text{Maximize} \quad \begin{pmatrix} 1 & 1 \\ 0 & 0 \end{pmatrix} \begin{pmatrix} x_1 \\ x_2 \end{pmatrix}$$

$$\text{subject to} \quad \begin{pmatrix} 1 & -1 \end{pmatrix} \begin{pmatrix} x_1 \\ x_2 \end{pmatrix} = 1$$

$$x_1, x_2 \geq 0.$$

In the value space $\mathbb{R}^2$ the value set Q consists of the vectors $(2x_1 - 1 \; 0)^T, x_1 \geq 1$. Hence every feasible solution is weakly maximal. For $Y = (1/2, 0)^T$ and $\lambda = (0, 1)^T$ we have

$$\lambda^T Y A = (0, 1) \times (1/2, 0)^T \times (1, -1) = (0, 0)$$

$$\lambda^T C = (0, 1) \begin{pmatrix} 1 & 1 \\ 0 & 0 \end{pmatrix} = (0, 0).$$

Consequently Y is a feasible solution of (VD3). However, there is no feasible solution x of the primal problem such that $Yb = Cx$. This is because for every feasible x we have $Cx = (2x_1 - 1\ 0)^T$ with $x_1 \geq 1$, while $Yb = (1/2, 0)^T$.

Extended dual problem

It is evident that Lemma 5.4.1 and Theorem 5.4.2 are not true without the assumption that b is nonzero. The extended dual problem is again our best recourse to amend this issue. We consider the problem (MOLP′) as defined in Sect. 5.3 in which the vector on the right hand side of the equality constraint is always nonzero, and so we may apply the above duality to obtain the following weak dual problem, denoted (VD3′):

$$\text{Minimize } Yb + Y_{m+1}$$
$$\text{subject to } \lambda^T Y A \geq \lambda^T C$$
$$\lambda^T Y_{m+1} \geq 0 \text{ for some } \lambda \in \mathbb{R}^k, \lambda \geq 0.$$

It is evident that a $k \times m$-matrix Y is a feasible solution of (VD3) for some $\lambda \geq 0$ if and only if the $k \times (m+1)$-matrix $(Y\ Y_{m+1})$ with $\lambda^T Y_{m+1} \geq 0$ is feasible for (VD3′). Applying the same method of the extended problem (VD2) we obtain the following conclusions related to (VD3):

(i) $Q^{D3} + \mathbb{R}_+^k \subseteq Q^{D3'} = Q^{***}$ and equality holds if $b \neq 0$;
(ii) Provided that $b \neq 0$, a feasible solution Y of (VD3) is minimal if and only if the solution $(Y\ 0)$ is minimal for (VD3′).

Kolumban's duality

We consider the following dual problem, denoted (KD):

$$\text{Minimize} \qquad z$$
$$\text{subject to } \langle b, u \rangle = \langle \lambda, z \rangle,$$
$$A^T u \geqq C^T \lambda,$$
$$\lambda \geq 0,$$

with variables $(u, \lambda, z) \in \mathbb{R}^m \times \mathbb{R}^k \times \mathbb{R}^k$. It is not linear because of the first constraint equality, but it fulfils a strong duality relation (see Corollary 5.4.5 below). Moreover, since the data of the problem are positively homogeneous with respect to (u, λ), we may replace the constraint $\lambda \geq 0$ by $\lambda \in \Delta$. Actually (KD) and (VD3) are equivalent in the sense that their value sets coincide.

Lemma 5.4.4 *Assume that Q is nonempty and let Q^{KD} be the value set of the dual problem (KD). Then*

$$Q^{KD} + \mathbb{R}_+^k = \mathbb{R}^k \setminus (Q - \text{int}(\mathbb{R}_+^k)).$$

Proof When $Q - \mathrm{int}\mathbb{R}^k_+$ coincides with $\mathbb{R}^k$, the scalar linear problem (P$_\lambda$)

$$\text{maximize } \langle \lambda, Cx \rangle$$
$$\text{subject to } Ax = b,$$
$$x \geqq 0,$$

is unbounded for every $\lambda \geq 0$. Hence its dual problem (D$_\lambda$)

$$\text{minimize } \quad \langle b, u \rangle$$
$$\text{subject to } A^T u \geqq C^T \lambda$$

has no feasible solution and so does the problem (KD). Conversely, if Q^{KD} is empty, then for all $\lambda \geq 0$, the problems (D$_\lambda$) have no feasible solution, which implies that the problems (P$_\lambda$) are unbounded. This is possible only when $Q - \mathrm{int}(\mathbb{R}^k_+)$ coincides with the whole space $\mathbb{R}^k$. Assume now that Q^{KD} is nonempty. Let z be any element of it and let (u, λ, z) be a feasible solution of (KD). Suppose to the contrary that z belongs to the set $Q - \mathrm{int}(\mathbb{R}^k_+)$, say $z = Cx - v$ for some feasible solution x of (MOLP) and $v \in \mathrm{int}(\mathbb{R}^k_+)$. Then x and u are feasible solutions of the problems (P$_\lambda$) and (D$_\lambda$). We have

$$\langle b, u \rangle = \langle \lambda, z \rangle = \langle \lambda, Cx \rangle - \langle \lambda, v \rangle < \langle \lambda, Cx \rangle,$$

which contradicts the duality between (P$_\lambda$) and (D$_\lambda$). Thus Q^{KD} is included in $\mathbb{R}^k \setminus (Q - \mathrm{int}(\mathbb{R}^k_+))$, and equality follows. Conversely, let $z \notin Q - \mathrm{int}(\mathbb{R}^k_+)$. If $z \in Q - \mathbb{R}^k_+$, then it is clear that z is a weakly maximal element of $Q - \mathbb{R}^k_+$. By scalarization (Theorem 4.3.1) there is some vector $\lambda \in \Delta$ such that z maximizes $\langle \lambda, . \rangle$ on $Q - \mathbb{R}^k_+$. Let x be a feasible solution of (MOLP) such that $z = Cx - v$ for some $v \in \mathbb{R}^k_+$. Then $\langle \lambda, z \rangle = \langle \lambda, Cx \rangle$ and x is an optimal solution of the problem (P$_\lambda$). Consequently, the dual problem (D$_\lambda$) admits also an optimal solution $u \in \mathbb{R}^m$ with $\langle b, u \rangle = \langle \lambda, Cx \rangle = \langle \lambda, z \rangle$, which shows that (u, λ, z) is a feasible solution of (KD). If $z \notin Q - \mathbb{R}^k_+$, then one may find some positive vector v such that $z - v$ belongs to $Q - \mathbb{R}^k_+$, but outside of $Q - \mathrm{int}(\mathbb{R}^k_+)$. Then, as we have already proven, $z - v$ belongs to Q^{KD} which yields $z \in Q^{KD} + \mathbb{R}^k_+$. $\square$

Below is a strong duality relation for (KD).

Corollary 5.4.5 *For the couple of primal and dual problems (MOLP) and (KD) the following assertions hold.*

(i) *A feasible solution x of (MOLP) is a weakly maximal solution if and only if there is a weakly minimal solution (u, λ, z) of (KD) such that $Cx = z$ and $\langle b, u \rangle = \langle \lambda, Cx \rangle$.*

(ii) *A feasible solution (u, λ, z) of (KD) is a weakly minimal solution if and only if there is a weakly maximal solution x of (MOLP) such that $Cx \geqq z$ and $\langle b, u \rangle = \langle \lambda, Cx \rangle$.*

Proof The first assertion is clear from scalarization. For the second assertion, let (u, λ, u) be a weakly minimal solution of (KD). Then, by Lemma 5.4.4, z is a weakly minimal element of the set Q^{***}. By Theorem 5.1.2 there is a positive vector v and a feasible solution x of (MOLP) such that $z = Cx - v$. Hence $Cx \geqq z$. This latter inequality and the weak duality between (P_λ) and (D_λ) yield $\langle b, u \rangle = \langle \lambda, Cx \rangle$. Hence it is a weakly minimal point of the set Q^{KD}, which implies also that (u, λ, z) is a weakly minimal solution. □

5.5 Lagrangian Duality

In order to develop a theory of Lagrangian duality we consider the problem (MOLP)

$$\text{Maximize} \quad Cx$$
$$\text{subject to} \quad Ax = b$$
$$x \geqq 0$$

and define the Lagrangian function associated with it to be a function of two variables: x from $\mathbb{R}^n_+$ and Y from the space of $k \times m$-matrices $L(\mathbb{R}^m, \mathbb{R}^k)$ given by

$$L(x, Y) = Cx + Yb - YAx.$$

It is clear that L is a linear function in each variable when the other is fixed. Before we can establish a connection between the Lagrangian function and (MOLP) and its duals, we introduce an important concept.

Saddle points

Given a point $(\overline{x}, \overline{Y})$ in $\mathbb{R}^n_+ \times L(\mathbb{R}^m, \mathbb{R}^k)$ we consider three relations issued from three orders in the space $\mathbb{R}^k$:

$$L(x, \overline{Y}) \leqq L(\overline{x}, Y) \tag{5.25}$$

$$L(x, \overline{Y}) \ngeqq L(\overline{x}, Y) \tag{5.26}$$

$$L(x, \overline{Y}) \not> L(\overline{x}, Y) \tag{5.27}$$

for every $x \geqq 0$ and $Y \in L(\mathbb{R}^m, \mathbb{R}^k)$. It is obvious that (5.25) implies (5.26) and (5.26) implies (5.27), but the converse is not true.

Definition 5.5.1 The point $(\overline{x}, \overline{Y})$ is called an *ideal saddle point* (respectively a *strong saddle point* and a *weak saddle point*) of the Lagrangian function L if it satisfies (5.25) (respectively (5.26) and (5.27)).

Of course, an ideal saddle point is a strong saddle point and a strong saddle point is a weak saddle point, but the converse is not true in general. Given a vector x in $\mathbb{R}^n$ and a $k \times m$-matrix Y we set

$$\Pi(x) = \begin{cases} \{L(x, Y') : Y' \in L(\mathbb{R}^m, \mathbb{R}^k)\} & \text{if } x \geq 0 \\ \emptyset & \text{else,} \end{cases}$$

$$\Theta(Y) = \{L(x', Y) : x' \geq 0\}.$$

Then the relations (5.25)–(5.27) are respectively equivalent to the following ones

$$\Pi(\overline{x}) \subseteq L(\overline{x}, \overline{Y}) + \mathbb{R}_+^k \text{ and } \Theta(\overline{Y}) \subseteq L(\overline{x}, \overline{Y}) - \mathbb{R}_+^k \qquad (5.28)$$

$$\left(\Theta(\overline{Y}) - \Pi(\overline{x})\right) \cap \mathbb{R}_+^k = \{0\} \qquad (5.29)$$

$$\left(\Theta(\overline{Y}) - \Pi(\overline{x})\right) \cap \text{int}(\mathbb{R}_+^k) = \emptyset. \qquad (5.30)$$

In the case $k = 1$, the three relations above are identical and so there is no distinction between ideal, strong and weak saddle points. In fact, saddle points of vector functions can be characterized by saddle points of real-valued functions.

Theorem 5.5.2 *Let $\overline{x}$ be a vector in $\mathbb{R}^n$ and $\overline{Y}$ a $k \times m$-matrix. The following assertions hold.*

(i) *The point $(\overline{x}, \overline{Y})$ is an ideal saddle point of L if and only if it is a saddle point of the real-valued function $\langle \lambda, L \rangle$ for all positive vectors $\lambda \in \mathbb{R}^k$.*
(ii) *The point $(\overline{x}, \overline{Y})$ is a strong saddle point of L if and only if it is a saddle point of the real-valued function $\langle \lambda, L \rangle$ for some strictly positive vector $\lambda \in \mathbb{R}^k$.*
(iii) *The point $(\overline{x}, \overline{Y})$ is a weak saddle point of L if and only if it is a saddle point of the real-valued function $\langle \lambda, L \rangle$ for some (nonzero) positive vector $\lambda \in \mathbb{R}^k$.*

Proof The first assertion is clear from (5.28). To prove (ii), we observe that the "if" part is evident. For the "only if" part, we assume that $(\overline{x}, \overline{Y})$ is a strong saddle point of $L(x, Y)$. It follows from (5.29) that

$$\text{pos}\left(\Theta(\overline{Y}) - \Pi(\overline{x})\right) \cap \mathbb{R}_+^k = \{0\}.$$

Taking the polar cone in the both sides of the latter equality and applying Corollary 2.3.21 yield

$$\left[\text{pos}\left(\Theta(\overline{Y}) - \Pi(\overline{x})\right)\right]^o - \mathbb{R}_+^k = \mathbb{R}^k.$$

Consequently, the polar cone of the cone $\text{pos}(\Theta(\overline{Y}) - \Pi(\overline{x}))$ contains a strictly positive vector λ satisfying

$$\langle \lambda, z' - z \rangle \leq 0 \text{ for all } z' \in \Theta(\overline{Y}), z \in \Pi(\overline{x}). \qquad (5.31)$$

Equivalently,

$$\langle \lambda, L(\overline{x}, Y) \rangle \geq \langle \lambda, L(x, \overline{Y}) \rangle \quad \text{for all } x \geq 0,\, Y \in L(\mathbb{R}^m, \mathbb{R}^k).$$

By this, $(\overline{x}, \overline{Y})$ is a saddle point of the real function $\langle \lambda, L(x, Y) \rangle$.

To prove the last assertion, we assume (5.30) and distinguish two cases: 1) $(\Theta(\overline{Y}) - \Pi(\overline{x})) \cap \mathbb{R}^k_+ = \{0\}$, and 2) $(\Theta(\overline{Y}) - \Pi(\overline{x})) \cap \mathbb{R}^k_+ \neq \{0\}$. The first case has already been treated in (ii). Moreover, since the set $\Theta(\overline{Y}) - \Pi(\overline{x})$ is convex, the second case can happen only when there is an index i among $1, \cdots, k$ such that the ith component of every element of $\Theta(\overline{Y}) - \Pi(\overline{x})$ is negative or zero. In other words, (5.31) holds with $\lambda = e^i$ the ith axis unit direction. We conclude as before that $(\overline{x}, \overline{Y})$ is a saddle point of the real function $\langle \lambda, L(x, Y) \rangle$.   □

To establish some more characterizations of saddle points we need the following lemma.

Lemma 5.5.3 *Given nonzero vectors $d \in \mathbb{R}^m$ and $\lambda \in \mathbb{R}^k$, for every vectors $w \in \mathbb{R}^m$ and $z \in \mathbb{R}^k$ with $\langle d, w \rangle = \langle \lambda, z \rangle$, there is a $k \times m$-matrix Y such that $Yd = z$ and $Y^T \lambda = w$. In particular, for any vector $u \in \mathbb{R}^m$, one has*

$$\{Yu : Y \in L(\mathbb{R}^m, \mathbb{R}^k)\} = \begin{cases} \mathbb{R}^k & \text{if } u \neq 0 \\ \{0\} & \text{else.} \end{cases}$$

Consequently the set $\{Yu : Y \in L(\mathbb{R}^m, \mathbb{R}^k)\}$ has minimal/maximal points if and only if $u = 0$.

Proof Since the vectors d and λ are nonzero, the values $\|d\|$ and $\|\lambda\|$ are nonzero. We define

$$Y = \frac{1}{\|\lambda\|^2} \lambda w^T + \frac{1}{\|d\|^2} z d^T - \frac{1}{\|\lambda\|^2 \|d\|^2} \lambda \lambda^T z d^T.$$

By the hypothesis that $w^T d = \lambda^T z$ we deduce

$$Yd = \frac{1}{\|\lambda\|^2} \lambda w^T d + \frac{1}{\|d\|^2} z d^T d - \frac{1}{\|\lambda\|^2 \|d\|^2} \lambda \lambda^T z d^T d$$

$$= \frac{1}{\|\lambda\|^2} \lambda \lambda^T z + \frac{1}{\|d\|^2} z \|d\|^2 - \frac{1}{\|\lambda\|^2 \|d\|^2} \lambda \lambda^T z \|d\|^2$$

$$= z$$

and

$$\lambda^T Y = \frac{1}{\|\lambda\|^2} \lambda^T \lambda w^T + \frac{1}{\|d\|^2} \lambda^T z d^T - \frac{1}{\|\lambda\|^2 \|d\|^2} \lambda^T \lambda \lambda^T z d^T$$

$$= w^T + \frac{1}{\|d\|^2} w^T d d^T - \frac{1}{\|\lambda\|^2 \|d\|^2} \|\lambda\|^2 w^T d d^T$$

$$= w^T$$

which shows that Y is a matrix we need. Now, let u be a vector from $\mathbb{R}^m$. If $u = 0$, then $Yu = 0$ for any matrix Y. If $u \neq 0$, then for every $w \in \mathbb{R}^m$ by choosing any nonzero vector λ and $z = \frac{\langle d,w \rangle}{\|\lambda\|^2} \lambda$, we have $\langle d, w \rangle = \langle \lambda, z \rangle$. It remains to apply the first part of the lemma to find Y such that $Yu = w$. Finally, the set $\{Yu : Y \in L(\mathbb{R}^m, \mathbb{R}^k)\}$ has a minimal/maximal point if and only if it reduces to $\{0\}$, that is when $u = 0$. $\square$

The next theorem has relevant implications.

Theorem 5.5.4 *Let $\bar{x}$ be a vector in $\mathbb{R}^n$ and $\bar{Y}$ a $k \times m$-matrix. The following statements hold.*

(i) *The point $(\bar{x}, \bar{Y})$ is an ideal saddle point of L if and only if the following conditions are satisfied*

$$A\bar{x} = b \tag{5.32}$$
$$C - \bar{Y}A \leq 0 \tag{5.33}$$
$$(C - \bar{Y}A)\bar{x} = 0. \tag{5.34}$$

(ii) *The point $(\bar{x}, \bar{Y})$ is a strong saddle point of L if and only if there is a strictly positive vector λ in $\mathbb{R}^k$ such that (5.32) and the following conditions are satisfied*

$$\lambda^T(C - \bar{Y}A) \leq 0 \tag{5.35}$$
$$\lambda^T(C - \bar{Y}A)\bar{x} = 0. \tag{5.36}$$

(iii) *The point $(\bar{x}, \bar{Y})$ is a weak saddle point of L if and only if there is a nonzero positive vector λ in $\mathbb{R}^k$ such that the conditions (5.32), (5.35) and (5.36) are satisfied.*

Proof Let $(\bar{x}, \bar{Y})$ be an ideal saddle point of the Lagrangian function L. Setting $Y = \bar{Y}$ in (5.25), we deduce

$$(C - \bar{Y}A)(x - \bar{x}) \leq 0 \tag{5.37}$$

for every positive vector x. Specifying $x = 2\bar{x}$ and then $x = \bar{x}/2$, we obtain (5.34). Thus, inequality (5.37) becomes $(C - \bar{Y}A)x \leq 0$ for all $x \geq 0$. This evidently implies (5.33). In order to obtain (5.32), we set $x = \bar{x}$ in (5.25) and deduce

$$(\bar{Y} - Y)(b - A\bar{x}) \leq 0 \text{ for all } Y.$$

This clearly yields (5.32) in view of Lemma 5.5.3. For the converse assume that the three conditions in (i) hold. Then for every positive vector x and for every matrix Y

we have

$$L(x, \overline{Y}) = \overline{Y}b + (C - \overline{Y}A)x$$
$$= \overline{Y}A\overline{x} + (C - \overline{Y}A)x + (C - \overline{Y}A)\overline{x}$$
$$\leqq C\overline{x}$$
$$\leqq L(\overline{x}, Y),$$

which shows that $(\overline{x}, \overline{Y})$ is an ideal saddle point of L.

To prove the second statement we apply Theorem 5.5.2. Let $(\overline{x}, \overline{Y})$ be a strong saddle point of L. There is a strictly positive vector λ from $\mathbb{R}^k$ such that it is a saddle point of the real-valued function $\langle \lambda, L \rangle$, that is

$$\lambda^T \left(\overline{Y}b + (C - \overline{Y}A)x \right) \leqq \lambda^T \left(Yb + (C - YA)\overline{x} \right) \tag{5.38}$$

for all $x \geqq 0$ and Y. Like the first part of the proof, by setting $Y = \overline{Y}$ in (5.38), we deduce

$$\lambda^T (C - \overline{Y}A)(x - \overline{x}) \leqq 0$$

for every $x \geqq 0$. This evidently implies (5.35) and (5.36). Similarly, by setting $x = \overline{x}$ in (5.38), we have

$$\lambda^T (\overline{Y} - Y)(b - A\overline{x}) \leqq 0$$

which yields (5.32). Conversely, assume that the conditions (5.32), (5.35) and (5.36) are satisfied for some strictly positive vector λ. Then for every positive vector x and matrix Y we have

$$\langle \lambda, L(x, \overline{Y}) \rangle = \lambda^T \left(\overline{Y}b + (C - \overline{Y}A)x \right)$$
$$= \lambda^T \left(\overline{Y}A\overline{x} + (C - \overline{Y}A)x + (C - \overline{Y}A)\overline{x} \right)$$
$$\leqq \lambda^T (C\overline{x} + Yb - YA\overline{x})$$
$$\leqq \langle \lambda, L(\overline{x}, Y) \rangle.$$

In view of Theorem 5.5.2, we conclude that $(\overline{x}, \overline{Y})$ is a strong saddle point of L. The third statement is proven by the same method.   □

In the previous chapter we proved that the set of maximal solutions of a multiobjective linear problem consists of faces of the feasible set and is arcwise connected. We now show that this nice structure is also valid for the set of saddle points.

Corollary 5.5.5 *The following properties are true.*

(i) *If (x, Y) is an ideal saddle point of L, then $Cx = Yb$ and this value is unique. Consequently, the set of ideal saddle points of L is a convex polyhedral set.*

(ii) *If (x, Y) is a strong saddle point (respectively weak saddle point) of L and if x is a relative interior point of a face F of the feasible set of (MOLP), then for* every $x' \in F$, the couple (x', Y) is a strong saddle point (respectively weak saddle point) of L.

(iii) *The sets of strong as well as weak saddle points of L are composed of convex polyhedral sets and are arcwise connected.*

Proof It follows from Theorem 5.5.4 that if (x, Y) and (x', Y') are ideal saddle points, then (x', Y) and (x, Y') are ideal saddle points too. The first part of (i) is then obtained from (5.34). Under (5.32) the system of equations and inequalities determining ideal saddle points in Theorem 5.5.4 are linear, hence the solution set is a convex polyhedral set.

For (ii) let λ be a vector associated with a strong saddle point (x, Y) which is defined by (ii) of Theorem 5.5.4. Then $C^T \lambda$ is a normal vector to the feasible set of (MOLP) at x, and for every vector x' in F we have $\lambda^T C x' = \lambda^T C x$ (see Theorem 3.1.3). This latter equality and the feasibility of x' show that conditions (5.32), (5.35) and (5.36) are satisfied for (x', Y). Hence (x', Y) is a strong saddle point.

To prove the last statement we observe that for a fixed vector λ, the system (5.32), (5.35) and (5.36) is linear in (x, Y), hence its solution set is a convex polyhedral set. Notice further that λ determines an efficient solution face of (MOLP), and since the number of such faces is finite, the set of strong saddle points being solutions of the above system with a finite number of λ is a finite union of convex polyhedral sets. To prove the arcwise connectedness of this set, let (x, Y) and (x', Y') be two strong saddle points of L. Because the set of maximal solutions of (MOLP) is arcwise connected, we find a finite number of maximal solutions $x^1 = x, \cdots, x^\ell = x'$ such that the segments $[x^i, x^{i+1}], i = 1, \cdots, \ell - 1$ are efficient. Let F_i be an efficient face containing $z^i = (x^i + x^{i+1})/2$ in its relative interior, and let Y^i and W^i be matrices with $Y^1 = Y$ and $Y^\ell = Y'$ for which (x^i, Y^i) and (z^i, W^i) are strong saddle points (see Theorem 5.5.15 for existence of such matrices). By (ii), the segments $[(x^i, Y^i), (x^i, W^i)], [(x^i, W^i), (x^{i+1}, W^i)]$ and $[(x^{i+1}, W^i), (x^{i+1}, Y^{i+1})]$ belong to the set of strong saddle points and joint (x^i, Y^i) with (x^{i+1}, Y^{i+1}) for $i = 1, \cdots, \ell - 1$. This proves that the set of strong saddle points is arcwise connected. The proof for the set of weak saddle points follows the same line.   □

Primal problem

We shall see now that the objective function Cx of (MOLP) can be obtained by minimizing the Lagrangian function over the set of matrices Y.

Corollary 5.5.6 *For every vector x, one has*

$$\mathrm{IMin}\big(\Pi(x)\big) = \mathrm{Min}\big(\Pi(x)\big) = \mathrm{WMin}\big(\Pi(x)\big) = \begin{cases} \{Cx\} & \text{if } Ax = b, x \geqq 0 \\ \emptyset & \text{else.} \end{cases}$$

Proof When $x \not\geqq 0$, by definition $\Pi(x)$ is empty. When $Ax - b \neq 0$, by Lemma 5.5.3 for every vector z in $\mathbb{R}^k$ there is a matrix Y such that $Y(b - Ax) = z$, which implies that $\Pi(x)$ has no weakly minimal point, hence no minimal point and no ideal point at all. Finally, if $Ax = b$ and $x \geq 0$, then $\Pi(x)$ consists of a singleton Cx and the corollary follows immediately.   □

In consequence, we may express (MOLP) as a max-min problem of the Lagrangian function

$$\text{Maximize Min}\big(\Pi(x)\big)$$
$$\text{subject to} \quad x \in \mathbb{R}^n.$$

When a maximal solution $\bar{x}$ exists with an associated matrix $\bar{Y}$, that is $C\bar{x} = L(\bar{x}, \bar{Y})$, the matrix $\bar{Y}$ is called a *multiplier* of (MOLP). In the classical framework the dimension of the multiplier vector is the number of the constraints of the initial problem, which is equal to m in our case. However, as we have k objective functions, the number of multiplier components is equal to km. This explains why multipliers of multiobjective problems are matrices.

We know now that (MOLP) is the max-min problem of the Lagrangian function. What happens if we switch max and min? It is to hope that the min-max problem will yield a dual problem for (MOLP) as in the case of the linear programming (Sect. 3.2). It is worthwhile to note that even if for the set $\Pi(x)$ three concepts of minimality coincide (Corollary 5.5.6), it is not so for the set $\Theta(Y)$. This peculiarity leads to different dual problems that we develop next.

Ideal dual problem

Given a matrix Y, we define a function of variable Y with values in $\mathbb{R}^k$ by

$$G_1(Y) = \sup \{L(x, Y) : x \geq 0\}.$$

Lemma 5.5.7 *For every matrix Y one has*

$$G_1(Y) = \begin{cases} \{Yb\} & \text{if } YA \geq C \\ \emptyset & \text{else.} \end{cases}$$

Proof When $YA - C \geq 0$, it is clear that $Yb \geq L(x, Y)$ for all $x \geq 0$. Hence $G_1(Y) = Yb$. Assume that $YA \ngeq C$. There exist a row Y^i of Y and a column a_j of A such that $c_{ij} - Y^i a_j > 0$, where c_{ij} is the ijth entry of C. Choose $x = te^j$ for $t \geq 0$ and e^j being the jth axis unit direction. Then the jth component of the vector $(C - YA)x$ tends to ∞ as t goes to ∞. By this, the set $G_1(Y)$ is empty.   □

The min-max problem of the Lagrangian function will take form

$$\text{Minimize } G_1(Y)$$
$$\text{subject to } Y \in L(\mathbb{R}^m, \mathbb{R}^k),$$

which is exactly the ideal dual problem we have studied in Sect. 5.2.

Theorem 5.5.8 *A point* (x, Y) *is an ideal saddle point of* L *if and only if* x *and* Y *are ideal solutions of (MOLP) and (VD1), in which case* $Cx = Yb$.

Moreover, if x *is an ideal solution of (MOLP), then there exists an ideal solution* Y *of (VD1) such that* (x, Y) *is an ideal saddle point of* L*; and if* Y *is an ideal solution of (VD1), then (MOLP) has at least* k *maximal solutions each of which forms with* Y *a weak saddle point of* L.

Proof Assume that $(\overline{x}, \overline{Y})$ is an ideal saddle point of L. In view of Theorem 5.5.4 both $\overline{x}$ and $\overline{Y}$ are feasible solutions. Moreover, for every feasible solutions x and Y one has

$$Cx = Cx + \overline{Y}b - \overline{Y}Ax \leqq C\overline{x} + Yb - YA\overline{x}$$

which yields $Cx \leqq C\overline{x}$ when setting $Y = \overline{Y}$ in the expression on the right hand side and $\overline{Y}b \leqq Yb$ when setting $x = \overline{x}$ in the expression on the left hand side. Thus, $\overline{x}$ and $\overline{Y}$ are ideal solutions. Conversely, assume that $\overline{x}$ and $\overline{Y}$ are ideal solutions of (MOLP) and (VD1) respectively. Then for every matrix Y we have

$$L(\overline{x}, \overline{Y}) = C\overline{x} + \overline{Y}b - \overline{Y}A\overline{x} = C\overline{x} + Y(b - A\overline{x}) \leqq L(\overline{x}, Y)$$

because $b - A\overline{x} = 0$. It remains to prove $L(x, \overline{Y}) \leqq L(\overline{x}, \overline{Y})$. To this end, we claim that (5.34) holds, that is,

$$(C - \overline{Y}A)\overline{x} = 0.$$

Indeed, since $\overline{x}$ and $\overline{Y}$ are ideal solutions, the feasible solution $\overline{x}$ is an optimal solution of the scalar problem (P_i):

$$\text{maximize} \quad \langle c^i, x \rangle$$
$$\text{subject to} \quad Ax = b$$
$$x \geqq 0$$

and the ith column $\overline{y}^i$ of $\overline{Y}^T$ is an optimal solution of the dual (D_i):

$$\text{minimize} \quad \langle b, y \rangle$$
$$\text{subject to} \quad A^T y \geqq c^i.$$

By duality for scalar problems, one has $\langle c^i, \overline{x} \rangle = \langle \overline{y}^i, b \rangle$. This being true for all $i = 1, \cdots, k$, we deduce $C\overline{x} = \overline{Y}b = \overline{Y}A\overline{x}$. Thus (5.34) is valid. Theorem 5.5.4 along with the feasibility of $\overline{x}$ and $\overline{Y}$ and (5.34) imply that $(\overline{x}, \overline{Y})$ is an ideal saddle point of L.

To prove the second part we assume x is an ideal solution of (MOLP). Then it is an optimal solution of the problems (P_i), $i = 1, \cdots, k$. By duality, the dual problems (D_i) have optimal solutions, say $y^i, i = 1, \cdots, k$ that satisfy $\langle c^i, x \rangle = \langle y^i, b \rangle$ for all $i = 1, \cdots, k$. Define Y to be the matrix whose rows are the transposes of

$y^1, \cdots, y^k$. It is clear that $Cx = Yb$ and (x, Y) is a saddle point of L. Now let Y be an ideal solution of (VD1). Then the columns $y^1, \cdots, y^k$ of Y^T are optimal solutions of the problems $(D_1), \cdots, (D_k)$ respectively. Again, by duality the primal problems $(P_1), \cdots, (P_k)$ have solutions, which imply that the value set of (MOLP) is bounded above. Hence each problem (P_i) admits optimal solutions among which there exists at least one maximal solution x^i of (MOLP). To prove that (x^i, Y) is a weak saddle point of L we define λ to be the ith coordinate unit vector in $\mathbb{R}^k$. Then

$$\lambda^T L(x, Y) = \langle y^i, b \rangle + \langle c^i - A^T y^i, x \rangle$$

which is the classical Lagrangian function associated with the problem (P_i), and deduce that (x^i, Y) is a saddle point of the real Lagrangian function $\langle \lambda, L \rangle$. By Theorem 5.5.2 it is a weak saddle point of L.   □

Note that when (VD1) has an ideal solution it is not necessary that L has ideal saddle points or strong saddle points. This is illustrated by the next example.

Example 5.5.9 Consider the following problem

$$\text{Maximize} \quad \begin{pmatrix} 1 & 0 \\ 0 & 1 \end{pmatrix} \begin{pmatrix} x_1 \\ x_2 \end{pmatrix}$$
$$\text{subject to} \quad x_1 + x_2 = 1$$
$$x_1, x_2 \geq 0.$$

The Lagrangian function associated with this problem takes the form

$$L(x, Y) = (1 - x_1 - x_2) \begin{pmatrix} y_1 \\ y_2 \end{pmatrix} + \begin{pmatrix} x_1 \\ x_2 \end{pmatrix},$$

where Y is a 2×1-matrix with the entries y_1 and y_2. It is clear that the matrix $\overline{Y}$ with $\overline{y}_1 = \overline{y}_2 = 1$ is an ideal solution of the dual problem

$$\text{Minimize} \begin{pmatrix} y_1 & y_2 \end{pmatrix}$$
$$\text{subject to} \begin{pmatrix} y_1 \\ y_2 \end{pmatrix} \begin{pmatrix} 1 & 1 \end{pmatrix} \geq \begin{pmatrix} 1 & 0 \\ 0 & 1 \end{pmatrix}.$$

However, the Lagrangian function L has no strong saddle point because the system (5.32) and (5.36) becomes

$$x_1 + x_1 = 1$$
$$(\lambda_1 \ \lambda_2) \begin{pmatrix} 0 & -1 \\ -1 & 0 \end{pmatrix} \begin{pmatrix} x_1 \\ x_2 \end{pmatrix} = 0$$
$$x_1, x_2 \geq 0; \ \lambda_1, \lambda_2 > 0$$

and is not solvable.

Strong dual problem

Given a matrix Y, we define a function of variable Y with set-values in $\mathbb{R}^k$ by

$$G_2(Y) = \text{Max}\big(\Theta(Y)\big) = \text{Max}\big\{L(x, Y) : x \geqq 0\big\}.$$

Our aim at the moment is to compute the value $G_2(Y)$ for a given matrix Y. We need some auxiliary lemmas.

Lemma 5.5.10 *Let M be a $k \times n$-matrix whose columns are denoted $M_1, \cdots, M_n$. Then*

$$\text{Max}\{Mx : x \geqq 0\} = \bigcup_{\lambda>0:\lambda^T M \leqq 0} \text{pos}\Big\{M_i : \lambda^T M_i = 0, i \in \{1, \cdots, n\}\Big\}.$$

Proof Since the set $\{Mx : x \geqq 0\}$ is the positive hull of the columns $M_1, \cdots, M_n$, it is a polyhedral cone. Therefore, it has maximal points if and only if its intersection with the positive orthant $\mathbb{R}^k_+$ consists of the zero vector only. In other words, that positive hull has maximal points if and only if there is a strictly positive vector λ such that $\lambda^T M_i \leqq 0$ for all $i = 1, \cdots, n$, in which case the set of maximal points is composed of faces spanned by the columns M_i that satisfy equation $\lambda^T M^i = 0$ or the zero vector if $\lambda^T M_i < 0$ for all $i = 1, \cdots, n$. $\qquad\square$

Lemma 5.5.11 *For every matrix Y one has*

$$G_2(Y) = Yb + \bigcup_{\lambda>0:\lambda^T (C-YA) \leqq 0} \text{pos}\Big\{(C - YA)_i : \lambda^T(C - YA)_i = 0,$$

$$i \in \{1, \cdots, n\}\Big\}$$

$$= Yb + \bigcup_{\lambda>0:\lambda^T (C-YA) \leqq 0} \Big\{(C - YA)u : u \geqq 0, \lambda^T(C - YA)u = 0\Big\}.$$

Proof If there is no $\lambda > 0$ such that $\lambda^T(C - YA) \leqq 0$, then the second term on the right hand side of the expression of $G_2(Y)$ is empty, and so $G_2(Y)$ is empty. This is the case when the positive hull of the columns of the matrix $C - YA$ contains a nonzero positive vector, and hence the set $\{L(x, Y) : x \geqq 0\} = \{Yb + (C - YA)x : x \geqq 0\}$ has no maximal point. When the set $\{L(x, Y) : x \geqq 0\}$ has maximal points, we may apply Lemma 5.5.10 to obtain

$$G_2(Y) = \text{Max}\{L(x, Y) : x \geqq 0\}$$
$$= \text{Max}\big\{Yb + (C - YA)x : x \geqq 0\big\}$$
$$= Yb + \bigcup_{\lambda>0:\lambda^T(C-YA)\leqq 0} \text{pos}\Big\{(C - YA)_i : \lambda^T(C - YA)_i = 0,$$

$$i \in \{1, \cdots, n\}\Big\}$$

as requested. To prove the second equality it suffices to observe that a vector z belongs to the cone pos $\{(C - YA)_i : \lambda^T(C - YA)_i = 0, i \in \{1, \cdots, n\}\}$ for some $\lambda > 0$ such that $\lambda^T(C - YA) \leq 0$ if and only if $z = (C - YA)u$ for some positive vector u whose components corresponding to $\lambda^T(C - YA)_i < 0$ are zero.   □

The min-max problem of the Lagrangian function with G_2 yields a dual problem, denoted (VD2)^L:

$$\text{Minimize } Yb + (C - YA)u$$
$$\text{subject to } \lambda^T YA \geq \lambda^T C$$
$$\lambda > 0, u \geq 0$$
$$\lambda^T(C - YA)u = 0.$$

Notice that once Y is given with $\lambda^T YA \geq \lambda^T C$ for some $\lambda > 0$, the polyhedral cone pos$\{(C - YA)_i : \lambda^T(C - YA)_i = 0, i \in \{1, \cdots, n\}\}$ is completely defined and contains all value vectors $(C - YA)u$.

We consider also the following maps:

$$G_2'(Y) = Yb + \bigcup_{\lambda > 0: \lambda^T(C - YA) \leq 0} \left\{ z \in \mathbb{R}^k : \langle \lambda, z \rangle \geq 0 \right\}$$

and

$$G_2''(Y) = Yb + \bigcup_{\lambda > 0: \lambda^T(C - YA) \leq 0} \left\{ z \in \mathbb{R}^k : \langle \lambda, z \rangle = 0 \right\}.$$

The dual problem obtained from G_2' is denoted (VD2'):

$$\text{Minimize } \quad Yb + z$$
$$\text{subject to } \quad \lambda^T YA \geq \lambda^T C$$
$$\langle \lambda, z \rangle \geq 0, \lambda > 0,$$

and the dual problem obtained from G_2'' is denoted (VD2''):

$$\text{Minimize } \quad Yb + z$$
$$\text{subject to } \quad \lambda^T YA \geq \lambda^T C$$
$$\langle \lambda, z \rangle = 0, \lambda > 0.$$

The dual problem (VD2') was given in Sect. 5.3 and called the *extended dual problem* of (MOLP). It is clear that if (Y, λ, z) is a feasible solution of (VD2''), then it is a feasible solution of (VD2'). Moreover, since for every $z \in \mathbb{R}^k$ with $\langle \lambda, z \rangle > 0$ there is some $z^0 \leq z$ such that $\langle \lambda, z^0 \rangle = 0$, we deduce easily that for every feasible solution (Y, λ, z) of (VD2') there is some vector $z^0 \leq z$ such that (Y, λ, z^0) is a feasible solution of (VD2″). Consequently, every minimal solution of (VD2′) is a minimal solution of (VD2″) and vice versa. In this sense the two problems (VD2′) and (VD2″) are equivalent. The problems (VD2)^L and (VD2″) apparently differ from each other. Namely, given a matrix Y, we have the following inclusions

$$Yb \in G_2(Y) \subseteq G_2''(Y) \subseteq G_2'(Y) \tag{5.39}$$

in which equality does not hold in general. Remember that the strong dual problem (VD2) is given by

$$\text{Minimize } Yb$$
$$\text{subject to } \lambda^T YA \geqq \lambda^T C$$
$$\text{for some } \lambda > 0.$$

The value sets of the problems (VD2), (VD2)^L and (VD2″) are denoted Q^{D2}, Q^L and Q'' respectively. Recall also $Q^{**} := \left(\mathbb{R}^k \setminus (Q - \mathbb{R}_+^k) \right) \cup \text{Max}(Q)$.

Lemma 5.5.12 *Assume that (MOLP) has maximal solutions. Then the following inclusions hold*

$$Q^{D2} \subseteq Q^L \subseteq Q'' \subseteq Q^{**}.$$

In addition, if $A \neq 0$ and $b = 0$, then $Q^L = Q''$.

Proof The first two inclusions of the lemma are clear. We prove the last one. Let $Yb + z$ be an element of Q'', say with $\langle \lambda, z \rangle = 0$ and $\lambda^T (C - YA) \leqq 0$ for some strictly positive vector λ. We claim that $Yb + z$ belongs to Q^{**}. Indeed, if this is not true, then one can find some feasible solution x of (MOLP) and a positive vector v such that

$$Yb + z = Cx - v.$$

Multiplying both side of the above equality by λ^T and using the fact that $Ax = b$ we obtain

$$\lambda^T (YA - C)x = -\lambda^T (v + z).$$

Since the value on the left hand side is nonnegative because $x \geq 0$ and $\lambda^T (C - YA) \leqq 0$, and λ is strictly positive, one must have $v = 0$, which implies that $Yb + z = Cx$. This latter vector being not a maximal element of Q, we find some feasible solution $\bar{x}$ of (MOLP) such that

$$YA\bar{x} + z = Yb + z \leq C\bar{x}.$$

Again, multiplying both sides of this inequality by λ^T we derive

$$\lambda^T (YA - C)\bar{x} < 0$$

which is a contradiction.

For the second part of the lemma it suffices to show that given any feasible solution (Y, λ, z) of (VD2″), there is a feasible solution (Y', λ, u) of $(VD2)^L$ such that

$$z = (C - Y'A)u. \tag{5.40}$$

To this end, consider the scalarized problem (P_λ) of maximizing $\lambda^T Cx$ under the constraint $Ax = 0$ and $x \geq 0$.

Claim 1. There exists a coordinate unit vector $e \geq 0$ in $\mathbb{R}^n$ such that $Ae \neq 0$ and the problem (Pe) below has the optimal value equal to 0:

$$\begin{array}{ll}
\text{maximize} & \lambda^T Cx \\
\text{subject to} & Ax = 0 \\
& x - te \geq 0 \\
& (x, t) \in \mathbb{R}^n \times \mathbb{R}.
\end{array}$$

In fact, suppose to the contrary that for every coordinate unit vector e^j with $Ae^j \neq 0$, the problem (Pe^j) has its optimal value strictly positive. Observe first that since $A \neq 0$, the index set $J = \{j \in \{1, \cdots, n\} : Ae^j \neq 0\}$ is nonempty. Then for $j \in J$, there is an optimal solution x^j such that $\lambda^T Cx^j > 0$. The vectors x^j, $j \notin J$ and e^j, $j \in J$ all belong to the kernel of the matrix A, hence they are linearly dependent, that is there are coefficients $\alpha_1, \cdots, \alpha_n$ not all zero such that $\sum_{j \in J} \alpha_j x^j + \sum_{j \notin J} \alpha_j e^j = 0$. Set

$$x = \sum_{j \in J} |\alpha_j| x^j.$$

We prove that x is a feasible solution of (P_λ) and $\lambda^T Cx > 0$, which is a contradiction due to the fact that x is also an asymptotic direction of the feasible set. Indeed, since the coordinate vectors e^j are linear independent, at least one of the coefficients α_j, $j \in J$ is non zero. Let J_+ be the set of indices j with $\alpha_j \geq 0$ and J_- the set of all j with $\alpha_j < 0$. Then the ith component $(x^j)_i$ of the vector x^j is positive for every $(j, i) \in (J_+ \times J_-) \cup (J_- \times J_+)$. Moreover, for every $i \notin J$, the ith component of x is nonnegative, while for $i \in J$,

$$\sum_{j \in J_-} \alpha_j (x^j)_i + \sum_{j \in J_+} \alpha_j (x^j)_i = 0,$$

which implies

$$x_i = \sum_{j \in J_-} (-\alpha_j)(x^j)_i + \sum_{j \in J_+} \alpha_j (x^j)_i = \left\{ \begin{array}{l} 2\sum_{j \in J_+} \alpha_j (x^j)_i \geq 0 \\ -2\sum_{j \in J_-} \alpha_j (x^j)_i \geq 0. \end{array} \right.$$

By this, $x \geq 0$. Moreover, $Ax = \sum_{j \in J} |\alpha_j| Ax^j = 0$ and

$$\lambda^T Cx = \sum_{j \in J} |\alpha_j| \lambda^T Cx^j > 0,$$

which contradicts the hypothesis.

Claim 2. Let y be a dual optimal solution of (Pe). There exists a matrix Y' such that $Y'^T \lambda = y$ and $Y'Au = Cu - z$. Indeed, since y is a dual optimal solution, we have $A^T y \geq C^T \lambda$ and $\langle A^T y - C^T \lambda, u \rangle = 0$. Consider the vectors $Au, y, Cu - z$ and λ in which Au and λ are nonzero and $\langle \lambda, Cu - z \rangle = \langle \lambda, Cu \rangle = \langle Au, y \rangle$ because $\langle \lambda, z \rangle = 0$. It remains to apply Lemma 5.5.3 to find a matrix Y' satisfying (5.40) as requested.   □

It should be noticed that when $A = 0$ and $b = 0$, it is not necessary that equality $Q^L = Q''$ be true as the next example shows.

Example 5.5.13 Consider the following problem

$$\text{Maximize} \quad \begin{pmatrix} -1 & 0 \\ 0 & -1 \end{pmatrix} \begin{pmatrix} x_1 \\ x_2 \end{pmatrix}$$
$$\text{subject to} \quad x_1, x_2 \geq 0.$$

The dual problems (VD2″) and (VD2)^L are respectively given by

$$\text{Mimimize} \quad \begin{pmatrix} z_1 \\ z_2 \end{pmatrix}$$
$$\text{subject to} \quad (0 \;\; 0) \geq (\lambda_1 \;\; \lambda_2) \begin{pmatrix} -1 & 0 \\ 0 & -1 \end{pmatrix}$$
$$\lambda_1 z_1 + \lambda_2 z_2 = 0$$
$$\lambda_1, \lambda_2 > 0$$

and

$$\text{Mimimize} \quad \begin{pmatrix} -1 & 0 \\ 0 & -1 \end{pmatrix} \begin{pmatrix} u_1 \\ u_2 \end{pmatrix}$$
$$\text{subject to} \quad (0 \;\; 0) \geq (\lambda_1 \;\; \lambda_2) \begin{pmatrix} -1 & 0 \\ 0 & -1 \end{pmatrix}$$
$$\lambda_1 u_1 + \lambda_2 u_2 = 0$$
$$u_1, u_2 \geq 0$$
$$\lambda_1, \lambda_2 > 0.$$

It is easy to see that $Q^L = \{0\}$ and $Q'' = \mathbb{R}^2 \setminus (-\mathbb{R}^2_+ \setminus \{0\})$ which proves that the inclusion $Q^L \subset Q''$ is strict.

Lemma 5.5.14 *Assume that (MOLP) has maximal solutions. Then*

$$\text{Max}(Q) = \text{Min}(Q^L).$$

Proof If $b \neq 0$, then in view of Lemma 5.3.2, we have

$$Q^{D2} + \mathbb{R}_+^k = \left(\mathbb{R}^k \setminus (Q - \mathbb{R}_+^k)\right) \cup \text{Max}(Q)$$

and deduce from Lemma 5.5.12 equalities

$$Q^{**} = Q^{D2} + \mathbb{R}_+^k = Q^L + \mathbb{R}_+^k = Q'' + \mathbb{R}_+^k$$

which shows that the problems (VD2), (VD2)^L and (VD2$''$) have the same minimal value set.

If $b = 0$ and $A \neq 0$, then by Lemma 5.5.12 we have $\text{Min}(Q^L) = \text{Min}(Q'')$. Since $\text{Min}(Q'')$ and $\text{Min}(Q^{**})$ coincide, and $\text{Min}(Q'') = \text{Max}(Q)$ according to Theorem 5.1.2 and Lemma 5.3.7, we deduce $\text{Max}(Q) = \text{Min}(Q^L)$.

Finally, if $A = 0$ and $b = 0$, then $C\bar{x}$ with $\bar{x} \geq 0$ is a maximal element of the set $\{Cx : x \geq 0\}$ if and only if there is a vector $\lambda > 0$ such that $0 = \lambda^T C\bar{x} \geq \lambda^T Cu$ for every $u \geq 0$, which implies that $\lambda^T C \leq 0$ and $C\bar{x}$ is a minimal element of (VD2)^L. Therefore, we have equality $\text{Max}(Q) = \text{Min}(Q^L)$ as well. $\square$

A relationship between saddle points of the Lagrangian function L and efficient solutions of the primal problem and the dual problems is given next.

Theorem 5.5.15 *The following statements hold true.*

(i) *A point (x, Y) is a strong saddle point of L if and only if there is $\lambda > 0$ such that x is an optimal solution of (P_λ) and maximal solution of (MOLP), (Y, λ, x) is a minimal solution of (VD2)^L and $Y^T\lambda$ is an optimal solution of (D_λ). Moreover, if x is a maximal solution of (MOLP) and (Y, λ, u) is a minimal solution of (VD2)^L with $\langle \lambda, Cx \rangle = \langle \lambda, Yb \rangle$, then (x, Y) is a saddle point of $\lambda^T L$ and hence a strong saddle point of L.*

(ii) *For every maximal solution x of (MOLP) there exists a minimal solution (Y, λ, u) of (VD2)^L such that (x, Y) is a saddle point of $\lambda^T L$, and hence a strong saddle point of L; and for every minimal solution (Y, λ, u) of (VD2)^L there is a maximal solution x of (MOLP) such that (x, Y) is a saddle point of $\lambda^T L$ and thus a strong saddle point of L.*

Proof Assume that (x, Y) is a strong saddle point of L. In view of Theorem 5.5.2 there is $\lambda > 0$ such that it is a saddle point of the real-valued function $\langle \lambda, L(., .) \rangle$, which implies that x is an optimal solution of (P_λ), and hence a maximal solution of (MOLP). Furthermore, $Y^T\lambda$ is an optimal solution of (D_λ) and $\lambda^T(Cx - Yb) = 0$ which imply that (Y, λ, x) is a feasible solution of (VD2)^L. Suppose (Y, λ, x) is not minimal for (VD2)^L. Since $Yb + (C - YA)x = Cx$, there exists a feasible solution (Y, λ', u) of (VD2)^L such that $Yb + (C - YA)u \leq Cx$. By multiplying each side of this inequality by λ'^T, we get $\lambda'^T Yb < \lambda'^T Cx$ and hence $\lambda'^T (YA - C)x < 0$ which is impossible. So (Y, λ, x) is a minimal solution of $(VD2)^L$.

Conversely, let x and (Y, λ, x) be efficient solutions of respectively (MOLP) and $(VD2)^L$. One directly obtains that (x, Y) satisfies conditions (5.32), (5.35) and (5.36) of Theorem 5.5.4 by which (x, Y) is a strong saddle point of L. For the second part of (i), (5.32), (5.35) and (5.36) are satisfied. We deduce that $(\overline{x}, \overline{Y})$ is a saddle point of $\lambda^T L$ and hence a strong saddle point of L.

For (ii), we assume that x is a maximal solution of (MOLP). Then there is $\lambda > 0$ such that x is an optimal solution of (P_λ) and y an optimal solution of the dual problem (D_λ). Choose Y such that $Y^T \lambda = y$ (Lemma 5.5.3). Then (5.32), (5.35) and (5.36) are satisfied. By Theorem 5.5.4 (x, Y) is a strong saddle point. Finally, let (Y, λ, u) be a minimal solution of $(VD2)^L$. By Lemma 5.5.14, there is a maximal solution x of (MOLP) such that $Cx = Yb + (C - YA)u$ and it is easy to check that (x, Y) is a saddle point of $\lambda^T L$ and thus a strong saddle point of L.   □

Weak dual problem

Given a matrix Y, we define a function of variable Y with set-values in $\mathbb{R}^k$ by

$$G_3(Y) = \text{WMax}(\Theta(Y)) = \text{WMax}\{L(x, Y) : x \geq 0\}.$$

Using the same method as for strong saddle points one may express $G_3(Y)$ by

$$G_3(Y) = Yb + \bigcup_{\lambda \geq 0 : \lambda^T (C - YA) \leq 0} \text{pos}\Big\{(C - YA)_i : \lambda^T (C - YA)_i = 0,$$

$$i \in \{1, \cdots, n\}\Big\}$$

$$= Yb + \bigcup_{\lambda \geq 0 : \lambda^T (C - YA) \leq 0} \Big\{(C - YA)u : u \geq 0, \lambda^T (C - YA)u = 0\Big\}.$$

and obtain a dual problem, denoted $(VD3)^L$:

$$\begin{aligned} & \text{Minimize } Yb + (C - YA)u \\ & \text{subject to } \lambda^T YA \geq \lambda^T C \\ & \qquad\qquad \lambda \geq 0, u \geq 0 \\ & \qquad\qquad \lambda^T (C - YA)u = 0. \end{aligned}$$

Two related dual problems denoted (VD3′) and (VD3″) are given respectively by

$$\begin{aligned} & \text{Minimize } Yb + z \\ & \text{subject to } \lambda^T YA \geq \lambda^T C \\ & \qquad\qquad \langle \lambda, z \rangle \geq 0 \\ & \qquad\qquad \lambda \geq 0, \end{aligned}$$

and

$$\text{Minimize } Yb + z$$
$$\text{subject to } \lambda^T Y A \geq \lambda^T C$$
$$\langle \lambda, z \rangle = 0$$
$$\lambda \geq 0.$$

The problem (VD3′) was obtained in Sect. 5.4 as an extended weak dual problem. The problems (VD3)^L, (VD3′) and (VD3″) are distinct, having the different value sets, which are denoted respectively $\hat{Q}^L$, $\hat{Q}'$ and $\hat{Q}''$. It is clear that (VD3′) and (VD3″) are equivalent, which means that their weakly minimal value sets coincide. We recall also that $Q^{(D3)} + \mathbb{R}_+^k \subseteq \hat{Q}' = Q^{***}$, and equality is true when $b \neq 0$ (Lemma 5.4.1).

Lemma 5.5.16 *The following inclusions hold true:*

$$\text{WMin}(\hat{Q}^L) \subseteq \text{WMin}(\hat{Q}'' + \mathbb{R}_+^k) = \text{WMax}(Q - \mathbb{R}_+^k). \tag{5.41}$$

Proof It is clear that $Q^{(D3)} + \mathbb{R}_+^k \subseteq \hat{Q}^L + \mathbb{R}_+^k \subseteq \hat{Q}'' + \mathbb{R}_+^k \subseteq \hat{Q}'$. When $b \neq 0$ these inclusions become equalities. We deduce from Theorem 5.1.2 that

$$\text{WMin}(\hat{Q}^L) \subseteq \text{WMin}(\hat{Q}'' + \mathbb{R}_+^k) = \text{WMin}(Q^{***}) = \text{WMax}(Q - \mathbb{R}_+^k).$$

If $b = 0$ and $A \neq 0$, the same argument of proof of Lemma 5.5.12 shows that $\hat{Q}^L = \hat{Q}''$. By this, the conclusion of the lemma remains true.

Finally, if $A = 0$ and $b = 0$, then, as in the proof of Lemma 5.5.14 we have $\text{WMin}(\hat{Q}^L) = \text{WMax}(Q) \subseteq \text{WMax}(Q - \mathbb{R}_+^k)$ as requested. The proof is complete.   □

The next result is an analogy of Theorem 5.5.15 for weak saddle points.

Theorem 5.5.17 *The following statements hold.*

(i) *A point (x, Y) is a weak saddle point of L if and only if there is $\lambda \geq 0$ such that x is an optimal solution of (P_λ) and weakly maximal solution of (MOLP), (Y, λ, x) is a weakly minimal solution of (LD3) and $Y^T \lambda$ is an optimal solution of (D_λ).*
Moreover, if x is a weakly maximal solution of (MOLP) and (Y, λ, u) is a weakly minimal solution of (LD3) with $\langle \lambda, Cx \rangle = \langle \lambda, Yb \rangle$, then (x, Y) is a saddle point of $\lambda^T L$ and hence a weak saddle point of L.
(ii) *For every weakly maximal solution x of (MOLP) there exists a weakly minimal solution (Y, λ, u) of (LD3) such that (x, Y) is a saddle point of $\lambda^T L$, and hence a weak saddle point of L; And for every weakly minimal solution (Y, λ, u) of (LD3) there is a weakly maximal solution x of (MOLP) such that (x, Y) is a saddle point of $\lambda^T L$ and thus a weak saddle point of L.*

Proof We apply the same method of Theorem 5.5.15 to prove the first item and the first part of the second item. We now proceed to prove the second part of (ii). Let (Y, λ, u) be a weakly minimal solution (LD3). If $b = 0$, then $x = 0$ is an optimal solution of (P_λ), while $\lambda^T Y$ is a feasible solution, hence an optimal solution of (D_λ). By this, (x, Y) is a saddle point of $\lambda^T L$. We apply (5.41) to find a feasible solution x of (MOLP) and a positive vector v such that

$$Yb + (C - YA)u = Cx - v.$$

We deduce that

$$\langle \lambda, Yb \rangle = \langle \lambda, Yb + (C - YA)u \rangle = \langle \lambda, Cx - v \rangle \leqq \langle \lambda, Cx \rangle.$$

Actually the last inequality is equality because x and $\lambda^T Y$ are feasible solutions of (P_λ) and (D_λ). Hence x and Y form a saddle point of $\lambda^T L$ as requested. □

Notice again that when (x, Y) is a weak saddle point of L, the matrix Y is not necessarily a minimal solution of $(VD3)^L$.

Scalarizing sets

Given a weak saddle point (x, Y) of L, according to Lemma 5.5.2, there is some positive vector $\lambda \in \mathbb{R}^k$ such that (x, Y) is a saddle point of the real valued function $\lambda^T L$. The set of all such $\lambda \geq 0$ is called a *scalarizing set* for the saddle point (x, Y) and denoted $\Lambda(x, Y)$. We recall that for a weakly maximal solution x of (MOLP), the set of scalarizing vectors $\Lambda(x)$ consists of $\lambda \geq 0$ such that x is an optimal solution of (P_λ). For a matrix Y we denote $\Lambda(Y)$ the set of vectors $\lambda \geq 0$ such that (Y, λ, u) is a weakly minimal solution of $(VD3)^L$ for some $u \geqq 0$. In this part we study the set $\Lambda(x, Y)$ by establishing some of its properties in relation with $\Lambda(x)$ and $\Lambda(Y)$. Below is an explicit formula of the scalarizing sets $\Lambda(x)$ and $\Lambda(Y)$ in terms of normal cones.

Lemma 5.5.18 *The following assertions hold:*

$$\Lambda(x) = N_Q(Cx) \cap \left(\mathbb{R}^k_+ \setminus \{0\} \right).$$

$$\Lambda(Y) = \bigcup_{\substack{u \geq 0 \\ Yb+(C-YA)u \in Q}} N_{co(Yb,Q)}\big(Yb + (C - YA)u\big)$$

$$\bigcap \left\{ \lambda \in \mathbb{R}^k_+ \setminus \{0\} : \lambda^T(C - YA) \leqq 0 \right\}.$$

Proof The first item follows from Theorem 4.2.6. We prove the second formula. Let $\lambda \in \Lambda(Y)$. There is $u \geq 0$ such that (Y, λ, u) is a weakly minimal solution of $(LD3)^L$. In particular $\lambda \geq 0$ and $\lambda^T(C - YA) \leqq 0$. Moreover, according to Theorem 5.5.17 (ii), there exists a feasible solution x of (MOLP) such that (x, Y) is a saddle point of $\lambda^T L$. We have then

$$Yb + (C - YA)x = Cx \in Q \quad \text{and}$$

$$\lambda^T Cx \geqq \lambda^T (Cx' + Yb - YAx') \text{ for every } x' \geqq 0.$$

It follows that $\lambda^T Cx \geq \lambda^T Yb$ (taking $x' = 0$ in the latter inequality) and $\lambda^T Cx \geq \lambda^T Cx''$ for every feasible solution x'' of (MOLP). Consequently,

$$\lambda^T Cx \geqq \lambda^T (tYb + (1-t)Cx'')$$

for every $t \in [0, 1]$ and $Cx'' \in Q$, which proves that λ belongs to the normal cone $N(co(Yb, Q), Yb + (C - YA)x)$.

For the converse inclusion, let λ be a positive k-vector such that $\lambda^T (C - YA) \leq 0$ and $\lambda \in N(co(Yb, Q), Yb + (C - YA)u)$ where $Yb + (C - YA)u = Cx \in Q$ for some $u \geq 0$ and $Cx \in Q$. We have

$$\lambda^T Cx = \lambda^T (Yb + (C - YA)u) \geqq \lambda^T Yb,$$

which implies $\lambda^T (C - YA)u \geq 0$. The latter inequality is actually equality because $\lambda^T (C - YA) \leq 0$ and $u \geq 0$. We deduce that (x, Y) is a saddle point of $\lambda^T L$ and by Theorem 5.5.17, $\lambda \in \Lambda(Y)$. $\qquad\qquad\square$

In the next result S_L and $\hat{S}_L$ denote respectively the set of weak saddle points and the set of exact weak saddle points of L (in the sense that $Yb = Cx$).

Theorem 5.5.19 *Let (x, Y) be a weak saddle point of L. Then the scalarizing set $\Lambda(x, Y)$ is a convex polyhedral cone without apex and satisfies*

$$\Lambda(x, Y) = \Lambda(x) \cap \Lambda(Y).$$

Moreover, the following relations hold true

(i) $\Lambda(x) = \bigcup_{(x,Y') \in S_L} \Lambda(x, Y') \supseteq \bigcup_{(x,Y') \in \hat{S}_L} \Lambda(x, Y')$. *Equality holds if $b \neq 0$.*
(ii) $\Lambda(Y) = \bigcup_{(x',Y) \in S_L} \Lambda(x', Y)$.

Proof Given a weak saddle point (x, Y), in view of Theorem 5.5.4 the set $\Lambda(x, Y)$ consists of the solutions to the homogenous linear system (5.35) and (5.36). Hence it is a convex polyhedral cone without apex because $\lambda \geq 0$. In view of (i) and (ii), to establish $\Lambda(x, Y) = \Lambda(x) \cap \Lambda(Y)$ it suffices to show that every $\lambda \in \Lambda(x) \cap \Lambda(Y)$ belongs to $\Lambda(x, Y)$. Indeed, since (Y, λ, u) is a weakly minimal solution of (LD3) for some $u \geq 0$, by Theorem 5.5.17 (ii) there is a weakly maximal solution x' of (MOLP) such that (x', Y) is a saddle point of $\lambda^T L$. Hence both of x and x' are optimal solutions of (P_λ) and $\lambda^T Cx = \lambda^T Cx' = \lambda^T Yb$. We deduce that (x, Y) is a saddle point of $\lambda^T L$.

We now proceed to prove (i). It clear that $\Lambda(x, Y') \subset \Lambda(x)$ for every Y' such that $(x, Y') \in S_L$. For the converse, let $\lambda \in \Lambda(x)$. Then x is an optimal solution of (P_λ), and hence (D_λ) admits some optimal dual solution y. Apply Lemma 5.5.3

to find some $\tilde{Y}$ such that $\tilde{Y}^T \lambda = y$. Then $(x, \tilde{Y})$ is a saddle point of $\lambda^T L$, that is $\lambda \in \Lambda(x, \tilde{Y})$. Moreover, when $b \neq 0$, in view of Lemma 5.5.3, it is possible to find $\tilde{Y}$ satisfying $\tilde{Y}b = Cx$ which shows that $(x, \tilde{Y})$ is an exact weak saddle point.

To establish (ii) we observe that if (x', Y) is a weak saddle point of L and $\lambda \in \Lambda(x', Y)$, then the argument of the proof of Theorem 5.5.15 (i) shows that (Y, λ, x') is a weakly minimal solution of (LD3), by which λ belongs to $\Lambda(Y)$. Conversely, if $\lambda \in \Lambda(Y)$, then $Y^T \lambda$ is an optimal solution of (D_λ). Let x' be an optimal solution of (P_λ). We have $\lambda^T Cx' = \lambda^T YA$ proving that (x', Y) is a saddle point of $\lambda^T L$, that is $\lambda \in \Lambda(x', Y)$. □

We recall that for a convex set G in $\mathbb{R}^k$, its relative interior is denoted ri(G).

Corollary 5.5.20 *Let x and x' be weakly maximal solutions of (MOLP) and G be a face of Q. Then the following assertions hold.*

(i) *$\Lambda(x, Y) \subseteq \Lambda(x', Y)$ if $Cx \in$ ri(G) and $Cx' \in G$, and equality is true when both Cx and Cx' are in ri(G).*
(ii) *$\Lambda(x, Y) = \Lambda(x', Y) = \Lambda(Y)$ for every $x, x' \in G$ if $Yb \in$ ri(G). In particular if $Yb = Cx$, then $\Lambda(x) = \Lambda(Y) = \Lambda(x, Y)$.*
(iii) *If $G \subset Max(Q)$ and both of Cx and Cx' are boundary points of G, then there exists Y such that $\Lambda(x, Y) \cap \Lambda(x', Y) \neq \emptyset$.*

Proof The inclusion in (i) is clear from the equality given in Theorem 5.5.19 and from the fact that $N_Q(Cx) \subseteq N_Q(Cx')$. When both Cx and Cx' are relative interior points of G equality is immediate from the inclusion.

To prove (ii) assume $Yb \in$ ri(G). Then co$(Yb, Q) = Q$ and for every $\lambda \in N_Q(Yb + (C - YA)u)$ with $\lambda^T (C - YA) \leqq 0$, $\lambda \geq 0$ and $Yb + (C - YA)u \in Q$, we have $\lambda^T (C - YA)u = 0$, which implies that $\lambda \in N_Q(Yb)$. It follows that

$$\Lambda(Y) = N_Q(Yb) \cap \{\lambda \in \mathbb{R}_+^k \setminus \{0\} : \lambda^T (C - YA) \leqq 0\}.$$

Since Yb is a relative interior point of G, we deduce the equalities in (ii) for every $x, x' \in G$. When $Cx = Yb$, we choose a face G of Q that contains Cx in its relative interior. Then $\Lambda(Y) = \Lambda(x, Y)$. Moreover, for every $\lambda \in \Lambda(x), \lambda \geq 0$, the vector $\lambda^T Y$ is an optimal solution of (D_λ), by which $\lambda^T (C - YA) \leqq 0$. In view of Lemma 5.5.3 we deduce $\Lambda(x) = \Lambda(x, Y)$.

For the last assertion it suffices to choose a feasible solution x'' of (MOLP) such that Cx'' is a relative interior point of G, find Y such that $\Lambda(x'', Y) \neq \emptyset$ (such a matrix Y exists in view of Theorem 5.5.17) and apply (i) to obtain the inclusion $\Lambda(x'', Y) \subseteq \Lambda(x, Y) \cap \Lambda(x', Y)$. □

A natural question arises whether for a weakly maximal solution x of (MOLP) there is some Y such that $\Lambda(x) = \Lambda(x, Y)$. An immediate answer is obtained from Theorem 5.5.4 (iii): such a matrix Y exists if and only if the system

$$\lambda^T (C - YA) \leqq 0,$$
$$\lambda^T (Cx - Yb) = 0 \text{ for all } \lambda \in \Lambda(x)$$

is consistent. In general it is difficult to check the consistency of this system because $\Lambda(x)$ is an infinite set. However, in some particular situations such a matrix Y can be easily obtained.

Corollary 5.5.21 *Let x be a weakly maximal solution of (MOLP). Each of the following conditions is sufficient for existence of Y such that $\Lambda(x) = \Lambda(x, Y)$:*

(i) x is a non-degenerate vertex of (MOLP);
(ii) the extreme directions of $\Lambda(\overline{x})$ are linearly independent.

Proof We prove the first assertion. Without loss of generality we may assume that the matrix A is composed of a basic square submatrix A_B and the remaining part A_N such that the basic part x_B of x is strictly positive and given by $A_B^{-1}b$, and the non-basic part x_N is zero. We set $Y = C_B A_B^{-1}$ where C_B is the submatrix of C corresponding to the basic variable x_B. Then $Yb = C_B A_B^{-1}b = C_B x_B = Cx$. It follows from Corollary 5.5.20 that $\Lambda(x) = \Lambda(x, Y)$.

For the second assertion, let $\Lambda(x) = \mathrm{pos}(\lambda_1, \cdots, \lambda_\ell)$, with $\ell \leq k$ where the family $\{\lambda_1, \cdots, \lambda_\ell\}$ is linearly independent. Since x is an optimal solution of (P_{λ_i}) for $i = 1, \cdots, \ell$, we may find dual optimal solutions y_{λ_i} of (D_{λ_i}) for $i = 1, \cdots, \ell$. Let Λ be the $k \times \ell$-matrix with the ith column equal to λ_i and Y_Λ the $\ell \times m$-matrix with the ith row equal to $y_{\lambda_i}^T$. Then it is easy to check that $Y = \Lambda(\Lambda^T \Lambda)^{-1} Y_\Lambda$ satisfies $\Lambda(x) = \Lambda(x, Y)$.   □

We notice that (ii) trivially holds for bi-criteria problems ($k = 2$) because $\Lambda(x)$ has either one extreme direction or two linearly independent extreme directions. Below is an example in which no Y exists such that $\Lambda(x) = \Lambda(x, Y)$ for a given weakly maximal solution x of (MOLP).

Example 5.5.22 Consider the following problem

$$\text{Maximize} \begin{pmatrix} 1 & -0.5 & 0 & 0 & 0 \\ -0.5 & 1 & 0 & 0 & 0 \\ -1 & -1 & 0.5 & 0 & 0 \end{pmatrix} \begin{pmatrix} x_1 \\ x_2 \\ x_3 \\ x_4 \\ x_5 \end{pmatrix}$$

subject to
$$x_1 + x_3 + x_4 = 1$$
$$0.5x_1 + 0.5x_2 + x_3 + x_5 = 1$$
$$x_1, x_2, x_3, x_4, x_5 \geq 0.$$

The feasible set is bounded and defined by its 5 vertices. One of these vertices is degenerate and a maximal solution of (MOLP), namely $x = (0, 0, 1, 0, 0)^T$. The other vertices are

$$x^1 = (1, 1, 0, 0, 0)^T, \quad x^2 = (1, 0, 0, 0, 0.5)^T,$$
$$x^3 = (0, 2, 0, 1, 0)^T, \quad x^4 = (0, 0, 0, 1, 1)^T.$$

They all are efficient solutions of (MOLP) except the last one which is merely a weakly efficient solution. By solving the system of 4 inequalities $\lambda^T C x \geq \lambda^T C x^i$, $i = 1, \cdots, 4$ we can show that $\Lambda(x)$ is the cone generated by the following vectors:

$$\lambda^1 = (5, 5, 2)^T, \ \lambda^2 = (8, 7, 3)^T, \ \lambda^3 = (3, 0, 2)^T, \ \lambda^4 = (0, 0, 1)^T, \ \lambda^5 = (0, 5, 4)^T.$$

Then we work out Theorem 5.5.4 to find Y such that $\Lambda(x) = \Lambda(x, Y)$. Using $\lambda^T (C - YA)x = 0$ for $\lambda \in \Lambda(x)$ we get

$$y_{31} + y_{32} - 0.5 = 0, \ \ y_{11} + y_{12} = 0, \ \ y_{21} + y_{22} = 0.$$

Computing the second and the fourth columns of $(\lambda^5)^T (C - YA)$, we get $5y_{21} + 4y_{31} = 0$. Similarly, the first and the fifth columns of $(\lambda^3)^T (C - YA)$, yield $6y_{11} - 5y_{21} - 2 = 0$; and the first and the second columns of $(\lambda^1)^T (C - YA)$ yield $8y_{11} = 1$. Finally, these necessary conditions for verifying $\Lambda(x) = \Lambda(x, Y)$ imply that $Y = \frac{1}{16} \begin{pmatrix} 2 & -2 \\ -4 & 4 \\ 5 & 3 \end{pmatrix}$. But then, one can see that $(\lambda^2)^T (C - YA) \nleq 0$, proving impossible to find Y such that $\Lambda(x) = \Lambda(x, Y)$ in this example.

Computing saddle points

In the previous section we proved that when (x, Y) is a saddle point of the Lagrangian function, the matrix Y is a maximal solution of a dual problem, which represents the rate of change of the value Cx with respect to the change of the constraint vector b (see Corollary 3.2.7). This explains the importance of computing Y when a minimal solution x of (MOLP) is known. We concentrate on the case of strong saddle points in this subsection and assume b nonzero.

Let $\bar{x}$ be a maximal vertex solution of (MOLP) and let λ be a strictly positive scalarizing vector (see Theorem 4.3.1), that is $\bar{x}$ solves the linear problem

$$\text{maximize} \ \langle C^T \lambda, x \rangle$$
$$\text{subject to} \ \ Ax = b$$
$$x \geq 0.$$

Without loss of generality we may assume that the last component b_m of b is nonzero. Since the optimal solution set of this problem is a face of the feasible polyhedron, in view of Theorem 3.1.3 it has vertices and extreme rays that we denote respectively by $x^1, \cdots, x^p$ and $x^{p+1}, \cdots, x^{p+q}$. There exist positive numbers $t_1, \cdots, t_{p+q}$ with $\sum_{i=1}^{p} t_i = 1$ such that

$$\bar{x} = t_1 x^1 + \cdots + t_{p+q} x^{p+q}.$$

Let B_i be the optimal basis associated with the optimal vertex $x^i, i = 1, \cdots, p$; and let (B_s, a_{i_s}) be a couple of basis and non-basic column that determine the extreme ray $x^s, s = p+1, \cdots, p+q$, that is the basic components of x^s are equal to $-B_s^{-1}a_{i_s}$, its i_sth component is one and the other non-basic components are zero (see Corollary 2.4.8). Define

$$Y^i = C_{B_i} B_i^{-1}, i = 1, \cdots, p,$$

where C_{B_i} is the submatrix of the matrix C corresponding to the basis B_i (see the proof of Corollary 5.5.21). Then

$$\lambda^T (Y^i A - C) \geq 0, i = 1, \cdots, p. \tag{5.42}$$

For $s = p+1, \cdots, p+q$, denote by $\overline{C}_{i_s}$ the i_sth column of the reduced cost matrix, that is

$$\overline{C}_{i_s} = C_{i_s} - C_{B_s} B_s^{-1} a_{i_s}$$

and define Y^s the $k \times m$-matrix whose first $(m-1)$ columns are all zero and the last column is $\frac{1}{b_m}\overline{C}_{i_s}$. Then Y^s satisfies the following relations

$$Cx^s = Y^s b = \overline{C}_{i_s} \tag{5.43}$$
$$\langle C^T \lambda, x^s \rangle = 0 \tag{5.44}$$
$$\lambda^T Y^s A = 0. \tag{5.45}$$

The first relation is direct from the definition of x^s and Y^s. The second equality is obtained from the fact that x^s is an extreme ray of the optimal solution face of the problem (MOLP). The last equality follows from (5.44) and from the fact that

$$Y^s A = \frac{1}{b_m}\overline{C}_{i_s} a_m.$$

Finally, let us define $\overline{Y}$ by

$$\overline{Y} = \sum_{i=1}^{p+q} t_i Y^i$$

and prove that it is what we are looking for. Indeed, in view of (5.42) and (5.45), we have

$$\lambda^T (\overline{Y}A - C) = \lambda^T (\sum_{i=1}^{p+q} t_i Y^i A - C)$$

$$= \lambda^T \sum_{i=1}^{p} t_i (Y^i A - C) + \lambda^T \sum_{i=p+1}^{p+q} t_i Y^i A$$

$$\geq 0.$$

Since λ is strictly positive, the latter inequality shows that $\overline{Y}$ is a feasible solution of (VD2). Moreover, by the definition of Y^i, one has

$$Cx^i = C_{B_i} B_i^{-1} b = Y^i b, \ i = 1, \cdots, p$$

which together with (5.43) produces

$$C\overline{x} = \sum_{i=1}^{p+q} t_i Cx^i = \sum_{i=1}^{p+q} t_i Y^i b = \overline{Y}b.$$

5.6 Parametric Duality

The correspondence we established in the previous chapter, between the efficient set of a polyhedron and the positive normal vectors suggests an approach to duality via polar cones.

Duality of polar cones

Let Γ be a convex polyhedral set in $\mathbb{R}^k$. The polar cone of Γ and the normal cone to Γ at a point $a \in \Gamma$ are denoted Γ° and $N_\Gamma(a)$ (Chap. 2). If F is a face of Γ, the normal cone to Γ at a relative interior point of F is denoted $N(F)$. We consider the support function of Γ given by

$$\delta^*(\xi) := \sup_{z \in \Gamma} \langle \xi, z \rangle \ \text{ for } \xi \in \mathbb{R}^k.$$

It is clear that δ^* is a polyhedral convex function, its effective domain is nonempty and its epigraph

$$\text{epi}(\delta^*) := \left\{ (\xi, t) \in \mathbb{R}^k \times \mathbb{R} : \delta^*(\xi) \leq t \right\}$$

is a convex polyhedral cone in the product space $\mathbb{R}^k \times \mathbb{R}$. Given a subset F of $\mathbb{R}^k$, the closed positive hull of the set $F \times \{-1\}$ in the space $\mathbb{R}^k \times \mathbb{R}$ is denoted $\hat{F}$. Here is a relationship between faces of Γ and those of $\hat{\Gamma}$. Below, when speaking about faces, we mean nonempty faces.

Lemma 5.6.1 *Let Γ be a convex polyhedral set in $\mathbb{R}^k$. Then the following assertions hold.*

(i) *A subset $F \subseteq \Gamma$ is a face of Γ if and only if $\hat{F}$ is a face of $\hat{\Gamma}$.*
(ii) *The face $\hat{\Gamma}_0 := \hat{\Gamma} \cap (\mathbb{R}^k \times \{0\})$ coincides with $\Gamma_\infty \times \{0\}$.*
(iii) *For every face G of $\hat{\Gamma}$, not lying in the hyperplane $\mathbb{R}^k \times \{0\}$, there is a face F of Γ such that $\hat{F} = G$.*

Proof Let F be a face of Γ. There is some nonzero vector u such that the function $\langle u, . \rangle$ determines F, that is,

$$\langle u, z \rangle \geq \langle u, z' \rangle \text{ for all } z \in F, z' \in \Gamma.$$

Inequality is strict when $z' \notin F$. Then the function $\langle u, . \rangle$ is constant on F, say with value β. Consider the linear function ϕ on $\mathbb{R}^k \times \mathbb{R}$ defined by

$$\phi(z, s) := \langle u, z \rangle + \beta s \text{ for every } \begin{pmatrix} z \\ s \end{pmatrix} \in \mathbb{R}^k \times \mathbb{R}.$$

Then for every $\begin{pmatrix} z \\ s \end{pmatrix} \in \hat{F}$ and $\begin{pmatrix} z' \\ s' \end{pmatrix} \in \hat{\Gamma}$ one has

$$\phi(z, s) \geq \phi(z', s') \tag{5.46}$$

which shows that $\hat{F}$ is a face of $\hat{\Gamma}$. Conversely, let F be a nonempty subset of Γ such that $\hat{F}$ is a face of $\hat{\Gamma}$. Suppose that $\hat{F}$ is determined by (5.46) with some nonzero vector $\begin{pmatrix} u \\ \beta \end{pmatrix}$ (defining ϕ). It is clear that $u \neq 0$, because otherwise β, satisfying $\beta s \geq \beta s'$ for all negative values of s and s', is zero. Now, setting $s = s' = -1$ in (5.46), we deduce that F is a face of Γ determined by the nonzero linear function $\langle u, . \rangle$. To prove the second assertion, we consider the linear function

$$\phi_0(z, s) = s \text{ for all } \begin{pmatrix} z \\ s \end{pmatrix} \in \mathbb{R}^k \times \mathbb{R}.$$

It determines the face $\hat{\Gamma}_0$ via inequality

$$0 = \phi_0(z, s) \geq \phi_0(z', s') \text{ for all } \begin{pmatrix} z \\ s \end{pmatrix} \in \hat{\Gamma}_0, \begin{pmatrix} z' \\ s' \end{pmatrix} \in \hat{\Gamma}.$$

By definition, a vector $\begin{pmatrix} z \\ s \end{pmatrix}$ belongs to $\hat{\Gamma}_0$ if and only if $s = 0$ and there are $z^i \in \Gamma$, $t_i > 0$ such that $\begin{pmatrix} z \\ s \end{pmatrix} = \lim_{i \to \infty} t_i \begin{pmatrix} z^i \\ -1 \end{pmatrix}$. By Theorem 2.3.11, z is a recession direction of Γ. The converse is evidently true by the definition of recession directions. Finally, if G is a face of $\hat{\Gamma}$, say defined by (5.46) with some nonzero vector $\begin{pmatrix} u \\ \beta \end{pmatrix}$, and does not lie in the hyperplane $\mathbb{R}^k \times \{0\}$, then it contains elements whose last component is strictly negative. Define $\hat{F} = G \cap (\mathbb{R}^k \times \{-1\})$. Then $\hat{F}$ is a nonempty set, and, in view of (5.46), one has

$$\langle u, z \rangle \geq \langle u, z' \rangle$$

for every $z \in F$ and $z' \in \Gamma$. Hence F is a face of Γ, and $G = \hat{F}$.   □

The following lemma explains a link between the epigraph of δ^* and the set Γ.

Lemma 5.6.2 *Let Γ be a convex polyhedral set. The following assertions hold.*

(i) *The epigraph of δ^* is the polar cone of the set $\hat{\Gamma}$.*
(ii) *If a nonempty set $F \subseteq \Gamma$ is a face of Γ, then $N(\hat{F})$ is a face of $\mathrm{epi}(\delta^*)$. Conversely, if a nonempty set F^* is a face of $\mathrm{epi}(\delta^*)$, then $N(F^*)$ is a face of $\hat{\Gamma}$. In particular, if F^* does not contain the ray $\{0\} \times \mathbb{R}_+$, then $N(F^*) \cap (\Gamma \times \{-1\})$ is a face of $\Gamma \times \{-1\}$.*
(iii) *If F and F' are faces of Γ with $F \subseteq F'$, then $N(\hat{F'})$ is a face of $N(\hat{F})$.*

Proof To prove (i), let $\begin{pmatrix} \xi \\ t \end{pmatrix} \in \mathrm{epi}(\delta^*)$ and $z \in \Gamma$. Then

$$\left\langle \begin{pmatrix} \xi \\ t \end{pmatrix}, \begin{pmatrix} z \\ -1 \end{pmatrix} \right\rangle = \langle \xi, z \rangle - t \leq \langle \xi, z \rangle - \delta^*(\xi) \leqq 0.$$

Hence $epi(\delta^*) \subseteq (\Gamma \times \{-1\})^\circ$. On the other hand every vector $\begin{pmatrix} \xi \\ t \end{pmatrix} \in (\Gamma \times \{-1\})^\circ$ satisfies inequality

$$\langle \xi, z \rangle - t \leqq 0$$

for all $z \in \Gamma$ and yields $t \geqq \delta^*(\xi)$. This proves that $\begin{pmatrix} \xi \\ t \end{pmatrix} \in \mathrm{epi}(\delta^*)$ and $\mathrm{epi}(\delta^*) = (\Gamma \times \{-1\})^\circ$ too. It remains to notice that, by homogeneity the epigraph of δ^* is the polar cone of $\hat{\Gamma}$.

For (ii), let F be a face of Γ. By Lemma 5.6.1, $\hat{F}$ is a face of $\hat{\Gamma}$, hence $N(\hat{F})$ is a face of $(\hat{\Gamma})^\circ = \mathrm{epi}(\delta^*)$. The converse is clear because $N(F^*)$ is a face of $(\mathrm{epi}(\delta^*))^\circ = (\hat{\Gamma})^{\circ\circ} = \hat{\Gamma}$, the set $\hat{\Gamma}$ being a closed convex cone. Now, assume that F^* contains no vector $\begin{pmatrix} 0 \\ s \end{pmatrix}$ with $s > 0$. Then $N(F^*)$ does not lie in $\mathbb{R}^k \times \{0\}$. By Lemma 5.6.1 there is a face F of Γ such that $\hat{F} = N(F^*)$. The proof of Lemma 5.6.1 shows that $N(F^*) \cap (\Gamma \times \{-1\}) = F \times \{-1\}$. The last assertion is evident because F is a face of F' if and only if $\hat{F}$ is a face of $\hat{F'}$ (Lemma 5.6.1).   □

In the rest of this subsection we consider $\Gamma = Q - \mathbb{R}^k_+$, where Q is a convex polyhedral set in $\mathbb{R}^k$. Notice that for this Γ, the effective domain of δ^* is entirely included in the positive orthant $\mathbb{R}^k_+$ and δ^* coincides with the support function of Q on this domain. Maximal and weakly maximal elements of Q are obtained from maximal and weakly maximal elements of Γ by the formulas

$$\mathrm{Max}(Q) = \mathrm{Max}(\Gamma)$$
$$\mathrm{WMax}(Q) = \mathrm{WMax}(\Gamma) \cap Q.$$

Furthermore, except for the trivial case when the domain of δ^* consists of one point at the origin of the space, the polyhedral cone epi(δ^*) is completely characterized by its intersection with the set $\Delta \times \mathbb{R}$ that we denote by Γ^Δ, where Δ is the standard simplex in $\mathbb{R}^k$. More precisely, for a nonzero vector $\xi = (\xi_1, \cdots, \xi_k)^T$, the vector $\begin{pmatrix} \xi \\ t \end{pmatrix}$ belongs to the epigraph of δ^* if and only if the vector $\begin{pmatrix} \xi \\ t \end{pmatrix} / \sum_{i=1}^k \xi_i$ belongs to Γ^Δ. For convenience let us make the following notations

$$N^\Delta(z) := N_{\hat{\Gamma}}\begin{pmatrix} z \\ -1 \end{pmatrix} \cap \Gamma^\Delta$$

$$N^\Delta(F) := N(\hat{F}) \cap \Gamma^\Delta,$$

for any element z and face F of Γ. Here are some relations between efficient elements of Γ and Γ^Δ.

Theorem 5.6.3 *An element $z \in Q$ is maximal (respectively weakly maximal) if and only if there is some λ from the relative interior of Δ (respectively from Δ) such that* $\begin{pmatrix} \lambda \\ \delta^*(\lambda) \end{pmatrix} \in N^\Delta(z)$, *in which case* $\begin{pmatrix} \lambda \\ \delta^*(\lambda) \end{pmatrix}$ *is a minimal element of Γ^Δ.*

Furthermore, an element $\begin{pmatrix} \lambda \\ t \end{pmatrix} \in \Gamma^\Delta$ is minimal if and only if there is some element $z \in Q$ such that $\begin{pmatrix} z \\ -1 \end{pmatrix} \in N_{epi(\delta^*)}\begin{pmatrix} \lambda \\ t \end{pmatrix}$ in which case $t = \delta^*(\lambda)$ and z is maximal when $\lambda > 0$ and weakly maximal when $\lambda \geq 0$.

Proof We begin with the following characterization of minimal elements of the set Γ^Δ:

$$\begin{pmatrix} \lambda \\ t \end{pmatrix} \in \mathrm{Min}(\Gamma^\Delta) \Leftrightarrow \begin{pmatrix} \lambda \\ t \end{pmatrix} \in \Gamma^\Delta, t = \delta^*(\lambda). \tag{5.47}$$

Indeed, if $\begin{pmatrix} \lambda \\ t \end{pmatrix} \in \mathrm{Min}(\Gamma^\Delta)$, then $\begin{pmatrix} \lambda \\ t \end{pmatrix} \in \Gamma^\Delta$ and $\begin{pmatrix} \lambda \\ t \end{pmatrix} \leq \begin{pmatrix} \lambda \\ \delta^*(\lambda) \end{pmatrix}$ and we deduce $t = \delta^*(\lambda)$. Conversely, for any $\begin{pmatrix} \lambda' \\ t' \end{pmatrix} \in \Gamma^\Delta$ with $\begin{pmatrix} \lambda \\ \delta^*(\lambda) \end{pmatrix} \geq \begin{pmatrix} \lambda' \\ t' \end{pmatrix}$ we have $\lambda = \lambda'$ and $\delta^*(\lambda) \geq t'$ which yields $\delta^*(\lambda) = t'$. By this $\begin{pmatrix} \lambda \\ \delta^*(\lambda) \end{pmatrix}$ is a minimal element.

Now assume that z is a maximal element of Q. By scalarization there is a strictly positive vector $\lambda \in \Delta$ such that $\langle \lambda, z \rangle = \delta^*(\lambda)$. This implies evidently that $\begin{pmatrix} \lambda \\ \delta^*(\lambda) \end{pmatrix}$ belongs to $N^\Delta(z)$. By (5.47), $\begin{pmatrix} \lambda \\ \delta^*(\lambda) \end{pmatrix}$ is a minimal element of Γ^Δ. Conversely, if $\begin{pmatrix} \lambda \\ \delta^*(\lambda) \end{pmatrix} \in N^\Delta(z)$ with $\lambda \in \Delta$ and $\lambda > 0$, then for every $z' \in Q$ one has

$$0 \geq \left\langle \begin{pmatrix} \lambda \\ \delta^*(\lambda) \end{pmatrix}, \begin{pmatrix} z' \\ -1 \end{pmatrix} - \begin{pmatrix} z \\ -1 \end{pmatrix} \right\rangle = \langle \lambda, z' - z \rangle,$$

which shows that z is a maximal element of Q.

To proceed to the second part, let $\begin{pmatrix} \lambda \\ t \end{pmatrix} \in \Gamma^\Delta$ be a minimal element with $\lambda > 0$.
By (5.47), $t = \delta^*(\lambda)$ and $\langle \lambda, . \rangle$ attains its maximum on Q at some point z with value $\delta^*(\lambda)$. Since $\lambda > 0$, by scalarization z is a maximal element of Q. Moreover, for every $\begin{pmatrix} \lambda' \\ t' \end{pmatrix} \in \text{epi}(\delta^*)$ one has

$$\left\langle \begin{pmatrix} z \\ -1 \end{pmatrix}, \begin{pmatrix} \lambda' \\ t' \end{pmatrix} - \begin{pmatrix} \lambda \\ \delta^*(\lambda) \end{pmatrix} \right\rangle = \langle z, \lambda' \rangle - t' - (\langle z, \lambda \rangle - \delta^*(\lambda))$$

$$= \langle z, \lambda' \rangle - t'$$

$$\leq \langle z, \lambda' \rangle - \delta^*(\lambda') \leq 0,$$

which shows that $\begin{pmatrix} z \\ -1 \end{pmatrix}$ is normal to $\text{epi}(\delta^*)$ at $\begin{pmatrix} \lambda \\ \delta^*(\lambda) \end{pmatrix}$. Conversely, let $\begin{pmatrix} z \\ -1 \end{pmatrix} \in N_{\text{epi}(\delta^*)} \begin{pmatrix} \lambda \\ t \end{pmatrix}$. Then for every $\begin{pmatrix} \lambda' \\ t' \end{pmatrix} \in \text{epi}(\delta^*)$ we have

$$0 \geq \left\langle \begin{pmatrix} z \\ -1 \end{pmatrix}, \begin{pmatrix} \lambda' \\ t' \end{pmatrix} - \begin{pmatrix} \lambda \\ t \end{pmatrix} \right\rangle = \langle z, \lambda' - \lambda \rangle - t' + t.$$

In particular this implies $t = \delta^*(\lambda)$. By (5.47), $(\lambda\ t)^T$ is a minimal element of Γ^Δ. By polarity, the vector $\begin{pmatrix} \lambda \\ \delta^*(\lambda) \end{pmatrix}$ belongs to the normal cone to $\hat{\Gamma}$ at $\begin{pmatrix} z \\ -1 \end{pmatrix}$, and in view of (i), the vector z is a maximal element of Q whenever λ is strictly positive. The proof for weakly efficient elements is done in the same way. $\square$

A relation between efficient faces of Γ and efficient faces of Γ^Δ can also be established.

Theorem 5.6.4 *If F is a face of weakly maximal (respectively maximal) elements of Γ, then $N^\Delta(F)$ is a face of minimal elements of Γ^Δ (respectively, having at least one point from $\text{int}(\mathbb{R}^k_+) \times \mathbb{R})$. Conversely, if F^Δ is a face of minimal elements of Γ^Δ (respectively, containing at least one element from $\text{int}(\mathbb{R}^k_+) \times \mathbb{R})$, then $N(F^\Delta) \cap (\mathbb{R}^k \times \{-1\})$ is a face of weakly maximal (respectively, maximal) elements of $\Gamma \times \{-1\}$.*

Proof Let F be a face of maximal elements of Γ. By Lemma 5.6.2 and Theorem 5.6.3 the normal cone $N(\hat{F})$ is a face of $\text{epi}(\delta^*)$ and contains some element $\begin{pmatrix} \lambda \\ \delta^*(\lambda) \end{pmatrix}$ with $\lambda \in \Delta$ and $\lambda > 0$. By this, $N^\Delta(F)$ is a face of Γ^Δ. We observe that $\begin{pmatrix} \lambda \\ t \end{pmatrix} \in N^\Delta(F)$ implies that $t = \delta^*(\lambda)$, hence it is a minimal element in view of (5.47).

Conversely, let F^Δ be a face of minimal elements of Γ^Δ containing some $\begin{pmatrix} \lambda \\ t \end{pmatrix}$ with $\lambda > 0$. Then $t = \delta^*(\lambda)$ and the closed cone generated by F^Δ is a face of epi(δ^*). Its normal cone $N(\text{cone}(F^\Delta))$ is then a face of the cone $\hat{\Gamma}$. Denote by $F \subseteq \Gamma$ such that $F \times \{-1\} = N(\text{cone}(F^\Delta)) \cap (\mathbb{R}^k \times \{-1\})$. Then F is a face of Γ. Moreover, for $z \in F$ and $z' \in \Gamma$ one has

$$0 \geq \left\langle \begin{pmatrix} \lambda \\ \delta^*(\lambda) \end{pmatrix}, \begin{pmatrix} z' \\ -1 \end{pmatrix} - \begin{pmatrix} z \\ -1 \end{pmatrix} \right\rangle = \langle \lambda, z' - z \rangle,$$

which shows that F is a face of maximal elements. The proof for weakly maximal elements uses the same argument. □

Parametric dual

It is time to derive a dual problem for the problem (MOLP) from the duality features of polar cones and normal cones. Set

$$Q = \{Cx : Ax = b, x \geq 0\}$$
$$\Gamma = Q - \mathbb{R}^k_+.$$

Let us compute the epigraph of the support function δ^*. A vector $\begin{pmatrix} \xi \\ t \end{pmatrix}$ with $\xi \geq 0$ belongs to the epigraph of δ^* if and only if

$$t \geq \delta^*(\xi) = \sup\{\langle \xi, Cx \rangle : Ax = b, x \geq 0\}.$$

Hence, $\begin{pmatrix} \xi \\ t \end{pmatrix}$ belongs to $\hat{\Gamma}^\circ$ if and only if $\xi \in \Delta$, $\delta^*(\xi)$ is finite, $t \geq \delta^*(\xi)$ and $\delta^*(\xi) = \inf\{\langle b, u \rangle : A^T u \geq C^T \xi\}$ in view of duality in linear programming. The minimal elements of Γ^Δ are then efficient values of the following dual problem (D*):

$$\text{Minimize } \begin{pmatrix} \lambda \\ \langle b, u \rangle \end{pmatrix}$$
$$\text{subject to } A^T u \geq C^T \lambda$$
$$\lambda \in \Delta.$$

This is a linear problem with values in the space $\mathbb{R}^{k+1}$ in which the parameter λ is considered as a variable. In order to have a primal problem in the same value space, it suffices to embed (MOLP) into $\mathbb{R}^{k+1}$ by considering the problem (P*):

$$\text{Maximize} \quad \begin{pmatrix} Cx \\ -1 \end{pmatrix}$$

$$\text{subject to} \quad Ax = b$$

$$x \geqq 0,$$

or the problem (P**):

$$\text{Maximize} \quad \begin{pmatrix} Cx \\ t \end{pmatrix}$$

$$\text{subject to} \quad Ax = b$$

$$x \geqq 0, t \in \mathbb{R}.$$

It is clear that the feasible sets as well as the maximal solution set of (MOLP) and (P*) coincide, while x is a weakly maximal solution of (MOLP) if and only if $\begin{pmatrix} x \\ t \end{pmatrix}$ with $t \in \mathbb{R}$ is a weakly maximal solution of (P**). Here are the main duality results between (MOLP) and (D*).

Corollary 5.6.5 (Weak duality) *If x is a feasible solution of (MOLP) and $\begin{pmatrix} \lambda \\ u \end{pmatrix}$ is a feasible solution of (D*), then*

$$\left\langle \begin{pmatrix} Cx \\ -1 \end{pmatrix}, \begin{pmatrix} \lambda \\ \langle b, u \rangle \end{pmatrix} \right\rangle \leqq 0. \tag{5.48}$$

Moreover, if in addition equality holds and $\lambda > 0$ (respectively $\lambda \geq 0$), then x is a maximal (respectively weakly maximal) solution of (MOLP) and $\begin{pmatrix} \lambda \\ u \end{pmatrix}$ is a minimal solution of (D).*

Proof If x and $\begin{pmatrix} \lambda \\ u \end{pmatrix}$ are feasible, then Cx belongs to Q and $\begin{pmatrix} \lambda \\ \langle b, u \rangle \end{pmatrix}$ belongs to the epigraph of δ^*. Inequality (5.48) follows from polarity between the epigraph of δ^* and $Q \times \{-1\}$. If equality holds, we obtain

$$\begin{pmatrix} Cx \\ -1 \end{pmatrix} \in N_{epi(\delta^*)} \begin{pmatrix} \lambda \\ \langle b, u \rangle \end{pmatrix}$$

$$\begin{pmatrix} \lambda \\ \langle b, u \rangle \end{pmatrix} \in N^{\triangle}(Cx).$$

By Theorem 5.6.3, the vector $\begin{pmatrix} \lambda \\ u \end{pmatrix}$ is a minimal solution of (D*) and x is a maximal or weakly maximal solution of (MOLP) depending on whether $\lambda > 0$ or $\lambda \geq 0$. $\quad \square$

When (MOLP) has a maximal solution, a strong duality relation can be obtained.

Corollary 5.6.6 (Strong duality) *Let x be a feasible solution of (MOLP). The following assertions are equivalent.*

(i) *x is a maximal (respectively weakly maximal) solution of (MOLP).*

(ii) *There is a feasible solution* $\begin{pmatrix} \lambda \\ u \end{pmatrix}$ *of (D*) with* $\lambda > 0$ *(respectively* $\lambda \geq 0$*) such that*

$$\left\langle \begin{pmatrix} Cx \\ -1 \end{pmatrix}, \begin{pmatrix} \lambda \\ \langle b, u \rangle \end{pmatrix} \right\rangle = 0,$$

in which case $\begin{pmatrix} \lambda \\ u \end{pmatrix}$ *is a minimal solution of (D*) and the vector* $\begin{pmatrix} Cx \\ -1 \end{pmatrix}$ *is normal to the polytope* Γ^Δ *at* $\begin{pmatrix} \lambda \\ \langle b, u \rangle \end{pmatrix}$.

(iii) *There is a feasible solution* $\begin{pmatrix} \lambda \\ u \end{pmatrix}$ *of (D*) with* $\lambda > 0$ *(respectively* $\lambda \geq 0$*) such that the vector* $\begin{pmatrix} \lambda \\ \langle b, u \rangle \end{pmatrix}$ *is normal to the polytope* $\Gamma \times \{-1\}$ *at* $\begin{pmatrix} Cx \\ -1 \end{pmatrix}$.

Proof If x is a maximal solution of (MOLP), then $z = Cx$ is a maximal element of Q. By Theorem 5.6.3 there is some $\lambda > 0$ from the relative interior of Δ such that $\begin{pmatrix} \lambda \\ \delta^*(\lambda) \end{pmatrix}$ belongs to $N^\Delta(z)$ and $\begin{pmatrix} \lambda \\ \delta^*(\lambda) \end{pmatrix}$ is a minimal element of Γ^Δ. Let u be an optimal solution of the dual problem (D_λ) (see this dual in the proof of Lemma 5.4.4). Then $\delta^*(\lambda) = \langle b, u \rangle = \langle \lambda, z \rangle$ and (ii) follows. Conversely, under (ii), $\begin{pmatrix} \lambda \\ \delta^*(\lambda) \end{pmatrix}$ belongs to $N^\Delta(z)$ where $\delta^*(\lambda) = \langle b, u \rangle = \langle \lambda, Cx \rangle$. By Theorem 5.6.3, Cx is a maximal element of Q, and x is a maximal solution of (MOLP). The equivalence between (ii) and (iii) is clear. A similar argument goes through for weakly maximal solutions.   □

In like manner, when (D*) has a minimal solution one may realize the strong duality relation given in the preceding corollary.

Corollary 5.6.7 (Strong duality) *Let* $\begin{pmatrix} \lambda \\ u \end{pmatrix}$ *be a feasible solution of (D*) with* $\lambda > 0$ *(respectively* $\lambda \geq 0$*). The following assertions are equivalent.*

(i) $\begin{pmatrix} \lambda \\ u \end{pmatrix}$ *is a minimal solution of (D*).*

(ii) *There is a feasible solution x of (MOLP) such that*

$$\left\langle \begin{pmatrix} Cx \\ -1 \end{pmatrix}, \begin{pmatrix} \lambda \\ \langle b, u \rangle \end{pmatrix} \right\rangle = 0,$$

in which case x is a maximal (respectively weakly maximal) solution of (P*) and the vector $\begin{pmatrix} \lambda \\ \langle b, u \rangle \end{pmatrix}$ *is normal to the polytope* $\Gamma \times \{-1\}$ *at* $\begin{pmatrix} Cx \\ -1 \end{pmatrix}$.

(iii) *There is a feasible solution x of (MOLP) such that the vector* $\begin{pmatrix} Cx \\ -1 \end{pmatrix}$ *is normal to the polytope* Γ^Δ *at* $\begin{pmatrix} \lambda \\ \langle b, u \rangle \end{pmatrix}$.

Proof The equivalence between (ii) and (iii) is clear. Let us prove the equivalence between the first two assertions. If $\begin{pmatrix} \lambda \\ u \end{pmatrix}$, say with $\lambda > 0$, is a minimal solution of (D*), then $\begin{pmatrix} \lambda \\ \langle b, u \rangle \end{pmatrix}$ is a minimal element of Γ^*. By Theorem 5.6.3 there exists a maximal element $z \in Q$ such that $\begin{pmatrix} z \\ -1 \end{pmatrix} \in N_{epi(\delta^*)} \begin{pmatrix} \lambda \\ \langle b, u \rangle \end{pmatrix}$ and $\delta^*(\lambda) = \langle b, u \rangle$. Let x be a feasible solution of (MOLP) such that $Cx = z$. Then x is a maximal solution and (ii) is fulfilled. Conversely, if (λ, u) satisfies (ii), then by Corollary 5.6.5, it is a minimal solution of (D*). The proof of the case $\lambda \geq 0$ follows the same line. $\quad\square$

Duality between faces of maximal values of (MOLP) and faces of minimal values of (D*) can also be deduced.

Corollary 5.6.8 *Let F be a face of* Γ. *Then it is a face of maximal (respectively weakly maximal) values of (MOLP) if and only if the set*

$$F^* := \bigcap_{z \in F} \left\{ \begin{pmatrix} \lambda \\ t \end{pmatrix} \in \Gamma^\Delta : \left\langle \begin{pmatrix} z \\ -1 \end{pmatrix}, \begin{pmatrix} \lambda \\ t \end{pmatrix} \right\rangle = 0 \right\}$$

in which at least one λ *is strictly positive (respectively* $\lambda \geq 0$*), is a face of minimal elements of* Γ^Δ.

   Similarly, given a face F^* *of* Γ^Δ, *it is a face of minimal values of (D*) if and only if the set*

$$F := \bigcap_{(\lambda^T, t)^T \in F^*} \left\{ z \in \Gamma : \left\langle \begin{pmatrix} z \\ -1 \end{pmatrix}, \begin{pmatrix} \lambda \\ t \end{pmatrix} \right\rangle = 0 \right\},$$

in which at least one λ *is strictly positive (respectively* $\lambda \geq 0$*), is a face of maximal (respectively weakly maximal) elements of* Γ.

Proof This follows from Theorem 5.6.4 and Lemma 5.6.2. $\quad\square$

Example 5.6.9 Consider the problem

$$\text{Maximize} \quad \begin{pmatrix} 2 & 1 \\ 2 & 1 \end{pmatrix} \begin{pmatrix} x_1 \\ x_2 \end{pmatrix}$$

$$\text{subject to} \quad \begin{pmatrix} 1 & 1 \\ 0 & 0 \end{pmatrix} \begin{pmatrix} x_1 \\ x_2 \end{pmatrix} = \begin{pmatrix} 1 \\ 0 \end{pmatrix}$$

$$x_1, x_2 \geqq 0.$$

By definition the dual (D*) is given by

$$\text{Minimize} \quad \begin{pmatrix} \lambda_1 \\ \lambda_2 \\ u_1 \end{pmatrix}$$

$$\text{subject to} \quad \begin{pmatrix} 1 & 0 \\ 1 & 0 \end{pmatrix} \begin{pmatrix} u_1 \\ u_2 \end{pmatrix} \geqq \begin{pmatrix} 2 & 2 \\ 1 & 1 \end{pmatrix} \begin{pmatrix} \lambda_1 \\ \lambda_2 \end{pmatrix}$$

$$\lambda_1, \lambda_2 \geqq 0, \lambda_1 + \lambda_2 = 1.$$

It reveals from the proof of Theorem 5.6.3 that for the dual (D*) a feasible solution $(\lambda_1, \lambda_2, u_1, u_2)^T$ is minimal if and only if u_1 takes the minimum value. Thus, taking the dual constraints into account the minimal values of (D*) are of form $(\lambda_1, \lambda_2, 2, u_2)^T$. The complementarity condition in Corollaries 5.6.6 and 5.6.7 yields

$$(2x_1 + x_2)(\lambda_1 + \lambda_2) = 2$$

which gives $x_1 = 1, x_2 = 0$ the unique maximal solution of (MOLP).

5.7 Exercises

5.7.1 Construct the dual problems (VD1), (VD2) and (VD3) for the following primal problem and discuss their feasible sets and value sets.

$$\text{Maximize} \quad \begin{pmatrix} 1 & 0 \\ 0 & 1 \end{pmatrix} \begin{pmatrix} x_1 \\ x_2 \end{pmatrix}$$

$$\text{subject to} \quad \begin{pmatrix} 1 & 1 \\ 0 & 0 \end{pmatrix} \begin{pmatrix} x_1 \\ x_2 \end{pmatrix} = \begin{pmatrix} 1 \\ 0 \end{pmatrix}$$

$$x_1, x_2 \geqq 0.$$

5.7.2 Construct a linear problem (MOLP) with b nonzero and $\sup(Q)$ finite such that for some $z \geqq \sup(Q)$ the system

$$Yb = z$$
$$YA \geqq C$$

has no solution.

5.7.3 Consider the primal linear problem:

$$\text{Maximize} \quad \begin{pmatrix} -1 & 0 \\ 0 & 1 \end{pmatrix} \begin{pmatrix} x_1 \\ x_2 \end{pmatrix}$$

$$\text{subject to} \quad \begin{pmatrix} 1 & 1 \\ 0 & 0 \end{pmatrix} \begin{pmatrix} x_1 \\ x_2 \end{pmatrix} = \begin{pmatrix} 0 \\ 0 \end{pmatrix}$$

$$x_1, x_2 \geqq 0.$$

Find the extended dual problems (VD1′), (VD2′) and (VD3′). Compare their optimal solutions.

5.7.4 Given a problem

$$\text{Maximize} \quad \begin{pmatrix} 1 & 0 \\ 0 & 1 \end{pmatrix} \begin{pmatrix} x_1 \\ x_2 \end{pmatrix}$$

$$\text{subject to} \quad \begin{pmatrix} 1 & 1 \\ 0 & 0 \end{pmatrix} \begin{pmatrix} x_1 \\ x_2 \end{pmatrix} = \begin{pmatrix} 1 \\ 0 \end{pmatrix}$$

$$x_1, x_2 \geq 0.$$

Check that the feasible vertices are maximal solutions and find the corresponding dual minimal solutions such that $Cx = Yb$.

5.7.5 Kornbluth's duality. Let R be an $m \times r$-matrix and let $\mu \in \mathbb{R}^r$ be a strictly positive vector. Prove that $\bar{x}$ is a maximal solution of the problem (MOLP) with $b = R\mu$:

$$\text{Maximize} \quad Cx$$
$$\text{subject to} \quad Ax = R\mu$$
$$x \geq 0$$

if and only if there exist a strictly positive vector $\lambda \in \mathbb{R}^k$ and a minimal solution $\bar{y} \in \mathbb{R}^m$ of the dual problem (called Kornbluth's dual)

$$\text{Minimize} \quad R^T y$$
$$\text{subject to} \quad A^T y \geq C^T \lambda$$

such that $\langle A^T \bar{y} - C^T \lambda, \bar{x} \rangle = 0$.

5.7.6 Symmetry of ideal duality. Express the ideal dual problem in form

$$-\text{Maximize} \quad -\left(b^T, -b^T, 0 \right) \begin{pmatrix} Y_+^T \\ Y_-^T \\ Z^T \end{pmatrix}$$

$$\text{subject to} \quad \left(A^T, -A^T, I \right) \begin{pmatrix} Y_+^T \\ Y_-^T \\ Z^T \end{pmatrix} = C^T$$

$$Y_+, Y_-, Z \geq 0$$

where I is the identity $m \times m$-matrix, Y_+ and Y_- are $k \times m$-matrix variables and Z is an $m \times m$-matrix slack variable. Find the dual problem (VD1) for this maximization problem and prove that it consists of m maximization problems each of which is identical to (MOLP). Generalize the method of ideal duality to the primal problem (MOLP) in which the variable x is a matrix.

5.7.7 Heyde, Löhne and Tammer's set-valued duality. Consider the following dual problem, denoted (HD):

$$\text{Minimize} \quad H(u, \lambda)$$
$$\text{subject to} \quad A^T u \geqq C^T \lambda,$$
$$\lambda \in \Delta,$$

where

$$H(u, \lambda) := \left\{ z \in \mathbb{R}^k : \langle b, u \rangle = \langle \lambda, z \rangle \right\}$$

and the set-valued minimization problem is understood in the sense that a feasible solution $(\bar{u}, \bar{\lambda})$ is weakly efficient if and only if the hyperplane $H(\bar{u}, \bar{\lambda})$ contains a weakly minimal point of the value set

$$Q^{HD} = \left\{ z \in \mathbb{R}^k : z \in H(u, \lambda), A^T u \geqq C^T \lambda, \lambda \in \Delta \right\}.$$

Prove the following properties

(a) a vector (u, λ) is a feasible solution of (HD) if and only if the vector (u, λ, z) with $z \in H(u, \lambda)$ is a feasible solution of (KD), and a feasible solution (u, λ) of (HD) is a weakly efficient solution if and only if (u, λ, z) with $z \in H(u, \lambda) \cap$ WMin(Q^{HD}) is a weakly minimal solution of (KD).
(b) The value sets of (HD) and (KD) coincide.
(c) A feasible solution (u, λ) is a weakly minimal solution of (HD) if and only if each of the following equivalent conditions holds:

   (1) $H(u, \lambda) \cap$ WMin$(Q^{HD}) \neq \emptyset$;
   (2) there is a feasible solution x of (MOLP) such that $\langle b, u \rangle = \langle \lambda, Cx \rangle$;
   (3) u is an optimal solution of (D_λ) given in the proof of Lemma 5.4.4.
(d) If feasible solutions x and (u, λ) satisfy $\langle b, u \rangle = \langle \lambda, Cx \rangle$, then they are weakly minimal solutions.

5.7.8 Galperin and Guerra's duality. This is a duality based on the duality for scalar problems (P_i) and the weak duality relation of the ideal duality approach. Assume that the feasible set X of the primal problem (MOLP) is bounded. Then the optimal values α_i's (Lemma 5.2.1) are finite, the vector $\alpha := (\alpha_1, \cdots, \alpha_k)^T$ is finite. For $\xi = (\xi_1, \cdots, \xi_k)^T \in -\mathbb{R}_+^k$ set

$$X_i(\xi_i) = \{x \in X : \langle c^i, x \rangle \geq \alpha_i + \xi_i\}$$

$$X(\xi) = \bigcap_{i=1}^{k} X_i(\xi_i)$$

$$\mathcal{B} = \{\xi \in \mathbb{R}^k : \alpha + \xi \in Q - \mathbb{R}_+^k\}.$$

The problem (MOLP) is said to be *balanced* if $X(0)$ is nonempty, the set $\mathrm{Max}(\mathcal{B})$ is called the *balance set*, and (MOLP) is said to be of general position if $X(\xi)$ is a singleton for all ξ from the balance set. Prove the following assertions.

(a) (MOLP) is balanced if and only if it has an ideal maximal solution.
(b) $\mathcal{B}$ is a subset of $-\mathbb{R}_+^k$ and

$$\alpha + \mathcal{B} = Q - \mathbb{R}_+^k.$$

(c) $C(X(\xi)) = Q \cap (\alpha + \xi + \mathbb{R}_+^k)$ for every $\xi \in -\mathbb{R}_+^k$.
(d) A vector ξ belongs to the balance set if and only if the vector $\alpha + \xi$ is a maximal value of (MOLP).
(e) If (MOLP) is in general position and if a vector ξ belongs to the balance set, then the following system has a unique solution

$$Cx = \alpha + \xi$$
$$Ax = b$$
$$x \geq 0, \ \xi \leq 0.$$

(f) The balance set is composed of faces determined by certain equalities over X.
(g) Consider the dual problems (D_i) (given in the proof of Lemma 5.2.1) and for feasible solutions x and y of (MOLP) and (D_i), $i = 1, \cdots, k$, define

$$\eta_i = \langle b, y \rangle - \alpha_i$$

$$\delta(x, y) = \sum_{i=1}^{k} \langle c^i, x \rangle - \langle b, y \rangle = \sum_{i=1}^{k} (\xi_i + \eta_i).$$

The amount $\delta(x, y)$ is called the *total duality gap*. Show that if (MOLP) is unbalanced, the total duality gap is strictly positive.

(h) Assume that (MOLP) is unbalanced. For every maximal solution vertex x^* of (MOLP), consider scalar subproblems $(P_i(x^*))$:

$$\text{Maximize} \quad \langle c^i, x \rangle$$
$$\text{subject to} \quad Ax = b$$
$$x \geq 0$$
$$Cx \leq Cx^*,$$

and their dual problems $(D_i(x^*))$, with optimal solutions, say $y^*(i)$, $i = 1, \cdots, k$. The vectors $\eta = (\eta_1, \cdots, \eta_k)^T$ with $\eta_i = \langle b, y^*(i)\rangle - \alpha_i$ form a set, called a *dual balance set*. Then the multiple objective problem (MOLP(x^*)) obtained from (MOLP) by adding the last constraint of $(P_i(x^*))$ is balanced (x^* is an ideal solution), and hence its corresponding ideal dual problem (VD1) has a solution Y^* whose rows are optimal solutions of the dual problems $(D_i(x^*))$, $i = 1, \cdots, k$. This Y^* is called a *dual cluster* corresponding to x^*. Show that $Cx^* = Y^*b$ which implies that the dual balance set is exactly the opposite of the balance set.

5.7.9 Kolumban's dual problem. Consider a dual system of two variables:

$$A^T u \geqq C^T \lambda$$
$$\lambda \geq 0.$$

A solution (u, λ) is said to be optimal if there are k real numbers $r_1, \cdots, r_k$ such that

(i) $\langle b, u\rangle = \sum_{i=1}^k r_i \lambda_i$
(ii) there is no solution of the dual system (u', λ') and real numbers $r'_1, \cdots, r'_k$ such that $r_i > r'_i, i = 1, \cdots, k$ and $\langle b, u'\rangle = \sum_{i=1}^k r'_i \lambda'_i$.

Prove that optimal solutions are exactly weakly minimal solutions of (KD).

5.7.10 Heyde and Lohne's geometric duality. Consider the problem (MOLP) defined in Sect. 5.1. The geometric dual problem, given by Heyde and Lohne, is defined as

$$\text{K-Minimize } \left(\lambda_1, \cdots, \lambda_{k-1}, \langle b, u\rangle\right)^T$$
$$\text{subject to} \quad \lambda \in \Delta,$$
$$A^T u \geqq C^T \lambda,$$

where K is the positive kth axis of $\mathbb{R}^k$ and K-minimization means minimizing with respect to the order $a \geq_K b$ if and only if $a - b \in K$, which implies that the last component of a is bigger or equal to the last component of b, while the other components are equal. Denote

$$\mathcal{D} = \{(\lambda_1, \cdots, \lambda_{k-1}, t)^T \in \mathbb{R}^k : t \geqq \langle b, u\rangle, A^T u \geqq C^T \lambda, \lambda \in \Delta\}$$

and for a face F^* of $\mathcal{D}$, set

$$\Phi(F^*) = \bigcap_{\xi \in F^*} \left\{z \in \mathbb{R}^k : \sum_{i=1}^{k-1} \xi_i y_i + (1 - \sum_{i=1}^{k-1} \xi_i) y_k - \xi_k = 0\right\} \bigcap \Gamma$$

where $\Gamma = Q - \mathbb{R}_+^k$.

(i) Define a projection $\Pi : \Gamma^* \to \mathcal{D}$ by

$$\Pi\big((\lambda_1, \cdots, \lambda_n), t\big) = (\lambda_1, \cdots, \lambda_{k-1}, t)$$

for every $(\lambda_1, \cdots, \lambda_k) \in \Delta$ and $t \in \mathbb{R}$. Show that Π is isomorphic and order-preserving, that is for $w, w' \in \Gamma^*$, one has $w \geqq w'$ if and only if $\Pi(w) \geqq_K \Pi(w')$.

(ii) Prove that a face of Γ^* consists of minimal elements if and only if its image by Π is a face of K-minimal elements of $\mathcal{D}$, and vice versa.

(iii) Deduce that the map Φ is an inclusion-reserving one-to-one map between the set of all faces of K-minimal elements of $\mathcal{D}$ and the set of all faces of weakly maximal elements of Γ.

5.7.11 Exact saddle points. A saddle point $(\overline{x}, \overline{Y})$ of L is said to be exact if $C\overline{x} = \overline{Y}b$.

(i) Show that all ideal saddle points of L are exact saddle points, but that is not the case for weak or strong saddle points of L.

(ii) Assume that b is a nonzero vector. Prove that for every maximal solution x of (MOLP) there exists a minimal solution $(Y, \lambda, 0)$ of $(\text{VD2})^L$ such that (x, Y) is an exact strong saddle point of L, and that for every minimal solution $(Y, \lambda, 0)$ of $(\text{VD2})^L$, there exist a maximal solution x of (MOLP) such that (x, Y) is an exact strong saddle point of L.

(iii) Is the conclusion of (ii) true if $b = 0$?

5.7.12 Consider the problem

$$\text{Maximize} \quad \begin{pmatrix} 1 & 0 \\ 0 & 1 \end{pmatrix} \begin{pmatrix} x_1 \\ x_2 \end{pmatrix}$$
$$\text{subject to} \quad x_1 + x_2 = 1$$
$$x_1, x_2 \geqq 0.$$

Show that there is some feasible solution x such that (x, Y) where $Y = (2, -1)^T$, is a strong saddle point of the Lagrangian function associated with this problem, but no feasible solution of type $(Y, \lambda, 0)$ of the Lagrangian dual problem $(\text{VD2})^L$ is minimal.

5.7.13 Consider an extended problem $(\text{MOLP})'$:

$$\text{Maximize} \ (C \ 0) \begin{pmatrix} x \\ t \end{pmatrix}$$
$$\text{subject to} \ \begin{pmatrix} A & 0 \\ 0 & 1 \end{pmatrix} \begin{pmatrix} x \\ t \end{pmatrix} = \begin{pmatrix} b \\ 1 \end{pmatrix}$$
$$\begin{pmatrix} x \\ t \end{pmatrix} \geqq 0$$

and the associated Lagrangian function

$$\mathcal{L}\left(\begin{pmatrix} x \\ t \end{pmatrix}, (Y\ z)\right) = Yb + z + (C - YA)x - tz$$

where x is from $\mathbb{R}^n_+$, Y is a $k \times m$-matrix, z is from $\mathbb{R}^k$ and t is from $\mathbb{R}_+$.

(i) Establish that the feasible sets of (MOLP) and (MOLP)$'$ coincide.
(ii) Construct the Lagrangian dual problem of (MOLP)$'$ and prove that its value set coincides with the value set Q'' of the problem (VD2$''$) (Sect. 5.5).
(iii) Apply (ii) of Exercise 5.7.11 to obtain a relationship between saddle points of the extended Lagrangian function $\mathcal{L}$ and efficient solutions of (MOLP) and (VD2$''$).

5.7.14 Wolfe-type duality. Consider a (MOLP) of the form

$$\text{Maximize} \quad Cx$$
$$\text{subject to} \quad Ax \leq b$$

and a dual problem, called a Wolfe-type dual and denoted (WD)

$$\text{Minimize } Cy - \mu^T(Ay - b)e$$
$$\text{subject to} \quad A^T\mu - C^T\lambda = 0 \tag{5.49}$$
$$\mu^T(Ay - b) \geq 0$$
$$\lambda^T e = 1 \tag{5.50}$$
$$\lambda \geq 0, \mu \geq 0. \tag{5.51}$$

Establish the following statements.

(i) For every feasible solutions x of (MOLP) and (λ, μ, y) of (WD) one has

$$Cx \not> Cy - \mu^T(Ay - b)e.$$

(ii) If (λ, μ, y) is a feasible solution of (WD) such that y is a feasible solution of (MOLP), then (λ, μ, y) is a weakly minimal solution of (WD) and y is a weakly maximal solution of (MOLP).
(iii) Assume that there are positive numbers α and δ such that

- $\mu^T b \leq \alpha$ *for all* (λ, μ) *satisfying* (5.49)–(5.51)
- the system $A'x \leq b'$ is consistent for all A' and b' with $\max\{\|A - A'\|, \|b - b'\|\} \leq \delta.$

Then for every weakly maximal solution x of (MOLP) there exists (λ, μ) such that (α, μ, x) is a weakly minimal solution of (WD).

Chapter 6
Sensitivity and Stability

In practice the objective function and the constraints of a multiobjective optimization problem may undergo a small perturbation or depend on a parameter. It is crucial to know how the efficient solution set and the efficient value set change. This question is particularly useful in computational error estimation when the decision makers have only partial information on the data. We shall analyze the multiobjective problem (MOLP) in which the objective matrix C, the constraint matrix A and the second term b continuously or smoothly depend on a parameter or are slightly perturbed.

6.1 Parametric Convex Polyhedra

We consider a convex polyhedron $\Gamma(\omega)$ depending on a parameter ω from a nonempty open set Ω in a finite dimensional Euclidean space. It is assumed that $\Gamma(\omega)$ is determined by a parametric system of inequalities

$$\langle a^i(\omega), x \rangle - b_i(\omega) \leqq 0, i = 1, \cdots, m \tag{6.1}$$

where $a^i, i = 1, \cdots, m$ are vector functions on Ω with values in $\mathbb{R}^n$ and $b_i, i = 1, \cdots, m$ are real functions on Ω. We distinguish two types of constraints: those that can be reduced to equalities and the remaining ones. For convenience we write (6.1) as

$$\langle a^i(\omega), x \rangle - b_i(\omega) \leqq 0, i = 1, \cdots, p; \tag{6.2}$$

$$\langle a^j(\omega), x \rangle - b_j(\omega) = 0, j = p+1, \cdots, p+q = m. \tag{6.3}$$

We are interested in the change of $\Gamma(\omega)$ in dependence on ω. To be more specific we consider the set-valued map $\Gamma : \Omega \rightrightarrows \mathbb{R}^n$ which associates with every parameter ω from Ω a subset $\Gamma(\omega)$ of the space $\mathbb{R}^n$ and study its stability or continuity with regard to the perturbation of ω. The solution sets to the systems (6.2) and (6.3) are

© Springer International Publishing Switzerland 2016   183
D.T. Luc, *Multiobjective Linear Programming*,
DOI 10.1007/978-3-319-21091-9_6

denoted respectively by $P(\omega)$ and $Q(\omega)$. The following notations will be also in use. For subsets of indices $I \subseteq \{1, \cdots, p\}$ and $J \subseteq \{1, \cdots, m\}$,

$$P_I(\omega) := \left\{x \in \mathbb{R}^n : \langle a^i(\omega), x\rangle - b_i(\omega) \leq 0, i \in I\right\}$$
$$\hat{P}_I(\omega) := \left\{x \in \mathbb{R}^n : \langle a^i(\omega), x\rangle - b_i(\omega) < 0, i \in I\right\}$$
$$P^I(\omega) := \left\{x \in \mathbb{R}^n : \langle a^i(\omega), x\rangle - b_i(\omega) \leq 0, i \in \{1, \cdots, p\}\backslash I\right\}$$
$$Q_J(\omega) := \left\{x \in \mathbb{R}^n : \langle a^j(\omega), x\rangle - b_j(\omega) = 0, j \in J\right\}$$

and $A_I(\omega)$ is the matrix whose rows are the transposes of $a^i(\omega), i \in I$, and $b_I(\omega)$ is the vector whose components are $b_i(\omega), i \in I$.

Continuity of set-valued maps

We consider a set-valued map $G : \Omega \rightrightarrows \mathbb{R}^n$. The graph of G is the set

$$\mathrm{gr}(G) = \left\{(\omega, x) \in \Omega \times \mathbb{R}^n : x \in G(\omega)\right\}.$$

Definition 6.1.1 We say that G is closed at $\omega_0 \in \Omega$ if the limit of every convergent sequence $\{(\omega_r, x^r)\}_{r \geq 1}$ in $\Omega \times \mathbb{R}^n$ with $x^r \in G(\omega_r)$ and $\lim_{r\to\infty} \omega_r = \omega_0$, belongs to the graph of G.

The map G is said to be lower semi-continuous at $\omega_0 \in \Omega$ if for every open set V in $\mathbb{R}^n$ with $G(\omega_0) \cap V \neq \emptyset$, there exists a neighborhood U of ω_0 in Ω such that $G(\omega) \cap V \neq \emptyset$ whenever $\omega \in U$. And G is upper semi-continuous at $\omega_0 \in \Omega$ if for every open set containing $G(\omega_0)$ there exists a neighborhood U of ω_0 in Ω such that $G(\omega) \subseteq V$ whenever $\omega \in U$.

A set-valued map is closed, lower semi-continuous or upper semi-continuous on Ω if it is so at any point of Ω (Figs. 6.1 and 6.2). Continuous maps are those that are simultaneously lower and upper semi-continuous.

By convention if a map has empty values on an open set, it is continuous there. Usual (single valued) functions are a particular case of set-valued maps. For them the concepts of closedness, lower semi-continuity or upper semi-continuity above connote the concept of continuity in the classical sense. The closed hull of a set-valued map G is the map $\mathrm{cl}G$ whose value at every ω is the closure of the set $G(\omega)$.

Below we list some useful properties of set-valued maps which are quite obvious from the definition.

(1) The union of two closed (respectively lower semi-continuous/upper semi-continuous) set-valued maps is closed (respectively lower semi-continuous/upper semi-continuous);
(2) The intersection of two closed maps is closed;
(3) If a map has closed values in a compact subset of the space $\mathbb{R}^n$, then its closedness is equivalent to its upper semi-continuity on that subset;

Fig. 6.1 *Upper* semi-continuity, but not *lower* semi-continuity at ω_0

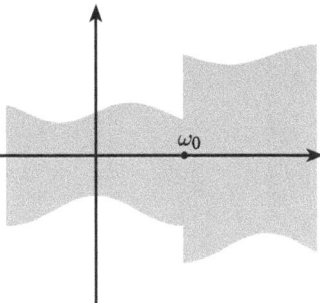

Fig. 6.2 *Lower* semi-continuity, but not *upper* semi-continuity at ω_0

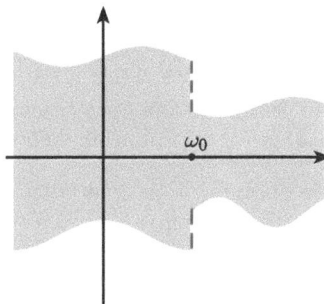

(4) A set-valued map G is lower semi-continuous at ω_0 if for any sequence $\{\omega_r\}_{r \geq 1}$ in Ω converging to ω_0 and for every element $x^0 \in G(\omega_0)$ there is a sequence $\{x^r\}_{r \geq 1}$ with $x^r \in G(\omega_r)$ converging to x^0;

(5) The closed hull of a lower semi-continuous map is lower semi-continuous;

(6) The intersection map $G_1 \cap G_2$ of two set-valued maps G_1 and G_2 is lower semi-continuous at $\omega_0 \in \Omega$ if the following conditions hold: a) G_1 is lower semi-continuous at ω; b) there is a subset $D \subseteq G_2(\omega_0)$ such that for every element x of D one can find a neighborhood W of ω_0 and a neighborhood U of x with $U \subseteq G_2(\omega)$ for all $\omega \in W$; c) $G_1(\omega_0) \cap G_2(\omega_0)$ coincides with the closure of the set $G_1(\omega_0) \cap D$.

Continuity of the map $\Gamma(\omega)$

We assume throughout this subsection that the vector functions $a^1(\omega), \cdots, a^m(\omega)$ and $b(\omega) := (b_1(\omega), \cdots, b_m(\omega))^T$ are continuous. The next example shows that the set-valued map $\Gamma(\omega)$ is neither upper semi-continuous, nor lower semi-continuous even though $a^1(\omega), \cdots, a^m(\omega)$ and $b(\omega)$ are continuous.

Example 6.1.2 Consider the following system in $\mathbb{R}^2$:

$$-\omega x_1 + x_2 \leqq 0$$
$$-x_1 \leqq 0$$
$$-x_2 \leqq 0$$

for $\omega \in (-1, 1) \subseteq \mathbb{R}$. At $\omega = 0$ the solution set $\Gamma(0)$ is the set $\mathbb{R}_+ \times \{0\}$. Let V be the set of vectors x with $-1 < x_1$ and $-1 < x_2 < 1$. It is an open set containing $\Gamma(0)$. For every $\omega > 0$ the set $\Gamma(\omega)$ contains the ray $x_2 = \omega x_1, x_1 \geqq 0$, and cannot be included in V. Hence Γ is not upper semi-continuous at $\omega = 0$. For $\omega < 0$ the set $\Gamma(\omega)$ consists of the singleton $\{0\}$ while $\Gamma(0)$ contains nonzero element. Consequently Γ is not lower semi-continuous at $\omega = 0$ either.

In the above example the map Γ is upper semi-continuous at no point $\omega \geqq 0$, but it is lower semi-continuous at every point except for $\omega = 0$. This generic property of lower semi-continuity is the main concern of the present subsection. Namely, we shall prove the following theorem.

Theorem 6.1.3 *Assume that the functions $a^1(\omega), \cdots, a^m(\omega)$ and $b(\omega)$ are continuous. Then the set-valued map $\Gamma(\omega)$ is closed. Moreover, for every open set $W_0 \subseteq \Omega$ there is an open subset W of W_0 such that $\Gamma(\omega)$ is lower semi-continuous on W. Consequently, the set on which $\Gamma(\omega)$ is not lower semi-continuous is nowhere dense.*

The proof of this result is based on a number of auxiliary statements that we present as lemmas.

Lemma 6.1.4 *Let ω_0 be a point in Ω and $J \subseteq \{1, \cdots, m\}$ be nonempty. If the family $\{a^j(\omega_0) : j \in J\}$ is linearly independent, then there is a neighborhood W of ω_0 in Ω such that the set-valued map $Q_J(\omega)$ is lower semi-continuous on W.*

Proof Without loss of generality, we may suppose that $J = \{1, \cdots, \ell\}$ for some $1 \leqq \ell \leqq m$. Denote by $A(\omega)$ the $\ell \times n$-matrix whose rows are the transposes of $a^1(\omega), \cdots, a^\ell(\omega)$. It is clear that $\ell \leqq n$ because at $\omega = \omega_0$ those rows are linearly independent, and so the matrix $A(\omega_0)$ possesses a nonsingular $\ell \times \ell$-submatrix, denoted $C(\omega_0)$. The complementary part of $C(\omega_0)$ in $A(\omega_0)$ is denoted $D(\omega_0)$. We may assume that $A(\omega)$ is written as $(C(\omega), D(\omega))$. The system of equalities determining $Q_J(\omega)$ is then

$$\big(C(\omega), D(\omega)\big) \begin{pmatrix} x_C \\ x_D \end{pmatrix} = b(\omega),$$

where x_C is the vector of the first ℓ components, and x_D is the vector of the $(n - \ell)$ remaining components of a vector x in $\mathbb{R}^n$. The set $Q_J(\omega_0)$ is given by

$$Q_J(\omega_0) = \left\{ x \in \mathbb{R}^n : x_C = \big(C(\omega_0)\big)^{-1}(b(\omega_0) - D(\omega_0)x_D), \; x_D \in \mathbb{R}^{n-\ell} \right\} \quad (6.4)$$

where $\left(C(\omega_0)\right)^{-1}$ is the inverse of the matrix $C(\omega_0)$. By the continuity hypothesis of the entries of $C(\omega_0)$, there is a neighborhood W of ω_0 such that $C(\omega)$ is nonsingular for all $\omega \in W$. Then, the formula (6.4) remain true for all $\omega \in W$. Now, let $\omega \in W$ and $x \in Q_J(\omega)$ with the components x_C and x_D as above. Let $\{\omega_r\}_{r \geq 1}$ be a sequence in W converging to ω. Set

$$x_C^r = \left(C(\omega_r)\right)^{-1}\left(b(\omega_r) - D(\omega_r)x_D\right).$$

It is clear that x^r belongs to $Q_J(\omega_r)$ and converges to x as r tends to ∞. Hence Q_J is lower semi-continuous on W. $\qquad\square$

Lemma 6.1.5 *Let W_0 be an open set in Ω and let $J \subseteq \{1, \cdots, m\}$ be nonempty. If the family $\{a^j(\omega) : j \in J\}$ is linearly dependent for every $\omega \in W_0$, then there is an open subset W of W_0 and a proper subset J' of J such that the family $\{a^j(\omega) : j \in J'\}$ is a maximal linearly independent subfamily of the family $\{a^j(\omega) : j \in J\}$ for every $\omega \in W$.*

Proof As in the proof of Lemma 6.1.4, we assume that $J = \{1, \cdots, \ell\}$. For all ω in W_0 the rank of the system $\{a^j(\omega) : j = 1, \cdots, \ell\}$ does not exceed $\ell - 1$. We may find a point ω_0 in W_0 such that the rank of the latter system is maximal, say equal to $r < \ell$ with $\{a^j(\omega_0) : j = 1, \cdots, r\}$ linearly independent. The $r \times n$-matrix $A(\omega_0)$ constituted from the rows transposes of the vectors $a^1(\omega_0), \cdots, a^r(\omega_0)$ has a nonsingular $r \times r$-submatrix $C(\omega_0)$. Let W be a neighborhood of ω_0 in W_0 on which $C(\omega)$ remains nonsingular. Then for every $\omega \in W$, the system $\{a^1(\omega), \cdots, a^r(\omega)\}$ has rank r and is a maximal linearly independent subsystem of $\{a^j(\omega) : j \in J\}$. $\qquad\square$

Lemma 6.1.6 *For every open set W_0 in Ω there is an open subset W of W_0 such that the set-valued map $Q(\omega)$ is lower semi-continuous on W.*

Proof If there exists a point ω_0 in W_0 such that the system $\{a^j(\omega_0) : j = p + 1, \cdots, m\}$ is linearly independent, then in view of Lemma 6.1.4 the map Q is lower semi-continuous on some neighborhood W of ω_0 in W_0. If that system is linearly dependent for all ω in W_0, one applies Lemma 6.1.5 to get an open subset W_1 of W_0 and a proper subset $J \subset \{p + 1, \cdots, m\}$ such that for all $\omega \in W_1$ the system $\{a^j(\omega) : j \in J\}$ is a maximal linearly independent subsystem of the system $\{a^j(\omega) : j = p + 1, \cdots, m\}$. We claim that there exists an open set W in W_1 such that either $Q(\omega)$ is empty for all $\omega \in W$, or $Q(\omega)$ coincides with $Q_J(\omega)$ for all $\omega \in W$. In fact, take any index j_0 from the index set $\{p + 1, \cdots, m\}\setminus J$. The system $\{a^j(\omega) : j \in J \cup \{j_0\}\}$ being linearly dependent, one is able to find coefficients $t_j(\omega) \in \mathbb{R}$ such that

$$a^{j_0}(\omega) = \sum_{j \in J} t_j(\omega)a^j(\omega),$$

where $t_j(\omega)$, $j \in J$ are uniquely determined because the system $\{a^j(\omega) : j \in J\}$ is linearly independent. Moreover, $t_j(\omega)$, $j \in J$ are continuous in some neighborhood of ω when the latter ω is fixed. If at some ω from W_1 one has

$$b_{j_0}(\omega) \neq \sum_{j \in J} t_j(\omega) b_j(\omega)$$

then, by continuity there is a neighborhood W of ω in W_1 such that

$$b_{j_0}(\omega') \neq \sum_{j \in J} t_j(\omega') b_j(\omega') \text{ for all } \omega' \in W.$$

Consequently, the set $Q(\omega')$ is empty for all $w' \in W$. Now, if

$$b_{j_0}(\omega) = \sum_{j \in J} t_j(\omega) b_j(\omega) \text{ for all } \omega \in W_1,$$

the j_0-th equality in the system determining Q is redundant, and so

$$Q_J(\omega) = Q_{J \cup \{j_0\}}(\omega) \text{ for all } w \in W_1.$$

Continuing this process for other indices from the set $\{p+1, \cdots, p+q\} \backslash J$, we arrive at the conclusion that either $Q(\omega)$ is empty for all $\omega \in W$, or $Q(\omega) = Q_J(\omega)$ for all $\omega \in W$. In both cases the map Q is lower semi-continuous on W in view of Lemma 6.1.4. $\qquad \square$

Lemma 6.1.7 *Let ω_0 be a point in Ω and $I \subseteq \{1, \cdots, p\}$ be nonempty. If the set $\hat{P}_I(\omega_0)$ is nonempty, then for every element x^0 of it, there exist a neighborhood U of x^0 in $\mathbb{R}^n$ and a neighborhood W of ω_0 in Ω such that U is included in $\hat{P}_I(\omega)$ for all $\omega \in W$. Consequently the set-valued map $P_I(\omega)$ is lower semi-continuous on W.*

Proof For every element x from $\hat{P}_I(\omega_0)$ we have strict inequality

$$\langle a^i(\omega_0), x \rangle - b_i(\omega_0) < 0 \text{ for all } i \in I.$$

One can find a positive number ε such that

$$\langle a^i(\omega_0) + \alpha^i, x + x' \rangle - (b_i(\omega_0) + \beta_i) < 0$$

for every index $i \in I$, for all vectors α^i and x' from $\mathbb{R}^n$ and real number β_i with $\max\{\|\alpha^i\|, \|x'\|, |\beta_i|\} < \varepsilon$. Since all $a^i(\omega)$ and $b_i(\omega)$ are continuous, there is a neighborhood W of ω_0 such that

$$\max\{\|a^i(\omega) - a^i(\omega_0)\|, |b_i(\omega) - b_i(\omega_0)|\} < \varepsilon \text{ for all } \omega \in W.$$

By choosing $U = \{x' \in \mathbb{R}^n : \|x' - x\| < \varepsilon\}$ we conclude that

$$\langle a^i(\omega), x' \rangle - b_i(\omega) < 0$$

for all $i \in I$, $x' \in U$ and $\omega \in W$, which implies that U lies in $\hat{P}_I(\omega)$ for all $\omega \in W$. The lower semi-continuity of P_I follows because it is the closure of $\hat{P}_I$. $\qquad\square$

Lemma 6.1.8 *For every open set W_0 in Ω there exist an open subset W of W_0 and a subset $I \subseteq \{1, \cdots, p\}$ such that for every $w \in W$ one has*

$$P(\omega) = P_I(\omega) \cap Q_J(\omega)$$
$$\hat{P}_I(\omega) \neq \emptyset,$$

where $J = \{1, \cdots, m\} \setminus I$.

Proof First we observe that $\hat{P}_i(\omega)$ is empty only if $a^i(\omega) = 0$ and $b_i(\omega) \leqq 0$, and that if at some point w_0, the set $\hat{P}_i(w_0)$ is nonempty, then $\hat{P}_i(\omega)$ is nonempty for all ω in a small neighborhood of w_0. Hence, without loss of generality, one may assume that

$$\hat{P}_i(\omega) \neq \emptyset \text{ for all } \omega \in W_0, i = 1, \cdots, p.$$

Let $I(\omega)$ denote a maximal subset of the index set $\{1, \cdots, p\}$ for which $\hat{P}_{I(\omega)}(\omega)$ is nonempty (with respect to the inclusion order), that is,

$$\hat{P}_{I(\omega)}(\omega) \cap \hat{P}_i(\omega) = \emptyset \text{ for } i \notin I(\omega).$$

Since the cardinal of $I(\omega)$ is less or equal to p, there is some point w_0 of W_0 such that

$$|I(w_0)| = \max\{|I(\omega)| : \omega \in W_0\}.$$

We claim that there is a neighborhood W of w_0 such that $I(\omega) = I(w_0)$ for all $\omega \in W$. In fact, as in the proof of Lemma 6.1.7, $\hat{P}_{I(w_0)}(w_0) \neq \emptyset$ implies that there is a neighborhood W of w_0 such that $\hat{P}_{I(w_0)}(\omega) \neq \emptyset$ for every $\omega \in W$. Hence $I(w_0)$ is a subset of $I(\omega)$ as soon as ω lies in W. Actually we have equality because the cardinality of $I(w_0)$ is already maximal. Set $I = I(w_0)$. On the one hand, for every $\omega \in W$ one has that

$$P_I(\omega) \cap Q_J(\omega) \subseteq P(\omega).$$

On the other hand, $x \in P(\omega)$ implies that

$$\langle a^j(\omega), x \rangle - b_j(\omega) = 0, \quad j \in J.$$

Indeed, if this were not true, for some $j \in J$ one would have

$$\langle a^j(\omega), x \rangle - b_j(\omega) < 0,$$

and would find a neighborhood U of x in $\mathbb{R}^n$ such that

$$\langle a^j(\omega), x'\rangle - b_j(\omega) < 0, \text{ for all } x' \in U.$$

Since $P_I(\omega)$ is the closure of $\hat{P}_I(\omega)$ that is nonempty, we can find some common point x' of U and $\hat{P}_I(\omega)$. Consequently, that x' belongs to the intersection of $\hat{P}_I(\omega)$ with $\hat{P}_j(\omega)$, which contradicts the maximality of I. In this way, we have $P(\omega) = P_I(\omega) \cap Q_J(\omega)$ with $\hat{P}_I(\omega)$ nonempty for all $\omega \in W$.   □

Here is the detailed proof we promised for Theorem 6.1.3.

Proof of Theorem 6.1.3 To prove that $\Gamma(\omega)$ is closed, we consider a sequence $\{(\omega_r, x^r)\}_{r \geq 1}$ of elements of the graph of Γ that converges to some (ω_0, x^0) with $\omega_0 \in \Omega$. By definition, the system (6.1) is satisfied at $\omega = \omega_r$ and $x = x^r$ for all $r \geq 1$. Since the functions a^i and b_i, $i = 1, \cdots, m$ are continuous, by passing the inequalities (6.1) with $\omega = \omega_r$ and $x = x^r$ to the limit as r tends to ∞ we deduce that (6.1) holds true for $\omega = \omega_0$ and $x = x^0$ too. Hence x^0 belongs to $\Gamma(\omega_0)$, by which Γ is closed.

The proof of the lower semi-continuity of $\Gamma(\omega)$ is based on the lemmas we have proven above. Let us write $\Gamma(\omega)$ as the intersection of $P(\omega)$ and $Q(\omega)$. In view of Lemma 6.1.8, there exists an open subset W_1 of W_0 such that

$$P(\omega) = P_I(\omega) \cap Q_J(\omega)$$
$$\hat{P}_I(\omega) \neq \emptyset$$

for all $\omega \in W_1$, where $J = \{1, \cdots, m\}\backslash I$. Express the map Γ by

$$\Gamma(\omega) = P(\omega) \cap Q(\omega)$$
$$= P_I(\omega) \cap \left[Q(\omega) \cap Q_J(\omega) \right]$$

for every $\omega \in W_1$. Apply Lemma 6.1.6 to the map $Q \cap Q_J$ we find an open subset W of W_1 on which the map $Q \cap Q_J$ is lower semi-continuous. It remains to apply Lemma 6.1.7 and the property 6) of set-valued maps to deduce the lower semi-continuity of the intersection map $P_I \cap (Q \cap Q_J)$ and the map Γ. The proof is complete.   □

Smooth representation of the map $\Gamma(\omega)$

In this subsection we assume that the vector functions $a^1(\omega), \cdots, a^m(\omega)$ and $b(\omega)$ are of class C^r (r-times continuously differentiable functions) with $r \geq 0$. Recall that $A_J(\omega)$ denotes the matrix whose rows are the transposes of the vectors a^j, $j \in J$ for a subset J of the index set $\{1, \cdots, m\}$ and A_J^{-1} denotes the inverse of the matrix A_J if it exists. We begin with the case where $\Gamma(\omega)$ has vertices.

Theorem 6.1.9 *Assume that Γ is lower semi-continuous on an open subset W of Ω and that $\Gamma(w_0)$ possesses at least one vertex for some $w_0 \in W$. Then there exists an open subset W_0 of W, and a finite number of subsets $J_1, \cdots, J_\kappa \subseteq \{1, \cdots, m\}$ with cardinality $|J_1| = \cdots = |J_\kappa| = n$ such that for every $w \in W_0$, the polyhedral convex set $\Gamma(w)$ possesses exactly κ vertices $v^1(w), \cdots, v^\kappa(w)$ defined by equations:*

$$v^i(w) = A_{J_i}^{-1}(w)\big(b_{J_i}(w)\big), i = 1, \cdots, \kappa.$$

Proof Let us denote by $\kappa(w)$ the number of vertices of $\Gamma(w)$. We show that for each $w_0 \in W$ there is a neighborhood W_0 of w_0 in W such that $\kappa(w) \geqq \kappa(w_0)$ for all $w \in W_0$. In fact, since $\kappa(w_0)$ is finite, there can be found a positive t such that all the vertices of $\Gamma(w_0)$ are located in the interior of the box

$$B := \big\{ x = (x_1, \cdots, x_n)^T \in \mathbb{R}^n : \max_{i=1,\cdots,n} |x_i| \leqq t \big\}.$$

We claim that around each vertex of $\Gamma(w_0)$ must at least one vertex of $\Gamma(w)$ be found once w is sufficiently close to w_0, thereby establishing the required inequality. Indeed, let v^0 be a vertex of $\Gamma(w_0)$. By the lower semi-continuity of Γ there is $x^w \in \Gamma(w)$ such that $\lim_{w \to w_0} x^w = v^0$. One can express $x^w = y^w + z^w$ where y^w is a convex combination of vertices of $\Gamma(w)$ and z^w is a direction of $\Gamma(w)$. It is evident that $\lim_{w \to w_0} z^w = 0$. Otherwise, one should choose a sequence $\{y^{w_i}\}_{\geq 1} \subseteq B$ converging to some $y^0 \in \Gamma(w_0)$, a sequence $\{z^{w_i}\}_{i \geq 1}$ converging to some nonzero direction z^0 of $\Gamma(w_0)$ as w_i tends to w_0, and arrive at a contradiction $v^0 = y^0 + z^0$ (remembering that v^0 is a vertex). Moreover, there is at least one vertex $v(w)$ in the convex combination y^w such that $\lim_{w \to w_0} v(w) = v^0$. Otherwise, since the number of vertices of $\Gamma(w)$ is bounded from above by $\binom{m}{n}$, there should exist a sequence $\{y^{w_i}\}_{i \geq 1}$ converging to a convex combination of some points of $\Gamma(w_0)$ which are distinct from v^0 and we again should arrive at a contradiction that the vertex v^0 is represented by a convex combination of other points of $\Gamma(w_0)$.

Let w_0 be a point which maximizes $\kappa(w)$ on W. Such a point exists because $\kappa(w)$ is bounded from above, as already said. Then there is a neighborhood $W_0 \subseteq W$ of w_0 such that $\kappa(w) = \kappa(w_0)$ for all $w \in W_0$. Let $v^i(w_0), i = 1, \cdots, \kappa(w_0)$ be the vertices of $\Gamma(w_0)$. We may assume that W_0 is so small that in a small neighborhood of each vertex $v^i(w_0)$ there is exactly one vertex, say $v^i(w)$ of $\Gamma(w)$. For every fixed index i, let

$$J(v^i(w)) = \big\{ j \in \{1, \cdots, p+q\} : \langle a^j(w), v^i(w) \rangle - b_j(w) = 0 \big\}.$$

Then $|J(v^i(w))| \geqq n$ and evidently, if an index j does not belong to $J(v^i(w))$, neither does it belong to $J(v^i(w'))$ when w' is sufficiently close to w. In other words $J(v^i(w'))$ is a subset of $J(v^i(w))$ for every w' in a sufficiently small neighborhood of ω. Choose a point ω_1 which minimizes the cardinality of $J(v^i(\omega))$ on W_0. Then one can find a neighborhood W_1 of ω_1 in W_0 such that

$$J(v^i(\omega_1)) = J(v^i(\omega)) \quad \text{for all } \omega \in W_1.$$

Pick any n indices from the set $J(v^i(\omega_1))$ with the property that the corresponding vectors $a^j(\omega_1)$ form a linearly independent system. There is a smaller neighborhood $W_2 \subseteq W_1$ where these vectors are still linearly independent. The vertex $v^i(\omega)$, $\omega \in W_2$, is then determined by the equation given in the theorem with $\kappa = \kappa(\omega_0)$. $\qquad \square$

To study the general case in which $\Gamma(\omega)$ may have no vertex, we denote by $L(\omega)$ the lineality space of $\Gamma(\omega)$ that is

$$L(\omega) = \big(\Gamma(\omega)\big)_\infty \cap \big(-\big(\Gamma(\omega)\big)_\infty\big).$$

Lemma 6.1.10 *Assume that Γ is nonempty valued and continuous on W. Then there exists an open subset $W_0 \subseteq W$ and a subset $J_0 \subseteq \{1, \cdots, m\}$ such that for every $\omega \in W_0$, $L(\omega)$ is the orthogonal space of the space spanned by the vectors $\{a^i(\omega) : i \in J_0\}$. Consequently, there can be found s functions $u^1, \cdots, u^s$ of class C^r on W_0 where $s = n - |J_0|$, such that $u^1(\omega), \cdots, u^s(\omega)$ form a basis of $L(\omega)$, for every $\omega \in W_0$.*

Proof Let $\omega_0 \in W$. We show first that there is a neighborhood W_0 of ω_0 in W such that

$$\dim L(\omega) \leqq \dim L(\omega_0) \quad \text{for all } \omega \in W_0.$$

In fact, since the lineality space $L(\omega)$ is defined by equations

$$\langle a^i(\omega), x \rangle = 0, \ i = 1, \cdots, m,$$

one has

$$\dim L(\omega) = n - \text{rank}\{a^1(\omega), \cdots, a^m(\omega)\}.$$

For a fixed $\omega_0 \in W$, one can find a neighborhood W_0 of ω_0 in W such that

$$\text{rank}\{a^1(\omega_0), \cdots, a^m(\omega_0)\} \leqq \text{rank}\{a^1(\omega), \cdots, a^m(\omega)\},$$

whenever $\omega \in W_0$. Hence $\dim L(\omega) \leqq \dim L(\omega_0)$ for every $\omega \in W_0$. By this, if we take $\omega_0 \in W_0$ with the property that

$$\dim L(\omega_0) = \min\{\dim L(\omega) : \omega \in W\},$$

then $\dim L(\omega_0) = \dim L(\omega)$ for every $\omega \in W_0$. Let $J_0 \subseteq \{1, \cdots, m\}$ be an index subset such that $\{a^i(\omega_0) : i \in J_0\}$ is a maximal linearly independent subsystem of the system $\{a^i(\omega_0) : i = 1, \cdots, m\}$. It is clear that

$$L(\omega_0) = \{x \in \mathbb{R}^n : \langle a^i(\omega_0), x \rangle = 0, i \in J_0\}$$

and that $\dim L(\omega_0) = n - |J_0|$. Set $s := \dim L(\omega_0)$. By taking a smaller neighborhood if necessary, we may assume that $\{a^i(\omega) : i \in J_0\}$ is still a linearly independent system whenever $\omega \in W_0$. Since $\dim L(\omega_0) = \dim L(\omega)$, one has

$$L(\omega) = \{x \in \mathbb{R}^n : \langle a^i(\omega), x \rangle = 0, i \in J_0\},$$

and hence $L(\omega)$ is the orthogonal space to the space spanned by $\{a^i(\omega) : i \in J_0\}$. Let $A_0(\omega_0)$ be a nonsingular $|J_0| \times |J_0|$-submatrix of $A_{J_0}(\omega_0)$. Without loss of generality, one may also assume that the matrix $A_0(\omega)$ is nonsingular for $\omega \in W_0$. To facilitate the writing, assume that $A_{J_0}(\omega) = (A_0(\omega)|B(\omega))$. Denote by $e^i \in \mathbb{R}^n$ the ith coordinate unit vector, $i = 1, \cdots, n$. It is evident that the vectors

$$u^i(\omega) = \begin{pmatrix} -(A_0(\omega))^{-1} B(\omega) e^{|J_0|+i} \\ 0 \end{pmatrix} + e^{|J_0|+i}, \ i = 1, \cdots, s$$

form a basis of $L(\omega)$ and are functions of class C^r on W_0. □

Now we are able to give the main representation theorem in the general case.

Theorem 6.1.11 *Assume as before that the functions $a^1(\omega), \cdots, a^{p+q}(\omega)$ and $b(\omega)$ are of class C^r. Then for every open set $W \subseteq \Omega$, there exists an open subset W_0 of W such that either Γ is empty valued on W_0, or one can find μ functions $v^1, \cdots, v^\mu$ of class C^r on W_0 and an integer $\kappa, 0 < \kappa \leq \mu$, such that*

$$\Gamma(\omega) = \left\{ x \in \mathbb{R}^n : x = \sum_{i=1}^{\mu} \lambda_i v^i(\omega), \sum_{i=1}^{\kappa} \lambda_i = 1, \lambda_i \geq 0, i = 1, \cdots, \mu \right\},$$

for every $\omega \in W_0$.

Proof By Theorem 6.1.3 we may assume that Γ is lower semi-continuous on W. If $\Gamma(\omega) = \emptyset$ for all $\omega \in W$, we are done. If not, by the lower semi-continuity there is a neighborhood in W where the values of Γ are nonempty. Therefore without loss of generality, it can be assumed that $\Gamma(\omega)$ is nonempty for all $\omega \in W$. Let $\omega_0 \in W$ be a point where $\dim L(\omega_0)$ is minimal among $\dim L(\omega), \omega \in W$. In virtue of Lemma 6.1.10, there is a neighborhood W_1 of ω_0 in W and s functions $u^1, \cdots, u^s$ of class C^r on W_1 such that $L(\omega)$ is generated by the vectors $u^1(\omega), \cdots, u^s(\omega)$ for each $\omega \in W_1$. Let

$$\Gamma_0(\omega) := \Gamma(\omega) \cap \{x \in \mathbb{R}^n : \langle u^i(\omega), x \rangle = 0, i = 1, \cdots, s\}.$$

In other words, $\Gamma_0(\omega) := \Gamma(\omega) \cap L^\perp(\omega)$, where $L^\perp(\omega)$ is the orthogonal space to $L(\omega)$. The proof of Corollary 2.3.16 shows that $\Gamma(\omega) = \Gamma_0(\omega) + L(\omega)$, and $\Gamma_0(\omega)$ has no lines which means that it has at least one vertex (see Corollary 2.3.14). Using Theorem 6.1.3, we may assume that the map Γ_0 is lower semi-continuous on W_1. By Theorem 6.1.9, one can find an open subset $W_2 \subseteq W_1$ and κ functions $v^1, \cdots, v^\kappa$ of class C^r on W_0 such that for every $\omega \in W_2$, the polyhedral set $\Gamma_0(\omega)$ has exactly κ vertices: $v^1(\omega), \cdots, v^\kappa(\omega)$. If for some $\bar\omega \in W_2$, the set $\Gamma_0(\bar\omega)$ has no nonzero asymptotic direction, that is, it is a polytope, then there is a neighborhood $\bar W$ of $\bar\omega$ in W_2 such that $\Gamma(\omega)$ is a polytope too, $\omega \in \bar W$. In this event, $v^1, \cdots, v^\kappa, u^1, \cdots, u^s, -u^1, \cdots, -u^s$ are the functions we need to represent Γ. In the other case, we fix any $\bar\omega$ in W_2. Since Γ is lower semi-continuous on W_2, one can find a neighborhood $\bar W$ of $\bar\omega$ in W_2 and a vector $d \in \mathbb{R}^n$ such that $\langle d, v\rangle > 0$ for every $v \in \big(\Gamma_0(\omega)\big)_\infty \backslash\{0\}$ and very $\omega \in \bar W$. Let us consider the convex polyhedron

$$\Gamma_1(\omega) := \big(\Gamma_0(\omega)\big)_\infty \cap \{x \in \mathbb{R}^n : \langle d, x\rangle = 1\},$$

for $\omega \in \bar W$. It is clearly bounded. In view of Theorem 6.1.9, we can find an open subset $W_0 \subseteq \bar W_2$ and l functions $v^{\kappa+1}, \cdots, v^{\kappa+l}$ of class C^r on W_0 such that for every $\omega \in W_0$, the polytope $\Gamma_1(\omega)$ has exactly l vertices $v^{\kappa+1}(\omega), \cdots, v^{\kappa+l}(\omega)$. It is evident that

$$(\Gamma_0(\omega))_\infty = \left\{ x \in \mathbb{R}^n : x = \sum_{i=\kappa+1}^{\kappa+l} \lambda_i v^i(\omega), \lambda_i \geq 0, i = \kappa+1, \cdots, \kappa+l \right\},$$

and also

$$\Gamma_0(\omega) = \left\{ x \in \mathbb{R}^n : x = \sum_{i=1}^{\kappa+l} \lambda_i v^i(\omega), \sum_{i=1}^{\kappa} \lambda_i = 1, \lambda_i \geq 0, i = 1, \cdots, \kappa+l \right\}.$$

The functions $v^1, \cdots, v^{\kappa+l}, u^1, \cdots, u^s, -u^1, \cdots, -u^s$ are those we look for. The number μ of functions to represent $\Gamma(\omega)$ is equal to $\kappa + l + 2s$. $\square$

We conclude this section by observing that if $\Gamma(\omega)$ possesses vertices, the lineality space $L(\omega)$ must be null. Moreover, all the vertices of $\Gamma(\omega)$ must be among $v^1, \cdots, v^\kappa$; while all its extreme directions must appear among $v^{\kappa+1}, \cdots, v^{\kappa+l}$. Non-extreme points and non-extreme directions of the system $\{v^1, \cdots, v^{\kappa+l}\}$ become superfluous in the representation of $\Gamma(\omega)$. Thus, in this case $\Gamma(\omega)$ may be represented by extreme points and extreme directions only.

6.2 Sensitivity

We consider the following parametric multiobjective linear problem denoted (MO LP$_\omega$):

$$\text{Maximize} \quad C(\omega)x$$
$$\text{subject to } x \in \Gamma(\omega)$$

where ω is a parameter from an open set Ω of a finite dimensional Euclidean space, $C(\omega)$ is a real $k \times n$-matrix, $\Gamma(\omega)$ is a polyhedral convex set, determined by the system (6.1). Throughout this section the entries of $C(\omega)$ and the entries of the system defining $\Gamma(\omega)$ are supposed to be of class C^r with $r \geq 0$. The *marginal function* or the *efficient value map* of the problem (MOLP$_\omega$) is defined by

$$\Phi(\omega) := \text{Max}\{C(\omega)x : x \in \Gamma(\omega)\}$$

and the *efficient solution map* is given by

$$S(\omega) := \{x \in \Gamma(\omega) : C(\omega)x \in \Phi(\omega)\}.$$

The aim of this section is to show that generically the marginal function and the efficient solution map are of class C^r in the sense that they are polyhedral sets, not necessarily convex, whose faces are convex combinations of vertices and extreme directions that are C^r functions of the parameter ω. Let us first develop some more details on representations of parametric polyhedra. Assume that W is an open set in the parameter space Ω on which the polyhedron $\Gamma(\omega)$ admits a representation as described in Theorem 6.1.11. Thus, $\Gamma(\omega)$ is written as the sum of a bounded polyhedron $P(\omega)$ and a polyhedral cone $Q(\omega)$ of the form

$$P(\omega) = \left\{ x \in \mathbb{R}^n : x = \sum_{i=1}^{\kappa} \lambda_i u^i(\omega), \ \lambda_i \geq 0, \ i = 1, \cdots, \kappa, \ \sum_{i=1}^{\kappa} \lambda_i = 1 \right\}$$

$$Q(\omega) = \left\{ x \in \mathbb{R}^n : x = \sum_{i=\kappa+1}^{\mu} \lambda_i u^i(\omega), \ \lambda_i \geq 0, i = \kappa+1, \cdots, \mu \right\},$$

where $u^1(\omega), \cdots, u^\mu(\omega)$ are of class C^r on W and for any fixed $\omega \in W$, $u^1(\omega), \cdots, u^\kappa(\omega)$ are vertices of $P(\omega)$, while $u^{\kappa+1}(\omega), \cdots, u^\mu(\omega)$ are extreme directions of $Q(\omega)$. Notice that κ and μ depend on ω.

Lemma 6.2.1 *Let $\Gamma(\omega)$ and W be specified as in Theorem 6.1.11. There are an open set $W_0 \subseteq W$, an integer $l \geq 1$ and $2l$ index sets*

$$I_1, \cdots, I_l \subseteq \{1, \cdots, \kappa\},$$
$$J_1, \cdots, J_l \subseteq \{\kappa+1, \cdots, \mu\}$$

such that for any $\omega \in W_0$, the polyhedral set $\Gamma(\omega)$ consists of exactly l faces, each face $\Gamma_i(\omega)$ of which can be written as $P_i(\omega) + Q_i(\omega)$ with $P_i(\omega)$ being a face of $P(\omega)$ generated by vertices $u^{i'}(\omega)$, $i' \in I_i$ and $Q_i(\omega)$ being a face of $Q(\omega)$ generated by directions $u^{j'}(\omega)$, $j' \in J_i$, $i = 1, ..., l$.

Proof Choose two arbitrary index sets I from $\{1, \cdots, \kappa\}$ and J from $\{\kappa+1, \cdots, \mu\}$, and choose $i_0 \in I$. Consider the following system of equations

$$\langle \xi, u^i(\omega) - u^{i_0}(\omega) \rangle = 0, \ i \in I \tag{6.5}$$

$$\langle \xi, u^j(\omega) \rangle = 0, \ j \in J. \tag{6.6}$$

In view of Theorem 6.1.11 there is an open set $W_1 \subseteq W$ such that either this system has no solution, or there are p functions $\xi^1(\omega), \cdots, \xi^p(\omega)$ of the class C^r on W_1 such that every solution of this system can be written in the form

$$\xi = \sum_{i=1}^{p} \lambda_i \xi^i(\omega), \ \lambda_i \geq 0, \ i = 1, \cdots, p, \ \sum_{i=1}^{p'} \lambda_i = 1, p' \leq p. \tag{6.7}$$

Since the number of index subsets of the set $\{1, \cdots, \mu\}$ is finite, we may assume that for arbitrary index sets $I \subseteq \{1, \cdots, \kappa\}$ and $J \subseteq \{\kappa + 1, \cdots, \mu\}$ either the system (6.5), (6.6) has no solution, or it has solutions written in the form (6.7). Of course, the number p and the functions $\xi_1, \cdots, \xi_p$ depend on I and J.

Now, choose a point $\omega_0 \in W_1$, that maximizes the number of faces of $\Gamma(\omega)$ over W_1. Such a point exists because the number of faces of $\Gamma(\omega)$ is bounded. As before, we have $\Gamma(\omega_0) = P(\omega_0) + Q(\omega_0)$. Let $\Gamma_0(\omega_0)$ be a face of $\Gamma(\omega_0)$. Then we have a decomposition

$$\Gamma_0(\omega_0) = P_0(\omega_0) + Q_0(\omega_0),$$

where $P_0(\omega_0)$ is a face of $P(\omega_0)$ and is generated by some vertices, say $u^i(\omega_0), i \in I_0 \subseteq \{1, \cdots, \kappa\}$, and $Q_0(\omega_0)$ is a face of $Q(\omega_0)$ and is generated by some directions, say $u^j(\omega_0), j \in J_0 \subseteq \{\kappa + 1, \cdots, \mu\}$. It is known that the set $\Gamma_0(\omega_0)$ is a face of $\Gamma(\omega_0)$ if and only if it is the set of all minima of some linear function $\langle \xi_0, y \rangle$ on $\Gamma(\omega_0)$. In other words, the system (6.5), (6.6) with $I = I_0$ and $J = J_0$ has ξ_0 as a solution satisfying the following inequalities

$$\langle \xi, u^i(\omega_0) - u^{i_0}(\omega_0) \rangle > 0, \ i \notin I_0 \tag{6.8}$$

$$\langle \xi, u^j(\omega_0) \rangle > 0, \ j \notin J_0 \tag{6.9}$$

Since u^i are continuous, we may assume that (6.8) and (6.9) are satisfied for all $\omega \in W_1$ and all ξ with $\| \xi - \xi_0 \| < \varepsilon$, where $\varepsilon > 0$ is a sufficiently small number. Moreover, since the system (6.5), (6.6) has solutions at ω_0, and by the way we have chosen W_1, it has solutions at every $\omega \in W_1$ and its solutions can be written by (6.7). Let

$$\xi_0 = \sum_{i=1}^{p} \lambda_i \xi_i(\omega_0).$$

Then $\xi_0(\omega) = \sum_{i=1}^{p} \lambda_i \xi^i(\omega)$ is a solution of (6.5), (6.6) with $I = I_0, J = J_0$. Now for the given $\varepsilon > 0$, one can find an open neighborhood $W_2 \subseteq W_1$ of ω_0 such that $\|\xi_0(\omega) - \xi_0\| < \varepsilon$ whenever $\omega \in W_2$. Thus, $\xi_0(\omega)$ satisfies the system (6.5), (6.6) and the inequalities (6.8), (6.9) with ω instead of ω_0. This shows that the set $\Gamma_0(\omega) = P_0(\omega) + Q_0(\omega)$, where $P_0(\omega)$ is the polyhedron generated by the vertices $u^i(\omega)$, $i \in I_0$ and $Q_0(\omega)$ is the polyhedral cone generated by the directions $u^j(\omega)$, $j \in J_0$, is a face of $\Gamma(\omega)$ whenever ω is in W_2. The above reasoning is valid for any face of $\Gamma(\omega_0)$. Hence we may find a neighborhood $W_0 \subseteq W_1$, index sets $I_1, \cdots, I_l \subseteq \{1, \cdots, \kappa\}$, $J_1, \cdots, J_l \subseteq \{\kappa+1, \cdots, \mu\}$, where l is the number of all faces of $\Gamma(\omega_0)$, to satisfy the requirements of the lemma. Since l is the biggest number of faces that $\Gamma(\omega)$ may have on W_1, actually $\Gamma(\omega)$, $\omega \in W_0$ has exactly l faces as stated in the lemma. $\square$

We shall make use of a terminology already mentioned in Chap. 3: a face of a polyhedral set in $\mathbb{R}^n$ is said to be an efficient solution face (respectively a weakly efficient solution face) if every point of it is efficient (resp. weakly efficient).

Lemma 6.2.2 *Let $\Gamma(\omega)$ and $W_0 \subseteq W$ be as in the preceding lemma. If $\Gamma'(\omega_0)$ is an efficient solution face of $\Gamma(\omega_0)$ for some $\omega_0 \in W_0$, then there is a neighborhood $W_1 \subseteq W_0$ of ω_0 such that $\Gamma'(\omega)$ is an efficient solution face of $\Gamma(\omega)$ for all $\omega \in W_1$.*

Proof It follows from Theorem 4.2.6 that the face $\Gamma'(\omega_0)$ is efficient if and only if there is a strictly positive vector λ in $\mathbb{R}^k$ such that $\Gamma'(\omega_0)$ coincides with the set of all maximizers of the linear function $\langle C^T(\omega)\lambda, x \rangle$ over $\Gamma(\omega_0)$. In other words, the system (6.5), (6.6) and the inequalities (6.8), (6.9) with $\omega = \omega_0$, $I_0 = I = I_i$, $J_0 = J = J_i$ have a common solution $C^T(\omega)\lambda$. Represent this solution in the form (6.7), we can find a small neighborhood $W_1 \subseteq W_0$ of ω_0 such that the above systems have a strictly positive vector $\lambda(\omega)$ as a solution for all $\omega \in W_1$. This shows that $\Gamma'(\omega)$ is an efficient solution face of $\Gamma(\omega)$ for all $\omega \in W_1$. $\square$

Lemma 6.2.3 *Let $\Gamma(\omega)$ and $W_0 \subseteq W$ be as in Lemma 6.2.1. If $\Gamma_i(\omega_0)$ is not a weakly efficient solution face of $\Gamma(\omega_0)$ for some $\omega_0 \in W_0$, then there exists a neighborhood $W_1 \subseteq W_0$ of ω_0 such that $\Gamma_i(\omega)$ is not a weakly efficient face of $\Gamma(\omega)$ for all $\omega \in W_1$.*

Proof If $\Gamma_i(\omega_0)$ is not a weakly efficient face of $\Gamma(\omega_0)$, then there is $y^0 \in \Gamma_i(\omega_0)$ and $x^0 \in \Gamma(\omega_0)$ such that $C(\omega_0)(x^0 - y^0)$ is strictly positive. By the continuity of $C(\omega)$ and Lemma 6.2.1, for ω sufficiently close to ω_0 there are $y^\omega \in \Gamma_i(\omega)$, and $x^\omega \in \Gamma(\omega)$ such that $\lim_{\omega \to \omega_0} y^\omega = y^0$, $\lim_{\omega \to \omega_0} x^\omega = x^0$ and $C(\omega)(y^\omega - x^\omega)$ remains strictly positive. This shows that the face $\Gamma_i(\omega)$ is not a weakly efficient solution face of (MOLP_ω). $\square$

Observe that in general, the conclusion of Lemma 6.2.2 is not true for weakly efficient faces, while the conclusion of Lemma 6.2.3 is not true if we replace "weakly efficient" by "efficient". This can be seen by the following example.

Example 6.2.4 For $\omega \in R$ we define $\Gamma(\omega) \subseteq \mathbb{R}^2$ by

$$\Gamma(\omega) = \{(x, y) \in \mathbb{R}^2 : y = \omega x\}.$$

For $\omega_0 = 0$, the whole set $\Gamma(\omega_0)$ is a weakly efficient face, but it is not an efficient face. There is no neighborhood V of ω_0 such that $\Gamma(\omega)$ continues to be a weakly efficient face or not to be an efficient face for all $\omega \in V$.

Now we are able to prove the main result of this section.

Theorem 6.2.5 *Let W be an open subset of the parameter space Ω on which the representation described in Theorem 6.1.11 is true. Then there exists an open subset W_0 of W such that either of the following statements holds.*

(i) *$S(\omega) = \emptyset$ for all $\omega \in W_0$.*
(ii) *There exists a number $l_* \geq 1$ and $2l_*$ index sets*

$$I_1, \cdots, I_{l_*} \subseteq \{1, \cdots, \kappa\}$$
$$J_1, \cdots, J_{l_*} \subseteq \{\kappa + 1, \cdots, \mu\}$$

such that $S(\omega)$ consists of exactly l_* faces of the set $\Gamma(\omega)$ each of which is generated by points $u^{i'}(\omega)$, $i' \in I_i$ and by directions $u^{j'}(\omega)$, $j' \in J_i, i = 1, \cdots, l_*$ for each $\omega \in W_0$.

Proof Without loss of generality we may assume that W is an open set on which Lemma 6.2.1 is true. If $S(\omega)$ is empty for all $\omega \in W$, we are done. If there is some point ω_0 in W such that $S(\omega_0)$ is nonempty, then by Lemma 6.2.2 it remains nonempty on some open neighborhood of ω_0. Therefore, we may assume that $S(\omega)$ is nonempty at every ω of W. Applying Lemma 6.2.1 to obtain an integer l and index sets $I_1, \cdots, I_l$ and $J_1, \cdots, J_l$ as described there. Since the efficient solution set $S(\omega)$ consists of faces along with the fact that the number of faces of Γ is unchanged and equal to l on W, we may find a point ω_* at which $S(\omega_*)$ has the largest number of efficient solution faces. For the sake of convenience we assume they are $\Gamma_1(\omega), \cdots, \Gamma_{l_*}(\omega)$ with $1 \leq l_* \leq l$. By Lemmas 6.2.2 and 6.2.3 there is an open subset W_0 of W on which these faces remain efficient. There are no more efficient solution faces beyond those ones by the choice of l_*. Thus, $S(\omega)$ consists of exactly l efficient solution faces given by the points $u^i(\omega)$ as described in Lemma 6.2.1. $\qquad\square$

Corollary 6.2.6 *Let W be an open subset of the parameter space Ω on which the representation described in Theorem 6.1.11 is true. Then there exist an open subset $W_0 \subseteq W$ such that either of the following holds*

(i) *$\Phi(\omega) = \emptyset$ for each $\omega \in W_0$;*
(ii) *There exists a number $l^* \geq 1$ and $2l^*$ index sets*

$$I_1, \cdots, I_{l^*} \subseteq \{1, \cdots, \kappa\}$$
$$J_1, \cdots, J_{l^*} \subseteq \{\kappa + 1, \cdots, \mu\}$$

such that $\Phi(\omega)$ consists of exactly l^* faces of the set $C(\omega)\big(\Gamma(\omega)\big)$ each of which is generated by points $v^{i'}(\omega)$, $i' \in I_i$ and by directions $v^{j'}(\omega)$, $j' \in J_i$, $i = 1, \cdots, l^*$ for each $\omega \in W_0$.

Proof Since $\Phi(\omega)$ is the image of the solution set $S(\omega)$ under the linear function $C(\omega)$ we may apply Theorem 6.2.5 to obtain an open subset W_0 in W on which Φ has only empty values, or only nonempty values. Moreover, $\Phi(\omega)$ is the efficient solution set of the problem of maximizing the identity function over the polyhedron $C(\omega)\big(\Gamma(\omega)\big)$. Hence the method of Theorem 6.2.5 goes through. $\qquad\square$

It is to remark that the numbers κ and μ of Theorem 6.2.5 and the ones of Corollary 6.2.6 are distinct. The next result deals with weakly efficient solutions.

Theorem 6.2.7 *The conclusions of Theorem 6.2.5 and Corollary 6.2.6 remain true for the set of weakly efficient solutions and the set of weakly efficient values.*

Proof We follow the argument of Theorem 6.2.5. Let W be an open subset on which the representation given in Theorem 6.1.11 holds. If the weakly efficient solution set is empty at some point $\omega \in W$, then in view of Lemma 6.2.3 it remains empty on a neighborhood of ω. We may therefore assume that the weak solution set is nonempty at every point of W and that the decomposition of $\Gamma(\omega)$ by its faces as described in Lemma 6.2.1 is true. Choose a point ω_* in W for which the number of weakly efficient faces of (MOLP$_{\omega_*}$) is the least. We assume that they are $\Gamma_1(\omega_*), \cdots, \Gamma_{l^*}(\omega_*)$ with $1 \leq l^* \leq l$. The number of faces that are not weakly efficient is the largest at ω_*. By Lemmas 6.2.3 there is an open subset W_0 of W on which these latter faces remain non-weakly efficient, and only they are non-weakly efficient. Hence the weakly efficient set of (MOLP$_\omega$) consists of exactly l^* faces $\Gamma_1(\omega), \cdots, \Gamma_{l^*}(\omega)$ with description in Lemma 6.2.1. The proof of the counterpart for Corollary 6.2.6 is by the same argument. $\qquad\square$

Below we present an example to illustrate the analysis described in this section.

Example 6.2.8 Let $\Gamma(\omega) = \Gamma$ be a triangle with three vertices $\begin{pmatrix} 0 \\ 0 \end{pmatrix}$, $\begin{pmatrix} 1 \\ 0 \end{pmatrix}$ and $\begin{pmatrix} 0 \\ 1 \end{pmatrix}$ in $\mathbb{R}^2$. Let C be a 2×2-matrix that depends on a parameter $\omega \in (-\pi, \pi)$:

$$C(\omega) = \begin{pmatrix} -\cos\omega & \sin\omega \\ -\sin\omega & -\cos\omega \end{pmatrix}.$$

We study the following parametric problem:

$$\begin{aligned} &\text{Maximize } C(\omega)x \\ &\text{subject to } x \in \Gamma. \end{aligned}$$

It is clear that the image $C(\omega)(\Gamma)$ is the triangle with vertices

$$u^0(\omega) = \begin{pmatrix} 0 \\ 0 \end{pmatrix}, u^1(\omega) = \begin{pmatrix} -\cos\omega \\ -\sin\omega \end{pmatrix} \text{ and } u^2(\omega) = \begin{pmatrix} \sin\omega \\ -\cos\omega \end{pmatrix}.$$

At the point $\omega_0 = 0$, the unique efficient value face is $\{u^0\}$. According to Lemma 6.2.2, there exists a neighborhood W_1 of ω_0 in $(-\pi, \pi)$ such that $\{u_0\}$ is an efficient value face of the problem for $\omega \in W_1$. In order to compute this neighborhood we exploit the proof of Lemma 6.2.2. Let us take the strictly positive vector $\lambda_0 = \begin{pmatrix} 1 \\ 1 \end{pmatrix}$ and consider the system:

$$\langle \lambda^0, u^i(\omega) - u^0(\omega) \rangle < 0, i = 1, 2.$$

This system holds for all $\omega \in (-\pi/4, \pi/4)$. Hence, the open interval $(-\pi/4, \pi/4)$ is a neighborhood on which $\{u_0\}$ is an efficient value face of the problem. Note that by choosing another strictly positive vector λ^0 one obtains another neighborhood that has the same property.

Observe further that $\omega = 0$ is not the point ω_* that has a neighborhood W_0 mentioned in the proof of Theorem 6.2.5. According to that proof, we choose a point $\omega_* \in W_1$ such that the number of efficient value faces of the problem is the largest at this point. In our case, any point of W_1 different from 0 will do. For instance, with $\omega_* = \pi/8$, the efficient value set of consists of the faces: $[u^0, u^2(\omega_*)]$, $\{u^0\}$ and $\{u^2(\omega_*)\}$. Since other faces of the value set $C(\omega)(\Gamma)$ at ω_* are not weakly efficient, by Lemma 6.2.3, they cannot be weakly efficient whenever ω is sufficiently close to ω_*. This implies that the number of efficient value faces of $C(\omega)(\Gamma)$ is the largest at this point. Consequently, in view of the proof of Theorem 6.2.5 one has $\Phi(\omega) = [u^0, u^1(\omega)]$ for ω sufficiently close to ω_*. In order to determine a neighborhood W_0 of the point ω_* where $\Phi(\omega)$ has this representation, we again invoke the proof of Theorem 6.2.5. We have to solve the system:

$$\langle \lambda, u^1(\omega) - u^0 \rangle = 0$$
$$\langle \lambda, u^2(\omega) - u^0 \rangle > 0,$$

with $\lambda = \begin{pmatrix} \lambda_1 \\ \lambda_2 \end{pmatrix}$, $\lambda_1 > 0$ and $\lambda_2 > 0$. It is evident that for each $\omega \in (0, \pi/4)$ the above system has a solution. Consequently, the interval $(0, \pi/4)$ is a neighborhood we wanted.

6.3 Error Bounds and Stability

We consider a linear system in matrix form that defines a convex polyhedron Γ in the space $\mathbb{R}^n$:

$$Ax - b \leqq 0, \tag{6.10}$$

where A is an $m \times n$-matrix whose rows are the transposes of $a^1, \cdots, a^m$ and b is an m-dimensional vector. It follows that an element x is a solution of the system (6.10) if and only if the distance of $Ax - b$ to the negative orthant $-\mathbb{R}_+^m$ is zero. When x is not a solution, we wish to know how far it is away from the solution set Γ once the distance from $Ax - b$ to $-\mathbb{R}_+^m$ is computed. This question is important in computation of solutions of (6.10). In fact, when solving the above system, at each iteration an error is committed. At the final stage the obtained solution may not solve the system, but it is possible to estimate an upper bound for the total error. Given a vector a in $\mathbb{R}^m$, the vector obtained from a by substituting all negative components of a by zero is denoted by a^+. It is clear that the distance $d(a, -\mathbb{R}_+^m)$ from a to the negative orthant $-\mathbb{R}_+^m$ is exactly the norm of the vector a^+.

Definition 6.3.1 A positive number α is called an error bound of the system (6.10) if

$$d(x, \Gamma) \leqq \alpha \|(Ax - b)^+\| \quad \text{for all } x \in \mathbb{R}^n.$$

Distance to a polyhedron

Given $I \subseteq \{1, \cdots, m\}$, the matrix A_I is composed of the rows $i, i \in I$ of the matrix A. Similarly the vector b_I has components $b_i, i \in I$. The cardinality of I is denoted $|I|$.

Lemma 6.3.2 *Assume that the system (6.10) is homogeneous that is $b = 0$ and let Γ° be the polar cone of the polyhedral cone Γ. Then for every x from the polar cone Γ° there exist an index set $I \subseteq \{1, \cdots, m\}$ and a nonzero positive vector λ from the space $\mathbb{R}^{|I|}$ such that the vectors $a^i, i \in I$ are linearly independent and*

$$d(x, \Gamma) = \|x\| = \frac{\langle \lambda, A_I x \rangle}{\|A_I^T \lambda\|}.$$

Proof The conclusion being trivial for $x = 0$, we consider the case $x \neq 0$. The equality $d(x, \Gamma) = \|x\|$ is clear because otherwise one would find some point $y \in \Gamma$ such that

$$\|x - y\|^2 < \|x\|^2$$

which implies

$$\|y\|^2 < 2\langle x, y \rangle.$$

The latter inequality is impossible because x belongs to the polar cone Γ° and hence the value on the right hand side must be nonpositive. On the other hand, by Theorem 2.3.19 the polar cone Γ° is the positive hull of the vectors $a^1, \cdots, a^m$. By

Caratheodory's Theorem 2.1.2 there exists an index set $I \subseteq \{1, \cdots, m\}$ and positive numbers $\lambda_i, i \in I$ such that the vectors $a^i, i \in I$ are linearly independent and

$$x = \sum_{i \in I} \lambda_i a^i.$$

Let λ be the vector whose components are $\lambda_i, i \in I$. We deduce that

$$\|x\| = \frac{\langle x, x \rangle}{\|x\|} = \frac{\langle \lambda, A_I x \rangle}{\|A_I^T \lambda\|}.$$

Finally, we note that $\|A_I^T \lambda\|$ is not zero because it is exactly the norm of the vector x.   □

It is clear from the proof of the lemma above that the index set I can be chosen among those indices i such that a^i are extreme directions of the polar cone Γ. Given $y \in \Gamma$, we denote by $I_0(y)$ the set of active indices i such that a^i are extreme directions of the positive hull $\mathrm{pos}\{a^i : i \in I(y)\}$.

Theorem 6.3.3 *Assume that the system (6.10) is consistent. Then for every $x \notin \Gamma$ there exist a point $y \in \Gamma$, an index set $I \subseteq \{1, \cdots, m\}$ and a positive vector λ from $\mathbb{R}^{|I|}$ such that*

(i) *$I \subseteq I_0(y)$;*
(ii) *the vectors $a^i, i \in I$ of the matrix A are linearly independent;*
(iii) *the distance from x to Γ is given by*

$$d(x, \Gamma) = \|x - y\| = \frac{\langle \lambda, A_I x - b_I \rangle}{\|A_I^T \lambda\|}.$$

Proof Because Γ is a closed set, there exists some point $y \in \Gamma$ such that

$$\|x - y\| = d(x, \Gamma). \qquad (6.11)$$

Denote by H_i the half-space defined by the inequality $\langle a^i, x \rangle - b_i \leqq 0$. We claim that

$$\|x - y\| = d\left(x, \bigcap_{i \in I(y)} H_i\right), \qquad (6.12)$$

where $I(y)$ is the active index set at y. Notice that y cannot be an interior point of Γ, so that the active index set $I(y)$ is nonempty. Suppose (6.12) is not true. The set Γ being included in the set $\cap_{i \in I(y)} H_i$, one may find some z from the latter intersection such that $\|x - z\| < \|x - y\|$. Consider the point $tz + (1 - t)y$ for $t \in [0, 1]$. For inactive indices $j \notin I(y)$ one has $\langle a^j, y \rangle - b_j < 0$, hence for $t > 0$ sufficiently small, $\langle a^j, tz + (1-t)y \rangle - b_j \leqq 0$ which means that $tz + (1-t)y$ belongs to H_j. For active indices $i \in I(y)$, it is clear that $tz + (1-t)y$ belongs to H_i. By this $tz + (1-t)y$ belongs to Γ. With such a value of t we deduce

$$\|x - (tz + (1-t)y)\| \leq t\|x - z\| + (1-t)\|x - y\| < \|x - y\|$$

which contradicts (6.11).

By translation the distance from x to the set $\cap_{i \in I(y)} H_i$ is equal to the distance from $x - y$ to the polyhedral cone $\cap_{i \in I(y)} H_i - y$ which is given by the system

$$\langle a^i, x \rangle \leqq 0, \ i \in I(y).$$

Since y realizes the distance from x to the set $\cap_{i \in I(y)} H_i$, the vector $x - y$ is normal to it at y. It follows that $x - y$ belongs to the polar cone of the cone $\cap_{i \in I(y)} H_i - y$. Applying Lemma 6.3.2 we find some index set $I \subseteq I_0(y)$ and a positive vector $\lambda \in \mathbb{R}^{|I|}$ such that (ii) holds and

$$\|x - y\| = \frac{\langle \lambda, A_I(x - y) \rangle}{\|A_I^T \lambda\|}$$

which completes the proof. $\qquad\qquad\square$

Let $\mathcal{I}$ denote the set of index subsets $I \subseteq \{1, \cdots, m\}$ satisfying two conditions:

(1) I is nonempty and $\{a^i : i \in I\}$ is a linearly independent family, and
(2) there is some $y \in \Gamma$ such that $I \subseteq I_0(y)$.

Theorem 6.3.4 *Assume that the system (6.10) is consistent. Then for every $x \notin \Gamma$ the distance from x to Γ is given by the formula*

$$d(x, \Gamma) = \max_{I \in \mathcal{I}} \ \max_{\lambda \in \mathbb{R}_+^{|I|} \setminus \{0\}} \frac{\langle \lambda, A_I x - b_I \rangle}{\|A_I^T \lambda\|}.$$

Proof Let $J \in \mathcal{I}$ and $\lambda \in \mathbb{R}_+^{|J|}$ with $\lambda \neq 0$. It is clear that the vector $A_J^T \lambda$ is nonzero because the vectors a^j, $j \in J$ are linearly independent. Let y be the vector given in Theorem 6.3.3. Since λ is positive and $A_J y - b_J \leq 0$, we have

$$\langle \lambda, A_J(x - y) \rangle \geqq \langle \lambda, A_J x - b_J \rangle.$$

Consequently,

$$d(x, \Gamma) \geqq \left\langle \frac{A_J^T \lambda}{\|A_J^T \lambda\|}, x - y \right\rangle$$

$$\geqq \frac{\langle \lambda, A_J(x - y) \rangle}{\|A_J^T \lambda\|}$$

$$\geqq \frac{\langle \lambda, A_J x - b_J \rangle}{\|A_J^T \lambda\|}$$

which, combining with (iii) of Theorem 6.3.3, yields the requested formula. □

Another expressions of the distance $d(x, \Gamma)$ are given below.

Corollary 6.3.5 *Assume that the system (6.10) is consistent. Then for every $x \notin \Gamma$ we have*

$$d(x, \Gamma) = \max_{I \in \mathcal{I}, \lambda \geq 0, \|A_I^T \lambda\| = 1} \sum_{i \in I} \lambda_i (\langle a^i, x \rangle - b_i)$$

$$= \max_{I \in \mathcal{I}, \lambda \geq 0, \sum_{i \in I} \lambda_i = 1} \frac{\sum_{i \in I} \lambda_i (\langle a^i, x \rangle - b_i)}{\|A_I^T \lambda\|}.$$

Proof It suffices to note that the function on the right hand side of the formula for $d(x, \Gamma)$ in the preceding theorem is positively homogeneous with respect to λ. Therefore the maximum over λ can be restricted either on λ with $\sum_{i \in I} \lambda_i = 1$ or on λ with $\|A_I^T \lambda\| = 1$ and yield the desired conclusion. □

Here are some constants, called Hoffman constants , which help to estimate non solutions of the system (6.10):

$$\kappa = \max_{I \in \mathcal{I}, \lambda \geq 0, \|A_I^T \lambda\| = 1} \sum_{i \in I} \lambda_i \qquad (6.13)$$

$$\sigma = \min_{I \in \mathcal{I}, \lambda \in \mathbb{R}_+^{|I|}, \sum_{i \in I} \lambda_i = 1} \|A_I^T \lambda\|. \qquad (6.14)$$

The next result is known as error bound of the system (6.10).

Corollary 6.3.6 *Assume that the system (6.10) is consistent. Then $\kappa = 1/\sigma$ and for every $x \in \mathbb{R}^n$ one has*

$$d(x, \Gamma) \leqq \kappa \max\{0, \langle a^i, x \rangle - b_i, i = 1, \cdots, m\}.$$

Proof Let I and λ realize the maximum in (6.13). Setting $\mu = \lambda / \sum_{i \in I} \lambda_i$ we have $\mu \geq 0$ and $\sum_{i \in I} \mu_i = 1$ and deduce

$$\sigma \leqq \|A_I^T \mu\| = \frac{\|A_I^T \lambda\|}{\sum_{i \in I} \lambda_i} = \frac{1}{\kappa}.$$

Conversely, let λ and I realize (6.14). We set $\eta = \lambda / \|A_I^T \lambda\|$ and deduce

$$\kappa \geq \sum_{i \in I} \eta_i = \frac{\sum_{i \in I} \lambda_i}{\|A_I^T \lambda\|} = \frac{1}{\sigma}$$

which yields equality as requested.

Further, if x belongs to Γ, then $d(x, \Gamma) = 0$ and the formula is evident. If x does not belong to Γ, we apply Theorem 6.3.3 to obtain the formula in (iii). The vectors λ and $A_I^T \lambda$ being nonzero, we may choose λ so that $\|A_I^T \lambda\| = 1$. Moreover,

$$\langle \lambda, A_I x - b_I \rangle = \sum_{i \in I} \lambda_i (\langle a^i, x \rangle - b_i)$$
$$\leq \left(\sum_{i \in I} \lambda_i \right) \max_{i=1, \cdots, m} (\langle a^i, x \rangle - b_i)$$

which yields the inequality of the corollary.   □

The max-norm $\|u\|_\infty$ of a vector $u \in \mathbb{R}^m$ is given by

$$\|u\|_\infty = \max\{|u_i| : i = 1, \cdots, m\}.$$

Using this norm we may give the error bound stated in Corollary 6.3.6 in the form

$$d(x, \Gamma) \leq \kappa \|(Ax - b)^+\|_\infty.$$

To express the error bound by the Euclidean norm we introduce a new constant

$$\beta := \min_{I \in \mathcal{I}, \lambda \in \mathbb{R}_+^{|I|} \cap S_{|I|}} \|A_I^T \lambda\|, \tag{6.15}$$

where $S_{|I|}$ denotes the unit sphere in $\mathbb{R}^{|I|}$. This value is strictly positive because otherwise $A_I^T \lambda = 0$ for some $\lambda \neq 0$ which contradicts the fact that A_I is of full rank.

Corollary 6.3.7 *Assume that the system (6.10) is consistent. Then*

$$d(x, \Gamma) \leq \frac{1}{\beta} \|(Ax - b)^+\|.$$

Proof The inequality is evident for $x \in \Gamma$. Consider $x \notin \Gamma$. Let λ, y and I be as in Theorem 6.3.3. Since λ is positive, we have

$$\frac{\langle \lambda, A_I x - b_I \rangle}{\|A_I^T \lambda\|} \leq \frac{\langle \lambda, (A_I x - b_I)^+ \rangle}{\|A_I^T \lambda\|}$$
$$\leq \frac{\|\lambda\|}{\|A_I^T \lambda\|} \|(A_I x - b_I)^+\|. \tag{6.16}$$

Observe also that

$$\|(A_I x - b_I)^+\| = d(A_I x - b_I, -\mathbb{R}_+^{|I|})$$
$$\leqq d(Ax - b, -\mathbb{R}_+^m).$$

Combining this with (6.16) and Theorem 6.3.4 we deduce that

$$d(x, \Gamma) \leqq \max_{I \in \mathcal{I}} \; \max_{\lambda \in \mathbb{R}_+^{|I|} \setminus \{0\}} \frac{\|\lambda\|}{\|A_I^T \lambda\|} \|A_I x - b_I\|$$
$$\leqq \frac{1}{\beta} d(Ax - b, -\mathbb{R}_+^m),$$

as requested.   □

Stability of linear systems

Given the system (6.10) we wish to compare it with a perturbed system

$$A'x - b' \leqq 0, \tag{6.17}$$

where A' is an $m \times n$-matrix and b' is an m-dimensional vector. The solution set of this system is denoted Γ'. To find a relation between Γ and Γ' we use the *Hausdorff distance* which is defined by the formula

$$h(\Gamma, \Gamma') = \max \left\{ \sup_{x \in \Gamma} d(x, \Gamma'); \; \sup_{x' \in \Gamma'} d(x', \Gamma) \right\}.$$

This value may be infinite when one of the two sets is unbounded.

Theorem 6.3.8 *Assume that the systems (6.10) and (6.17) are consistent and that $A = A'$. Then*

$$h(\Gamma, \Gamma') \leqq \kappa \|b' - b\|_\infty$$

where κ is given by (6.13).

Proof Let $x \in \Gamma$. In view of Theorem 6.3.3 there exist a point $y \in \Gamma'$, an index set $I \subseteq I(y)$ and $\lambda \in \mathbb{R}_+^{|I|}$ such that

$$d(x, \Gamma') = \|x - y\| = \frac{\langle \lambda, A_I x - b_I' \rangle}{\|A_I^T \lambda\|}.$$

Since $A_I x \leqq b_I$ and λ is positive, we deduce

$$d(x, \Gamma') \leqq \frac{\langle \lambda, b_I - b'_I \rangle}{\|A_I^T \lambda\|}$$
$$\leqq \kappa \|b'_I - b_I\|_\infty$$
$$\leqq \kappa \|b' - b\|_\infty.$$

The same argument works for $x' \in \Gamma'$ and leads to the inequality stated in the theorem. □

When the matrix A is also perturbed, an estimate for the Hausdorff distance between Γ and Γ' can be given when both of them are bounded.

Theorem 6.3.9 *Assume that the systems (6.10) and (6.17) are consistent and that both Γ and Γ' are contained in a ball of radius $r > 0$. Then*

$$h(\Gamma, \Gamma') \leqq \max\{\kappa, \kappa'\}(\|b' - b\|_\infty + r\|A' - A\|_\infty),$$

where κ' is given by (6.13) with A' substituting A.

Proof Let $x \in \Gamma$ and $y \in \Gamma'$ be as in the proof of the preceding theorem. We have

$$d(x, \Gamma') \leqq \frac{\langle \lambda, A'_I x - b'_I \rangle}{\|A'^T_I \lambda\|}$$
$$\leqq \frac{\langle \lambda, (A'_I - A_I)x + A_I x - b'_I \rangle}{\|A'^T_I \lambda\|}$$
$$\leqq \frac{\langle \lambda, (A'_I - A_I)x \rangle}{\|A'^T_I \lambda\|} + \frac{\langle \lambda, b_I - b'_I \rangle}{\|A'^T_I \lambda\|}$$
$$\leqq \kappa'(r\|A' - A\|_\infty + \|b' - b\|_\infty).$$

A similar inequality can be obtained for $d(x', \Gamma)$, $x' \in \Gamma'$ and completes the proof. □

When one of Γ and Γ' is bounded and the other is unbounded, the Hausdorff distance between them is clearly unbounded. However, if both of them are unbounded, it is not excluded that the Hausdorff distance between them is bounded. When it is the case an estimate for $h(\Gamma, \Gamma')$ similar to the one of Theorem 6.3.9 is available.

Lemma 6.3.10 *Assume that the systems (6.10) and (6.17) are consistent and both sets Γ and Γ' are unbounded. Then the Hausdorff distance between Γ and Γ' is finite if and only if the rows of A and A' generate the same positive hull.*

Proof According to the representation theorem (Theorem 2.4.9) the polyhedra Γ and Γ' can be decomposed by sums of bounded polyhedra and polyhedral cones, say $\Gamma = P + Q$ and $\Gamma' = P' + Q'$ with P and P' bounded polyhedra and Q and Q' polyhedral cones. The cones Q and Q' are determined respectively by the homogeneous systems $Ax \leqq 0$ and $A'x \leqq 0$. It is clear that the Hausdorff distance between Γ and Γ' is finite if and only if Q and Q' coincide, which means that the two above mentioned homogeneous systems share the same solution set. By Theorem 2.3.19 the polar cones of Q and Q' are positive hulls of the rows of A and A' respectively. Since $Q = Q'$ if and only if their polar cones coincide (Corollary 2.3.22), the conclusion of the theorem follows at once.   □

Corollary 6.3.11 *Assume that the systems (6.10) and (6.17) are consistent and that the rows of A and A' generate the same positive hulls. Then there exists a positive number $r > 0$ such that*

$$h(\Gamma, \Gamma') \leqq \max\{\kappa, \kappa'\}\big(\|b' - b\|_\infty + r\|A' - A\|_\infty\big),$$

where κ' is given by (6.13) with A' substituting A.

Proof Let Q denote the asymptotic cone of Γ, which, by the hypothesis, is also the asymptotic cone of Γ'. In view of Theorem 2.3.9, $\Gamma = P + Q$ and $\Gamma' = P' + Q$ with P and P' bounded polyhedra. Define r to be a positive number such that P and P' are contained in the ball of radius r. Let $x \in \Gamma$, say $x = u + v$ for some $u \in P$ and $v \in Q$. Let $y \in \Gamma'$ that realizes the distance $d(x, \Gamma')$ and $I \subseteq I(y)$ the index set as given in Theorem 6.3.3. We obtain

$$
\begin{aligned}
d(x, \Gamma') &\leqq \frac{\langle \lambda, A'_I(u + v) - b'_I \rangle}{\|A'^T_I \lambda\|} \\
&\leqq \frac{\langle \lambda, (A'_I - A_I)u + A'_I v + A_I u - b'_I \rangle}{\|A'^T_I \lambda\|} \\
&\leqq \frac{\langle \lambda, (A'_I - A_I)u \rangle}{\|A'^T_I \lambda\|} + \frac{\langle \lambda, b_I - b'_I \rangle}{\|A'^T_I \lambda\|} \\
&\leqq \kappa'(r\|A' - A\|_\infty + \|b' - b\|_\infty).
\end{aligned}
$$

To pass from the second inequality to the third one we have used the fact that λ is a positive vector, $A_I u \leqq b_I$, and $A'_I v \leqq 0$ because v is an asymptotic direction of Γ'.   □

Regular system

The solvability of the system (6.10) expresses the fact that the vector b belongs to the set $A(\mathbb{R}^n) + \mathbb{R}^m_+$. If b is on the boundary of that set, then there is a vector b' outside of that set as close to b as we wish. For such a b' and $A' = A$ the perturbed system (6.17) has no solution. On the other hand, if b is an interior point of the set

$A(\mathbb{R}^n) + \mathbb{R}^m_+$, then any b' sufficiently close to b still belongs to that set, and hence the perturbed system (6.17) with $A' = A$ is solvable.

Definition 6.3.12 The system $Ax - b \leqq 0$ is said to be regular if b is an interior point of the set $A(\mathbb{R}^n) + \mathbb{R}^m_+$.

The theorem below shows that regularity is necessary and sufficient for stability.

Theorem 6.3.13 *The system (6.10) is regular if and only if there is some $\eta > 0$ such that for every couple (A', b') with $\max(\|A - A'\|, \|b - b'\|) < \eta$ the system (6.17) has a solution.*

Proof If the system is irregular, then it is clear that the system (6.17) has no solution if $A' = A$ and b' is outside the set $A(\mathbb{R}^n) + \mathbb{R}^m_+$ and close to b as we have already discussed. Assume that the system is regular. We proceed by contradiction. If there are couples (A_ν, b_ν) converging to (A, b) such that the systems $A_\nu x - b_\nu \leqq 0$ have no solutions, then by Corollary 2.2.4 (a version of Farkas' theorem) there are nonzero vectors $\xi_\nu \geq 0$ such that

$$\langle b_\nu, \xi_\nu \rangle = -1$$
$$A_\nu^T \xi_\nu = 0.$$

We may assume that $\xi_\nu/\|\xi_\nu\|$ converges to some $\xi \neq 0$. Then $A^T \xi = 0$ and $\langle b, \xi \rangle \leqq 0$. Since the vector ξ is positive, the above equality yields that

$$\langle Ax + v, \xi \rangle \geqq 0 \text{ for all } x \in \mathbb{R}^n, v \in \mathbb{R}^m_+.$$

This shows b cannot be an interior point of the set $A(\mathbb{R}^n) + \mathbb{R}^m_+$ and contradicts the hypothesis.   □

Stability of multiobjective linear problems

Consider a multiobjective linear problem (MOLP)

$$\text{Maximize} \quad Cx$$
$$\text{subject to } Ax \leqq b$$

where C is a real $k \times n$-matrix, A is a real $m \times n$-matrix and b is an m-vector, and its dual (VD2)

$$\text{Minimize } Yb$$
$$\text{subject to } \lambda^T Y A = \lambda^T C$$
$$\lambda^T Y \geqq 0 \text{ for some } \lambda \in \mathbb{R}^k, \lambda > 0,$$

where Y is a real variable $k \times m$-matrix. We remember that the columns of the matrix C^T are denoted $c^1, \cdots, c^k$.

Definition 6.3.14 The constraint of (VD2) is said to be regular if the following conditions hold:

$$c^i \in A^T(\mathbb{R}^m_+), \ i = 1, \cdots, k;$$
$$c^1 + \cdots + c^k \in \text{int}(A^T(\mathbb{R}^m_+)).$$

It is plain to see that the constraint of (VD2) is regular if and only if for every vector $\lambda > 0$ the vector $C^T\lambda$ belongs to the interior of the set $A^T(\mathbb{R}^m_+)$. In this section $S(A, b, C)$ denotes the set of efficient solutions of (MOLP) and $S^*(A, b, C)$ the set of efficient solutions of (VD2).

Theorem 6.3.15 *Assume that b is nonzero. The following assertions are equivalent.*

  (i) *The constraints of (MOLP) and (VD2) are regular.*
 (ii) *The set of efficient solutions of (MOLP) is nonempty and bounded, and for every fixed vector $\lambda > 0$ of $\mathbb{R}^k$, the set of vectors $\lambda^T Y$ with Y being efficient solutions of (VD2) and satisfying the constraint $\lambda^T Y A = \lambda^T C$ is bounded.*
(iii) *There exists a positive number ε such that for any A', b', C' with $\max(\|A - A'\|, \|b - b'\|, \|C' - C\|) < \varepsilon$ the primal and dual problems (MOLP')*

$$\text{Maximize } C'x$$
$$\text{subject to } A'x \leqq b'$$

and (VD2*)

$$\text{Minimize } Yb'$$
$$\text{subject to } \lambda^T Y A' = \lambda^T C'$$
$$\lambda^T Y \geqq 0 \text{ for some } \lambda \in \mathbb{R}^k, \lambda > 0.$$

have efficient solutions.

Proof We prove implication from (i) to (ii) first. Assume (i) holds. For every vector $\lambda > 0$ we consider the scalarized problem (P$_\lambda$)

$$\text{Maximize } \langle \lambda, Cx \rangle$$
$$\text{subject to } Ax \leqq b$$

and its dual (D$_\lambda$)

$$\text{Minimize } \langle u, b \rangle$$
$$\text{subject to } u^T A = \lambda^T C$$
$$u \geqq 0.$$

Since both primal and dual problems have feasible solutions, they have optimal solutions. By scalarization (MOLP) has efficient solutions and in view of Theorem 5.3.5 the dual (VD2) has feasible solutions and efficient solutions too. We prove that the set of efficient solutions of (MOLP) is bounded. Indeed, if not there is some vector $\lambda > 0$ such that the set of optimal solutions of (P_λ) is unbounded. One finds some nonzero vector v with $Av \leqq 0$ such that

$$\langle \lambda, Cv \rangle = \lambda^T C v = 0. \tag{6.18}$$

Since $Av \leqq 0$, one also has that

$$\langle v, A^T u \rangle = \langle Av, u \rangle \leqq 0$$

for all $u \geq 0$. Moreover, as (VD2) is regular, the vector $C^T \lambda$ is in the interior of the set $\{A^T u : u \in \mathbb{R}^m_+\}$ which together with (6.18) implies that $v = 0$, a contradiction. Once the set of efficient solutions of (MOLP) is bounded, the set of optimal solutions of (P_λ) is bounded for all $\lambda > 0$. By symmetry the set of optimal solutions of (D_λ) is bounded too. Since the set $S^*(A, b, C)$ is included in the union of those matrices Y for which λY is an optimal solution of (D_λ) with $\lambda > 0$, we conclude that the set $\{\lambda^T Y : Y \in S^*(A, b, C)\}$ is bounded.

Next we prove that (ii) implies (iii). Assume (ii) and suppose the contrary that there are $(A_\nu, b_\nu, C_\nu), \nu = 1, 2, \cdots$ converging to (A, b, C) for which at least either the problem $(MOLP_\nu)$

$$\text{Maximize } C_\nu x$$
$$\text{subject to } A_\nu x \leqq b_\nu$$

or the dual problem $(VD2_\nu)$

$$\text{Minimize } Y b_\nu$$
$$\text{subject to } \lambda^T Y A_\nu = \lambda^T C_\nu$$
$$\lambda^T Y \geqq 0 \text{ for some } \lambda \in \mathbb{R}^k, \lambda > 0$$

is not solvable. Actually both of the problems are not solvable according to the duality relation given in Theorem 5.3.5. We may then assume that either $(MOLP_\nu)$, $\nu = 1, 2, \cdots$ are all infeasible, or they are feasible and their objective functions are unbounded from above. The first case does not happen because (ii) implies that (MOLP) is regular and so $(MOLP_\nu)$ is feasible when ν is sufficiently large. In the second case there exist vectors v^ν such that $A_\nu v^\nu \leqq 0$ and $C_\nu v^\nu \geq 0$. Without loss of generality we may assume that v^ν converges to some nonzero vector v which satisfies $Cv \geq 0$ and $Av \leqq 0$. Let x be any efficient solution of (MOLP). Then for every positive number t, we have $C(x + tv) \geq Cx$ and $A(x + tv) \leqq b$, which shows that the set of efficient solutions of (MOLP) is unbounded. This contradiction proves that (ii) implies (iii).

We proceed to the final implication from (iii) to (i). Under (iii), the problems (MOLP') are feasible. In view of Theorem 6.3.13 the constraint of (MOLP) is regular. We argue that the constraint of (VD2) is regular too. Indeed, if not, there exists $\lambda > 0$ such that the vector $C^T \lambda$ is not an interior point to the image $A^T(\mathbb{R}^m_+)$. Then, we may choose a matrix C' as close to C as we wish such that $C'^T \lambda$ does not belong

to $A^T (\mathbb{R}^m_+)$. Consequently, there is no matrix Y satisfying the constraint of (VD2')
with $A' = A$, $b' = b$, and so (VD2') is infeasible, a contradiction. $\qquad\qquad\square$

We remark that in (ii) of Theorem 6.3.15 the set of efficient solutions of (VD2) is
not necessarily bounded.

Example 6.3.16 We consider a linear problem of type (MOLP) with

$$C = \begin{pmatrix} 1 & 0 \\ 0 & 1 \end{pmatrix}, \quad A = \begin{pmatrix} 0 & 1 \\ 1 & 1 \\ 1 & 0 \end{pmatrix}, \quad b = \begin{pmatrix} 1 \\ 1 \\ 1 \end{pmatrix}.$$

The dual (VD2) is given by

$$\text{Minimize} \begin{pmatrix} y_1 & y_2 & y_3 \\ y_4 & y_5 & y_6 \end{pmatrix} \begin{pmatrix} 1 \\ 1 \\ 1 \end{pmatrix}$$

$$\text{subject to } (\lambda_1, \lambda_2) \begin{pmatrix} y_1 & y_2 & y_3 \\ y_4 & y_5 & y_6 \end{pmatrix} \begin{pmatrix} 0 & 1 \\ 1 & 1 \\ 1 & 0 \end{pmatrix} = (\lambda_1, \lambda_2) \begin{pmatrix} 1 & 0 \\ 0 & 1 \end{pmatrix}$$

$$(\lambda_1, \lambda_2) \begin{pmatrix} y_1 & y_2 & y_3 \\ y_4 & y_5 & y_6 \end{pmatrix} \geq 0$$
$$\text{for some } \lambda_1 > 0, \lambda_2 > 0.$$

It is clear that the set of efficient solutions of (MOLP) consists of the vectors $\begin{pmatrix} x_1 \\ x_2 \end{pmatrix} \in$
$\mathbb{R}^2$ satisfying

$$x_1 + x_2 = 1, \quad x_1 \geq 0 \text{ and } x_2 \geq 0.$$

The constraint of (MOLP) is regular because the vector b lies in the interior of the
set $A(\mathbb{R}^2) + \mathbb{R}^3_+$ which evidently contains $\mathbb{R}^3_+$. Moreover, the set $A^T (\mathbb{R}^3_+$ contains
$\mathbb{R}^2_+$, and so it contains also the column vectors of the matrix C^T, and the sum of the
two columns lies in its interior. By this (VD2) is regular.

Let us fix a vector $\lambda = \begin{pmatrix} 1 \\ 1 \end{pmatrix}$ and find dual efficient solutions that satisfy the
system

$$\lambda^T Y A = \lambda^T C$$
$$\lambda^T Y \geq 0$$
$$Yb = C\bar{x}$$

where $\bar{x} = \begin{pmatrix} 1 \\ 0 \end{pmatrix}$ is an efficient solution of (MOLP) that solves (P_λ). Here is a system
for the entries of Y:

$$y_2 + y_3 + y_5 + y_6 = 1$$
$$y_1 + y_2 + y_4 + y_5 = 1$$
$$y_1 + y_4 \geq 0$$
$$y_2 + y_5 \geq 0$$
$$y_3 + y_6 \geq 0$$
$$y_1 + y_2 + y_3 = 1$$
$$y_4 + y_5 + y_6 = 0.$$

We deduce from this system the matrices Y efficient solutions of (VD2) in the form

$$Y = \begin{pmatrix} t & s & 1-t-s \\ -t & -s & t+s \end{pmatrix}$$

with real numbers t and s. It is clear that the set of these Y is unbounded. However, the set of λY for all Y above is bounded, which is actually a singleton $(0, 0, 1)^T$.

Weak sharp maxima

We consider the problem (MOLP) as in the preceding subsection. The feasible solution set is given by $\Gamma = \{x \in \mathbb{R}^n : Ax \leq b\}$. In this subsection we wish to compare the distance from a feasible solution x to a given set of efficient solutions with the distance between their images by C.

Definition 6.3.17 A nonempty set of efficient solutions F of (MOLP) is called a set of weak sharp maxima if there exists a strictly positive number r such that

$$d(x, F) \leq r\, d(Cx, C(F)) \text{ for all } x \in \Gamma. \tag{6.19}$$

We note that not every set of efficient solutions is a set of weak sharp maxima. For instance when C is not injective on the efficient solution set of (MOLP), say $Cx = Cx'$ for two distinct efficient solutions x and x', the set $F = \{x\}$ is not a set of weak sharp maxima for, $d(Cx, Cx') = 0$ and $d(x, x') > 0$.

Lemma 6.3.18 Let $F_1, \cdots, F_p$ be nonempty sets of weak sharp maxima. Then their union $F_1 \cup \cdots \cup F_p$ is a set of weak sharp maxima too.

Proof Let r_i be a strictly positive constant with which (6.19) is satisfied for F_i, $i = 1, \cdots, p$. Set $r = \max\{r_i : i = 1, \cdots, p\}$. For every $x \in \Gamma$ there is some $j \in \{1, \cdots, p\}$ such that

$$d(Cx, C(F_1 \cup \cdots \cup F_p)) = d(Cx, C(F_j)).$$

From (6.19) we obtain

$$d(x, F_1 \cup \cdots \cup F_p) \leqq d(x, F_j)$$
$$\leqq r_j d(Cx, C(F_j))$$
$$\leq rd(Cx, C(F_1 \cup \cdots \cup F_p)),$$

which proves that $F_1 \cup \cdots \cup F_p$ is a set of weak sharp maxima.   □

Theorem 6.3.19 *Every efficient face of (MOLP) is a set of weak sharp maxima. Consequently, the efficient solution set of (MOLP) is a set of weak sharp maxima too.*

Proof Let E be an efficient face of (MOLP). Let $\bar{x}$ be a relative interior point of E. In view of Theorem 4.3.1 there is a strictly positive vector $\lambda \in \mathbb{R}^k$ with $\|\lambda\| = 1$ such that $\bar{x}$ solves the scalarized problem

$$\text{maximize } \langle \lambda, Cx \rangle$$
$$\text{subject to } x \in \Gamma$$

and E is its optimal solution set. It is clear that E is the solution set to the system

$$Ax \leqq b$$
$$-\lambda^T Cx \leqq -\lambda^T C\bar{x}.$$

Given $x \in \Gamma$, let $x^0 \in C(E)$ such that $d(Cx, C(E)) = \|Cx - Cx^0\|$. Then $\langle \lambda, Cx^0 \rangle = \langle \lambda, C\bar{x} \rangle$. In view of Corollary 6.3.6 there exists a strictly positive number r such that

$$d(x, E) \leq r \left\| \left[\begin{pmatrix} A \\ -\lambda^T C \end{pmatrix} x - \begin{pmatrix} b \\ -\lambda^T C\bar{x} \end{pmatrix} \right]^+ \right\|$$
$$\leqq r| - \lambda^T Cx + \lambda^T C\bar{x}|$$
$$\leqq r| - \lambda^T Cx + \lambda^T Cx^0|$$
$$\leqq rd(Cx, C(E)).$$

Because $-\lambda^T C\bar{x} = -\lambda^T C\bar{x}'$ for all $\bar{x}$ and $\bar{x}'$ from E, we deduce that r_E is independent upon $x \in \Gamma$. Thus, E is a set of weak maxima.

The last conclusion of the theorem follows from the first statement and from the fact that the efficient solution set of (MOLP) consists of finite number of efficient faces (Theorem 4.3.8).   □

The same argument of proof shows that weakly efficient faces and the weakly efficient solution set of (MOLP) are sets of weak sharp maxima.

6.4 Post-optimal Analysis

Among efficient solutions of (MOLP) those that are sensitive to perturbations in the data are often not useful in practical application. Therefore it is important to determine less sensitive efficient solutions and conditions under which a given efficient solution is still efficient when the objective matrix C undergoes a perturbation, an addition or removal of some of its rows.

Extension and reduction of the objective matrix

We assume throughout this paragraph that $\overline{x}$ is a fixed feasible solution of (MOLP). We now add some objective row vectors to the objective matrix C by considering the following extended problem, denoted (MOLP)^D:

$$\text{Maximize } \begin{pmatrix} C \\ D \end{pmatrix} x$$
$$\text{subject to } Ax \leqq b,$$

where D is a $k' \times n$ matrix. Two problems associated to the matrix D are also considered. They are denoted respectively by (MOLP1) and (MOLP2) and given below:

$$\text{Maximize } Dx$$
$$\text{subject to } Ax \leqq b$$
$$Cx \geqq C\overline{x}$$

and

$$\text{Maximize } Dx$$
$$\text{subject to } Ax \leqq b$$
$$Cx = C\overline{x}.$$

The problem (MOLP) is considered as a reduction of the problem (MOLP)^D by removing the rows of the matrix D. Here is a relationship between the problems (MOLP), (MOLP)^D, (MOLP1) and (MOLP2).

Theorem 6.4.1 *Given a feasible solution $\overline{x}$ of (MOLP), the following hold.*

(i) *If $\overline{x}$ is an efficient solution of (MOLP)^D, then it is an efficient solution of (MOLP1).*

(ii) *If $\overline{x}$ is an efficient solution of (MOLP1), then it is an efficient solution of (MOLP2).*

(iii) *If $\bar{x}$ is an efficient solution of (MOLP) and (MOLP2), then it is an efficient solution of (MOLP)^D.*

(iv) *If $\bar{x}$ is an efficient solution of (MOLP)^D and $D\bar{x} \leq Dx$ for all feasible solutions of (MOLP1), then it is an efficient solution of (MOLP).*

Proof To prove (i) let x be a feasible solution of (MOLP1) that satisfies $Dx \geqq D\bar{x}$. Then it is also a feasible solution of (MOLP)^D and satisfies $\begin{pmatrix} C \\ D \end{pmatrix} x \geqq \begin{pmatrix} C \\ D \end{pmatrix} \bar{x}$. Since $\bar{x}$ solves (MOLP)^D, the latter inequality is equality. In particular $Dx = D\bar{x}$ and shows that $\bar{x}$ is an efficient solution of (MOLP1).

For (ii) it suffices to observe that the feasible set of (MOLP2) is a subset of the feasible set of (MOLP1). Hence any feasible solution of (MOLP2) which solves (MOLP1) is also an efficient solution of (MOLP2).

To prove (iii) let x be a feasible solution of (MOLP)^D satisfying $\begin{pmatrix} C \\ D \end{pmatrix} x \geqq \begin{pmatrix} C \\ D \end{pmatrix} \bar{x}$. Then x is a feasible solution of (MOLP) and by hypothesis $Cx = C\bar{x}$. Consequently x is a feasible solution of (MOLP2). It follows from the hypothesis that $Dx = D\bar{x}$. Thus, we have $\begin{pmatrix} C \\ D \end{pmatrix} x = \begin{pmatrix} C \\ D \end{pmatrix} \bar{x}$, by which $\bar{x}$ solves (MOLP)^D.

Finally, let x be a feasible solution of (MOLP) and satisfy $Cx \geqq C\bar{x}$. Then it is a feasible solution of (MOLP1). By hypothesis $Dx \geqq D\bar{x}$. We have then $\begin{pmatrix} C \\ D \end{pmatrix} x \geqq \begin{pmatrix} C \\ D \end{pmatrix} \bar{x}$, which in fact is equality because $\bar{x}$ solves (MOLP)^D. In particular $Cx = C\bar{x}$, by which $\bar{x}$ solves (MOLP). $\qquad\square$

Corollary 6.4.2 *If $D = YC$ for some $k' \times k$-matrix Y, then every efficient solution of (MOLP) is efficient solution of (MOLP)^D. Conversely, if $D = YC$ where the entries of Y are all positive or zero, then every efficient solution of (MOLP)^D is efficient solution of (MOLP).*

Proof Assume that $\bar{x}$ is an efficient solution of (MOLP). For every feasible solution x of (MOLP2) we have $Cx = C\bar{x}$ and deduce $Dx = D\bar{x}$, which implies that $\bar{x}$ is also an efficient solution of (MOLP2). By Theorem 6.4.1 (iii), $\bar{x}$ is an efficient solution of (MOLP)^D. Conversely, if $\bar{x}$ is an efficient solution of (MOLP)^D, then for every feasible solution of (MOLP1) one has $Cx \geqq C\bar{x}$, implying $Dx \geqq D\bar{x}$. By Theorem 6.4.1 (iv), $\bar{x}$ is an efficient solution of (MOLP). $\qquad\square$

Convex combinations of two objective matrices

When the matrix D has the same dimension as C we wish to know whether an efficient solution of (MOLP) is still efficient for the problem whose objective function is a convex combination of C and D:

$$\text{Maximize } \big(\alpha C + (1 - \alpha)D\big)x$$
$$\text{subject to } Ax \leq b$$

with some $\alpha \in (0, 1)$. This problem is denoted by (MOLP_α) which coincides with (MOLP) when $\alpha = 1$.

Theorem 6.4.3 *Assume that $\bar{x}$ is an efficient solution of (MOLP) and (MOLP_0), and that there are a vector $d \in \mathbb{R}^n$ and real numbers of the same sign $p_1, \cdots, p_k$ such that the columns of C^T and D^T satisfy $d^i = c^i + p_i d, i = 1, \cdots, k$. Then $\bar{x}$ is an efficient solution of (MOLP_α) for all $\alpha \in [0, 1]$.*

Proof According to Theorem 4.2.6 there are strictly positive numbers $\lambda_1, \cdots, \lambda_k$, $\mu_1, \cdots, \mu_k$ such that the vectors

$$v^1 = \sum_{i=1}^{k} \lambda_i c^i$$

$$v^2 = \sum_{i=1}^{k} \mu_i d^i$$

belong to the normal cone to the feasible set of (MOLP) at $\bar{x}$. Without loss of generality we may assume that at least one of p_i is nonzero. Let $\alpha \in (0, 1)$. Define t and γ_i to be real numbers

$$t = \frac{\alpha \langle \mu, p \rangle}{\alpha \langle \mu, p \rangle + (1 - \alpha) \langle \lambda, p \rangle}$$
$$\gamma_i = t\lambda_i + (1 - t)\mu_i, i = 1, \cdots, k.$$

It is clear that $0 < t < 1$ and $\gamma_i, i = 1, \cdots, k$ are strictly positive. We also have

$$
\begin{aligned}
(1 - t)\langle \mu, p \rangle &= (1 - \alpha) \frac{\langle \lambda, p \rangle \langle \mu, p \rangle}{\alpha \langle \mu, p \rangle + (1 - \alpha \langle \lambda, p \rangle}\\
&= (1 - \alpha)(t\langle \lambda, p \rangle + (1 - t)\langle \mu, p \rangle)\\
&= (1 - \alpha)\langle \gamma, p \rangle.
\end{aligned}
\tag{6.20}
$$

Let us compute the image of the vector γ under $(\alpha C + (1 - \alpha)D)^T$:

$$
\begin{aligned}
(\alpha C + (1 - \alpha)D)^T \gamma &= \sum_{i=1}^{k} \left(\alpha c^i + (1 - \alpha)d^i \right) \gamma_i\\
&= \sum_{i=1}^{k} c^i \gamma_i + \sum_{i=1}^{k} (1 - \alpha)\gamma_i p_i d\\
&= \sum_{i=1}^{k} \left(t\lambda_i c^i + (1 - t)\mu_i c^i \right) + (1 - \alpha)\langle \gamma, p \rangle d.
\end{aligned}
$$

The latter equality and (6.20) yield

$$(\alpha C + (1 - \alpha)D)^T \gamma = t \sum_{i=1}^{k} \lambda_i c^i + (1 - t) \sum_{i=1}^{k} (\mu_i c^i + \mu_i p_i d)$$

$$= t v^1 + (1 - t) v^2.$$

Since the normal cone to the feasible set at $\bar{x}$ is convex, the above equality shows that the vector $(\alpha C + (1 - \alpha)D)^T \gamma$ is a normal vector at $\bar{x}$. It remains to apply Theorem 4.2.6 to conclude that $\bar{x}$ is an efficient solution of the problem (MOLP_α).   □

Under the hypothesis of Theorem 6.4.3 the matrix $D - C$ has at most rank one. In the next examples we shall see that the conclusion is no more available when $D - C$ has a rank greater than one, or when it has rank one, but the coefficients p_i are of different signs.

Example 6.4.4 We consider the following two linear problems denoted respectively by (MOLP1) and (MOLP2):

$$\text{Maximize } \begin{pmatrix} 0 & 0 & 1 \\ 1 & 1 & 0 \end{pmatrix} \begin{pmatrix} x_1 \\ x_2 \\ x_3 \end{pmatrix}$$

$$\text{subject to } \begin{pmatrix} 0 & 1 & 0 \\ 1 & 1 & 1 \\ 1 & 0 & 0 \\ 0 & 0 & 1 \end{pmatrix} \begin{pmatrix} x_1 \\ x_2 \\ x_3 \end{pmatrix} \leqq \begin{pmatrix} 1 \\ 1 \\ 1 \\ 1 \end{pmatrix}$$

and

$$\text{Maximize } \begin{pmatrix} 0 & 2 & 1 \\ 3/2 & 1/2 & 1 \end{pmatrix} \begin{pmatrix} x_1 \\ x_2 \\ x_3 \end{pmatrix}$$

$$\text{subject to } \begin{pmatrix} 0 & 1 & 0 \\ 1 & 1 & 1 \\ 1 & 0 & 0 \\ 0 & 0 & 1 \end{pmatrix} \begin{pmatrix} x_1 \\ x_2 \\ x_3 \end{pmatrix} \leqq \begin{pmatrix} 1 \\ 1 \\ 1 \\ 1 \end{pmatrix}.$$

The objective functions of these problems are denoted respectively by C and D. We choose $\bar{x} = \frac{1}{3}(1, 1, 1)^T$. It is a feasible solution of the two problems. The active index set at this solution is $I(\bar{x}) = \{2\}$ and the normal cone to the feasible set at it is the cone $\text{pos}\{(1, 1, 1)^T\}$. For $\lambda = \begin{pmatrix} 1 \\ 1 \end{pmatrix}$ and $\mu = \begin{pmatrix} 1/3 \\ 2/3 \end{pmatrix}$ we have

$$C^T \lambda = \begin{pmatrix} 1 \\ 1 \\ 1 \end{pmatrix} \text{ and } D^T \mu = \begin{pmatrix} 1 \\ 1 \\ 1 \end{pmatrix}$$

which belong to the normal cone to the feasible set at $\bar{x}$. By Theorem 4.2.6 the solution is efficient for both (MOLP1) and (MOLP2). Let us now consider a problem with a convex combination of C and D:

$$\text{Maximize} \left[\tfrac{1}{2} \begin{pmatrix} 0 & 0 & 1 \\ 1 & 1 & 0 \end{pmatrix} + \tfrac{1}{2} \begin{pmatrix} 0 & 2 & 1 \\ 3/2 & 1/2 & 1 \end{pmatrix} \right] \begin{pmatrix} x_1 \\ x_2 \\ x_3 \end{pmatrix}$$

$$\text{subject to} \begin{pmatrix} 0 & 1 & 0 \\ 1 & 1 & 1 \\ 1 & 0 & 0 \\ 0 & 0 & 1 \end{pmatrix} \begin{pmatrix} x_1 \\ x_2 \\ x_3 \end{pmatrix} \leqq \begin{pmatrix} 1 \\ 1 \\ 1 \\ 1 \end{pmatrix}.$$

We observe that $\bar{x}$ no longer solves this problem obtained by a convex combination. This is because for any strictly positive vector λ in $\mathbb{R}^2$ the vector $\left(\tfrac{1}{2}C + \tfrac{1}{2}D\right)^T \lambda$ is a normal vector to the feasible set at $\bar{x}$ if and only if the components of λ satisfy the system

$$\frac{5}{4}\lambda_2 = \lambda_1 + \frac{3}{4}\lambda_2 = \lambda_1 + \frac{1}{2}\lambda_2,$$

which evidently has no solution. In view of Theorem 4.2.6 the solution $\bar{x}$ is not efficient.

Example 6.4.5 In this example we consider two linear problems denoted (MOLP1) and (MOLP2) as follows

$$\text{Maximize} \begin{pmatrix} 3 & 1 & 2 \\ 0 & 1 & 2 \end{pmatrix} \begin{pmatrix} x_1 \\ x_2 \\ x_3 \end{pmatrix}$$

$$\text{subject to} \begin{pmatrix} 2 & 1 & 2 \\ 1 & 2 & 2 \\ 1 & 0 & 0 \\ 0 & 1 & 0 \\ 0 & 0 & 1 \end{pmatrix} \begin{pmatrix} x_1 \\ x_2 \\ x_3 \end{pmatrix} \leqq \begin{pmatrix} 6 \\ 6 \\ 3 \\ 3 \\ 3 \end{pmatrix}$$

and

$$\text{Maximize} \begin{pmatrix} 3 & -2 & 2 \\ 0 & 4 & 2 \end{pmatrix} \begin{pmatrix} x_1 \\ x_2 \\ x_3 \end{pmatrix}$$

$$\text{subject to} \begin{pmatrix} 2 & 1 & 2 \\ 1 & 2 & 2 \\ 1 & 0 & 0 \\ 0 & 1 & 0 \\ 0 & 0 & 1 \end{pmatrix} \begin{pmatrix} x_1 \\ x_2 \\ x_3 \end{pmatrix} \leqq \begin{pmatrix} 6 \\ 6 \\ 3 \\ 3 \\ 3 \end{pmatrix}.$$

The objective matrices of these problems are denoted by C and D. Choose a feasible solution $\bar{x} = (1, 1, 3/2)^T$. The active index set at $\bar{x}$ is given by $I(\bar{x} = \{1, 2\}$, and the normal cone to the feasible set at it is the cone $\mathrm{pos}\{(2, 1, 2)^T, (1, 2, 2)^T\}$. Hence it is an efficient solution for both (MOLP1) and (MOLP2) because with $\lambda = \begin{pmatrix} 2/3 \\ 1/3 \end{pmatrix}$ and $\mu = \begin{pmatrix} 1/3 \\ 2/3 \end{pmatrix}$ we have

$$C^T \lambda = \begin{pmatrix} 2 \\ 1 \\ 2 \end{pmatrix} \text{ and } D^T \mu = \begin{pmatrix} 1 \\ 2 \\ 2 \end{pmatrix}$$

which both belong to the normal cone to the feasible set at $\bar{x}$. For the linear problem whose objective function is the convex combination

$$\frac{1}{2}C + \frac{1}{2}D = \begin{pmatrix} 3 & -1/2 & 2 \\ 0 & 5/2 & 2 \end{pmatrix}$$

the feasible solution $\bar{x}$ is not efficient because the system

$$3\lambda_1 = 2\mu_1 + \mu_2$$
$$-\frac{1}{2}\lambda_1 + \frac{5}{2}\lambda_2 = \mu_1 + 2\mu_2$$
$$2\lambda_1 + 2\lambda_2 = 2\mu_1 + 2\mu_2$$
$$\lambda_1, \lambda_2 > 0, \quad \mu_1, \mu_2 \geqq 0$$

which characterize the efficiency of $\bar{x}$ (Theorem 4.2.6), has no solutions. We notice that the rows of the objective matrices C and D are linked by equalities

$$d^1 = c^1 + d$$
$$d^2 = c^2 - d$$

where $d = (0, -3, 0)^T$. This proves that the conclusion of Theorem 6.4.3 is not valid when the coefficients $p_1, \cdots, p_k$ are of different sign.

Robust efficient solutions

In line with change of the objective function we are particularly interested in the efficient solutions which remain efficient when the objective matrix are slightly perturbed.

Definition 6.4.6 Let $\bar{x}$ be an efficient solution of (MOLP). It is said to be robust if there is a positive ε such that it is an efficient solution of the problem (MOLP')

$$\text{Maximize } C'x$$
$$\text{subject to } Ax \leqq b$$

for every C' with $\|C - C'\| < \varepsilon$.

The concept of robustness should be formulated for weakly efficient solutions in the same manner. However, it is quite useless because weakly efficient solutions are never robust. In fact, let $\bar{x}$ be a weakly efficient solution of (MOLP) which is not efficient. Let x be a feasible solution such that $Cx \geq C\bar{x}$. Denote by J the set of indices i such that $\langle c^i, x \rangle = \langle c^i, \bar{x} \rangle$. It is clear that J is nonempty. Given any $\varepsilon > 0$, there are vectors c'^j with $\|c'^j - c^j\| < \varepsilon/n$, for all $j \in J$ such that $\langle c'^j, x \rangle > \langle c'^j, \bar{x} \rangle$, $j \in J$. Then for the matrix C' obtained from C by substituting c^j by c'^j for $j \in J$ satisfies $\|C' - C\| < \varepsilon$ and $C'x > C\bar{x}$, showing that $\bar{x}$ is no more a weakly efficient solution of (MOLP').

It is clear that robust solutions exist. For instance when the feasible solution set is a singleton, it is a robust solution whatever the objective function be. We observe also that not every problem having efficient solutions has robust solutions. This is illustrated by the next example.

Example 6.4.7 We consider the following linear problem:

$$\text{Maximize } \begin{pmatrix} 1 & 1 \\ 1 & 1 \end{pmatrix} \begin{pmatrix} x_1 \\ x_2 \end{pmatrix}$$
$$\text{subject to } x_1 + x_2 \leqq 0$$
$$x_2 \leqq 0.$$

Then the efficient solution set is composed of vectors $(x_1, x_2)^T$, solutions to the system $x_1 + x_2 = 0$ and $x_2 \leqq 0$. It is clear that no element from this solution set is robust.

To prove a necessary and sufficient condition for robust efficient solutions we need the following lemmas.

Lemma 6.4.8 *Let $a^1, \cdots, a^k$ be given in $\mathbb{R}^n$. Then a vector x belongs to the relative interior of the convex hull of the family $\{a^1, \cdots, a^k\}$ if and only if there exist strictly positive numbers $\lambda_1, \cdots, \lambda_k$ with $\sum_{i=1}^k \lambda_i = 1$ such that*

$$x = \lambda_1 a^1 + \cdots + \lambda_k a^k.$$

Proof Let x be a relative interior point of $P := \text{co}\{a^1, \cdots, a^k\}$ and $x = \lambda_1 a^1 + \cdots + \lambda_k a^k$ for some positive numbers $\lambda_1, \cdots, \lambda_k$ with $\sum_{i=1}^k \lambda_i = 1$. If all coefficients λ_i are strictly positive, we are done. If not, say $\lambda_j = 0$ for some $j \in \{1, \cdots, k\}$. Because x is a relative interior point, there exists some $\delta > 0$ such that $y := x + \delta(x - a^j) \in P$.

Let $y = \alpha_1 a^1 + \cdots + \alpha_k a^k$ for some positive numbers $\alpha_1, \cdots, \alpha_k$ with $\sum_{i=1}^{k} \alpha_i = 1$. Then $2x + \delta x = x + y + \delta a^j$ implying

$$x = \frac{1}{2+\delta}(x + y + \delta a^j)$$

$$= \frac{1}{2+\delta}\left(\sum_{i=1}^{k}(\lambda_i + \alpha_i)a^i + \delta a^j\right).$$

The latter expression is a new convex combination for x, in which the number of strictly positive coefficients increases at least by one. Continue this procedure to arrive at a final convex combination in which all coefficients are strictly positive.

Conversely, assume $x = \lambda_1 a^1 + \cdots + \lambda_k a^k$ with $\lambda_1, \cdots, \lambda_k$ being strictly positive numbers and $\sum_{i=1}^{k} \lambda_i = 1$. Let y be any element of P, say $y = \alpha_1 a^1 + \cdots + \alpha_k a^k$ for some positive numbers $\alpha_1, \cdots, \alpha_k$ with $\sum_{i=1}^{k} \alpha_i = 1$. Then there is some $\delta > 0$ sufficiently small such that $\lambda_i + \delta(\lambda_i - \alpha_i) > 0, i = 1, \cdots, k$. We deduce

$$x + \delta(x - y) = \sum_{i=1}^{k}\left(\lambda_i + \delta(\lambda_i - \alpha_i)\right)a^i$$

which belongs to P. Hence x is a relative interior point of P.   □

Lemma 6.4.9 *Let P be a polytope in $\mathbb{R}^n$ and $a \notin P$. Then the relative interior of the convex hull $\mathrm{co}\{a, P\}$ consists of convex combinations $ta + (1-t)p$ for $p \in \mathrm{ri}(P)$ and $0 < t < 1$.*

Proof Because P is a polytope, it is the convex hull of its vertices by Corollary 2.3.8, say $P = \mathrm{co}\{a^1, \cdots, a^k\}$ for some $a^1, \cdots, a^k \in \mathbb{R}^n$. Then

$$\mathrm{co}\{a, P\} = \mathrm{co}\{a, a^1, \cdots, a^k\}.$$

In view of Lemma 6.4.8, the relative interior of $\mathrm{co}\{a, P\}$ consists of x such that

$$x = \lambda a + \sum_{i=1}^{k}\lambda_i a^i$$

with $\lambda > 0, \lambda_i > 0, i = 1, \cdots, k$ and $\sum_{i=1}^{k} \lambda_i = 1$. Set

$$p = \sum_{i=1}^{k}\frac{\lambda_i}{\sum_{i=1}^{k}\lambda_i}a^i.$$

Then, again due to Lemma 6.4.8, p is a relative interior point of P. We have then $x = \lambda a + (1 - \lambda)p$ with $\lambda \in (0, 1)$ and $p \in \mathrm{ri}(P)$ as requested.   □

Lemma 6.4.10 *Let $d^1, \cdots, d^k$ be nonzero vectors of $\mathbb{R}^n$. Then a vector d belongs to the relative interior of the cone $pos\{d^1, \cdots, d^k\}$ if and only if there exist strictly positive numbers $\lambda_1, \cdots, \lambda_k$ such that*

$$d = \lambda_1 d^1 + \cdots + \lambda_k d^k.$$

Proof We proceed by induction on k. For $k = 1$ it is clear that d belongs to the relative interior of $pos\{d^1\}$ if and only if $d = \lambda_1 d^1$ for some $\lambda_1 > 0$. Suppose that for a fixed $k \geq 1$ the conclusion of the lemma is true for any $d^i, i = 1, \cdots, k$ and d in $\mathbb{R}^n$. Let $d^1, \cdots, d^{k+1}$ be given and let $d \in \mathbb{R}^n$. To prove the "only if" part, we assume that d is a relative interior point of the cone $pos\{d^1, \cdots, d^{k+1}\}$. We distinguish two cases: (a) the vector d^{k+1} belongs to the cone $pos\{d^1, \cdots, d^k\}$, and (b) the vector d^{k+1} is outside of that cone. In the first case there are some positive numbers $t_1, \cdots, t_k$ not all zero such that

$$d^{k+1} = t_1 d^1 + \cdots + t_k d^k. \tag{6.21}$$

Then, if d is a relative interior point of $pos\{d^1, \cdots, d^{k+1}\}$, it is also a relative interior point of $pos\{d^1, \cdots, d^k\}$. By induction, there are strictly positive numbers $\mu_1, \cdots, \mu_k$ such that

$$d = \mu_1 d^1 + \cdots + \mu_k d^k. \tag{6.22}$$

Choose $\lambda_{k+1} > 0$ so small that $\lambda_i := \mu_i - \lambda_{k+1} t_i > 0$ for all $i = 1, \cdots, k$. It follows from (6.21) and (6.22) that

$$d = (\mu_1 - \lambda_{k+1}t_1)d^1 + \cdots + (\mu_k - \lambda_{k+1}t_k)d^k + \lambda_{k+1}(t_1 d^1 + \cdots + t_k d^k)$$
$$= \lambda_1 d^1 + .. + \lambda_k d^k + \lambda_{k+1} d^{k+1}$$

with all λ_i strictly positive as requested.

In the case (b), we observe without difficulty that

$$pos\{d^1, \cdots, d^{k+1}\} = \bigcup_{t \geq 0} t \, co\{0, d^1, \cdots, d^{k+1}\}$$

$$ri\left(pos\{d^1, \cdots, d^{k+1}\}\right) = \bigcup_{t > 0} t \, ri\left(co\{0, d^1, \cdots, d^{k+1}\}\right).$$

Hence there exist $t > 0$ such that d is a relative interior point of the set

$$t co\{0, d^1, \cdots, d^{k+1}\} = co\{t d^{k+1}, P\}$$

with $P = co\{0, t d^1, \cdots, t d^k\}$. Apply Lemma 6.4.9 to obtain a strict positive number $s \in (0, 1)$ and a relative interior point u of P such that

$$d = s t d^{k+1} + (1 - s)u.$$

By induction, there are strictly positive numbers $\mu_1, \cdots, \mu_k$ such that

$$u = \mu_1 d^1 + \cdots + \mu_k d^k.$$

Setting $\lambda_i = (1 - s)\mu_i$ for $i = 1, \cdots, k$ and $\lambda_{k+1} = st$ we deduce $d = \lambda_1 d^1 + \cdots + \lambda_{k+1} d^{k+1}$ with all λ_i strictly positive.

Conversely, assume that d is a strictly positive combination of $d^1, \cdots, d^{k+1}$, say $d = \lambda_1 d^1 + \cdots + \lambda_{k+1} d^{k+1}$ with $\lambda_i > 0$ for all $i = 1, \cdots, k + 1$. By induction the vector $u := \lambda_1 d^1 + \cdots + \lambda_k d^k$ belongs to the relative interior of the cone $\text{pos}\{d^1, \cdots, d^k\}$, hence it is a relative interior point of the polytope tP for some $t > 0$. In view of Lemma 6.4.9 the point $(u + \lambda_{k+1} d^{k+1})/(1 + \lambda_{k+1})$ is a relative interior point of the set $\text{co}\{d^{k+1}, tP\}$, and so d is a relative interior point of $\text{pos}\{d^1, \cdots, d^{k+1}\}$.   □

In the next theorem "span(.)" stands for the subspace spanned by the vectors in the parentheses. We adopt a convention that the span and the positive hull of the empty set are the trivial set $\{0\}$.

Theorem 6.4.11 *Assume that all rows of the matrix C are nonzero. An efficient solution $\overline{x}$ of the problem (MOLP) is robust if and only if the following conditions hold*

(i) $\text{ri}(\text{pos}\{a^i : i \in I(\overline{x})\}) \cap \text{ri}(\text{pos}\{c^j : j = 1, \cdots, k\}) \neq \emptyset$,
     where $I(\overline{x})$ is the active index set at $\overline{x}$, a^i is the ith column of A^T and c^j is the jth column of C^T;
(ii) $\text{span}\{a^i : i \in I(\overline{x})\} + \text{span}\{c^1, \cdots, c^k\} = \mathbb{R}^n$.

Proof We consider first the case $I(\overline{x}) = \emptyset$, that is, $\overline{x}$ is an interior point of the feasible solution set. In view of Theorem 4.3.1 it is an efficient solution if and only if there exists a strictly positive k-vector λ such that $C^T \lambda$ is the zero vector. By Lemma 6.4.10 the zero vector belongs to the relative interior of the cone $\text{pos}\{c^j : j = 1, \cdots, k\}$ and (i) holds. Moreover, we have also

$$\text{pos}\{c^j : j = 1, \cdots, k\} = \text{span}\{c^j : j = 1, \cdots, k\}.$$

If $\overline{x}$ is robust, then $\text{span}\{c^j : j = 1, \cdots, k\} = \mathbb{R}^n$, because otherwise there would exist some nonzero vector $u \in \mathbb{R}^n$ such that

$$\langle u, c^j \rangle = 0 \text{ for all } j = 1, \cdots, k$$

and a small perturbation $c^j + \epsilon u$, $j = 1, \cdots, k$ with $\epsilon > 0$ would result $0 \notin \text{ri}(\text{pos}\{c^j : j = 1, \cdots, k\})$, which proves that $\overline{x}$ is not efficient for the perturbed problem. Conversely, if $\text{span}\{c^j : j = 1, \cdots, k\} = \mathbb{R}^n$, then the matrix C has rank equal to n. It is clear that if C' is closed to C, then C' has rank n too. By this, the zero vector still belongs to the positive cone generated by the rows of C', and therefore $\overline{x}$ remains an efficient solution of the perturbed problem, or in other word $\overline{x}$ is robust.

We turn now to discuss the case $I(\overline{x}) \neq \emptyset$. Let $\overline{x}$ be a robust solution of (MOLP). Suppose that the first necessary condition (i) does not hold. In view of the separation theorem (Theorem 2.3.10) there is a unit vector v such that

$$\langle v, a^i \rangle \geq 0 \geq \langle v, c^j \rangle \text{ for all } i \in I(\overline{x}), j = 1, \cdots, k.$$

For any $\varepsilon > 0$, by defining $c'^j = c^j - \frac{\varepsilon}{2n^2}v, j = 1, \cdots, k$ we obtain a matrix C', whose transpose is composed of the columns $c'^1, \cdots, c'^k$ satisfying $\|C' - C\| < \varepsilon$ and

$$\langle v, c'^j \rangle < 0, \quad j = 1, \cdots, k.$$

Consequently, the cones $\mathrm{pos}\{a^i : i \in I(\overline{x})\}$ and $\mathrm{pos}\{c'^j : j = 1, \cdots, k\}$ have only the zero vector in their intersection. This is impossible because $\overline{x}$ is an efficient solution of (MOLP') with C' sufficiently close to C (Theorem 4.2.6).

We proceed to (ii) by contradiction. Assume that (ii) does not hold. Let v be a unit vector orthogonal to the proper subspace $\mathrm{span}\{a^i : i \in I(\overline{x})\} + \mathrm{span}\{c^1, \cdots, c^k\}$. Define $c'^j = (\varepsilon/n)v, j = 1, \cdots, k$ when a small positive ε is given. We claim

$$\mathrm{pos}\{a^i : i \in I(\overline{x})\} \cap \mathrm{pos}\{c'^j : j = 1, \cdots, k\} = \{0\}.$$

Indeed, let d be a nonzero vector belonging to the intersection on the left hand side of the latter equality. Then there are positive numbers $t_i, i \in I(\overline{x})$ and $s_j, j = 1, \cdots, k$ such that

$$d = \sum_{i \in I(\overline{x})} t_i a^i = \sum_{j=1}^{k} s_j(c^j + \varepsilon v)$$

which yields

$$\left(\varepsilon \sum_{j=1}^{k} s_j \right) v = \sum_{i \in I(\overline{x})} t_i a^i - \sum_{j=1}^{k} s_j c^j.$$

It follows from the choice of v that $\sum_{j=1}^{k} s_j = 0$ and hence $s_j = 0$ for all $j = 1, \cdots, k$ because they are all nonnegative. Thus $d = 0$ and we arrive at the same contradiction as for (i).

For the sufficient condition we assume that $\overline{x}$ is not robust, which means that there is a sequence of matrices C_α converging to C such that $\overline{x}$ is not an efficient solution of the problems with the objective matrices C_α. In view of Theorem 4.2.6 and Lemma 6.4.10 the cone $\mathrm{pos}\{a^i : i \in I(\overline{x})\}$ does not meet the relative interior of the cone $\mathrm{pos}\{c_\alpha^j : j = 1, \cdots, k\}$, where $c_\alpha^1, \cdots, c_\alpha^k$ are the columns of the matrix C_α^T. According to Theorem 2.3.10 we find unit vectors v_α separating these cones.

By picking a subsequence if necessary we may assume that v_α converges to some nonzero vector v. Then it is clear that v separates the cone $\mathrm{pos}\{a^i : i \in I(\bar{x})\}$ and the cone $\mathrm{pos}\{c^j : j = 1, \cdots, k\}$, that is

$$\langle v, a^i \rangle \geq 0, \ i \in I(\bar{x})$$
$$\langle v, c^j \rangle \leq 0, \ j = 1, \cdots, k.$$

There are two possible cases about this separation:

 (i) both of the two systems of inequalities are systems of equalities; and
(ii) at least one inequality is strict.

The first case shows that the vectors a^i and c^j are all on the hyperplane orthogonal to v, by which condition (ii) cannot be true. In the second case, say a strict inequality holds for c^j, some j among $1, \cdots, k$. Then the relative interior of the cone $\mathrm{pos}\{c^j : j = 1, \cdots, k\}$ must lie in the open half-space given by $\langle v, x \rangle < 0$. Consequently condition (i) is not true. This achieves the proof.   □

In the next corollary we obtain another characterization of robust solutions.

Corollary 6.4.12 *An efficient solution $\bar{x}$ of (MOLP) is robust if and only if*

$$\mathrm{Ker}(C) \cap \mathrm{cone}(\Gamma - \bar{x}) = \{0\}.$$

Proof Let $\bar{x}$ be a robust solution. Assume the contrary that there is some feasible solution $y \neq \bar{x}$ such that $y - \bar{x} \in \mathrm{Ker}(C)$. Then by (ii) of Theorem 6.4.11,

$$\langle a^i, y - \bar{x} \rangle \neq 0 \text{ for some } i \in I(\bar{x})$$

because

$$[\mathrm{span}\{a^i : i \in I(\bar{x})\}]^\perp \cap \mathrm{Ker}(C) = \{0\}.$$

This implies that the vector $y - \bar{x}$ separates $\mathrm{ri}\big(\mathrm{pos}\{a^i : i \in I(\bar{x})\}\big)$ and $\mathrm{ri}\big(\mathrm{pos}\{c^j : j = 1, \cdots, k\}\big)$ which contradicts (i) of Theorem 6.4.11.

For the converse, suppose that $\bar{x}$ is not robust. We find a sequence of matrices C_ν converging to C and a sequence of feasible solutions y^ν such that $C_\nu(y^\nu - \bar{x}) \geq 0$. We may assume without loss of generality that the sequence of vectors $(y^\nu - \bar{x})/\|y^\nu - \bar{x}\|$ converges to some unit vector u from $\mathrm{cone}(\Gamma - \bar{x})$ as ν tends to infinity. Then, $Cu \geq 0$. But since $\bar{x}$ is efficient, we deduce that $Cu = 0$, which contradicts the hypothesis.   □

We notice that the second condition of Theorem 6.4.11 evidently holds when either the matrix C has rank n or the normal cone at $\bar{x}$ has a nonempty interior. Here are some useful particular cases under such conditions.

Corollary 6.4.13 *If the rank of the objective matrix C is equal to n, then every efficient solution of (MOLP) is robust.*

Proof As already noticed the second condition of Theorem 6.4.11 is always satisfied because the cone pos$\{c^j : j = 1, \cdots, k\}$ has a nonempty interior. Moreover, if x is an efficient solution of the problem, then, in view of Theorem 4.2.6 the normal cone to the feasible set at x, which is exactly the cone pos$\{a^i : i \in I(x)\}$ meets the interior of the cone pos$\{c^j : j = 1, \cdots, k\}$. This implies the first condition of Theorem 6.4.11, by which x is robust. $\square$

Corollary 6.4.14 *A feasible vertex x of (MOLP) is robust if and only if the cone* pos$\{c^j : j = 1, \cdots, k\}$ *meets the interior of the normal cone to the feasible set at x.*

Proof If the vertex x is robust, then the first condition of Theorem 6.4.11 holds in which the interior of the cone pos$\{a^i : i \in I(x)\}$ is nonempty, and so its intersection with the cone pos$\{c^j : j = 1, \cdots, k\}$ is nonempty. Conversely, when x is a vertex, the second condition of Theorem 6.4.11 is satisfied trivially, again, because the cone pos$\{a^i : i \in I(x)\}$ has a nonempty interior. Moreover, if the cone pos$\{c^j : j = 1, \cdots, k\}$ has a nonempty intersection with the interior of the cone pos$\{a^i : i \in I(x)\}$, then so does its relative interior. By Theorem 4.2.6, x is robust. $\square$

We deduce a generic property of robustness when the number of criteria is larger then the number of variables. We need the concept of Lebesgue measure in $\mathbb{R}^n$. By an open interval in $\mathbb{R}^n$ we mean the box

$$L = L_1 \times \cdots \times L_n$$

where $L_1, \cdots, L_n$ are finite open intervals in $\mathbb{R}$. Each open interval $L_s = (a_s, b_s)$ in $\mathbb{R}$ has a length $\ell(L_s) = b_s - a_s$. The volume of L is defined to be

$$vol(L) = \ell(L_1) \times \cdots \times \ell(L_n).$$

Given a nonempty set Q in $\mathbb{R}^n$, the Lebesgue outer measure of Q is defined by

$$\mu(Q) = \inf \left\{ \sum_\nu vol(L^\nu) : L^\nu \text{ is open interval in } \mathbb{R}^n \text{ such that } Q \subseteq \bigcup_\nu L^\nu \right\}.$$

We say that Q is of measure zero if $\mu(Q) = 0$ and it is of full measure if its complement is of measure zero.

Corollary 6.4.15 *Assume that $m \geq n$. Then the set of perturbations of C for which efficient solutions are robust is of full measure.*

Proof Since the set of matrices of full rank is of full measure, when $m \geq n$, the set of matrices of rank n is of full measure too. Then the conclusion follows from the first part of Corollary 6.4.14. $\square$

When a solution is robust, by definition it is efficient for any small perturbation of the matrix C. We are now interested in a question in which direction we may perturb C such that a given efficient, but not robust solution remains efficient along that perturbation. In other words, we wish to find a $k \times n$-matrix D such that an efficient solution $\bar{x}$ of (MOLP) solves the problem, denoted (P_{tD})

$$\text{Maximize } (C + tD)$$
$$\text{subject to } Ax \leqq b$$

for all $t > 0$ sufficiently small. If this is true we call $\bar{x}$ robust in direction D. Of course such a direction always exists, for instance $D = C$ itself, which is a trivial one, and so we look for nontrivial directions. Here is a sufficient condition of such directions.

Corollary 6.4.16 *Assume that $\bar{x}$ is efficient solution of (MOLP) and λ is a strictly positive vector such that $C^T \lambda$ belongs to the cone $\mathrm{pos}\{a^i : i \in I(\bar{x})\}$. Then $\bar{x}$ is robust in any direction D whose rows $(d^j)^T$, $j = 1, \cdots, k$ belong to the cone $\mathrm{cone}\{\mathrm{pos}\{a^i : i \in I(\bar{x})\} - C^T \lambda\}$. Moreover, if $\bar{x}$ is an efficient vertex solution of (MOLP) and there is a nonzero vector $u \in \mathbb{R}^n$ such that*

$$\langle a^i, u \rangle \geqq \langle C^T \lambda, u \rangle, \ i \in I(\bar{x})$$
$$\langle c^j, u \rangle \leqq \langle C^T \lambda, u \rangle, \ j = 1, \cdots, k,$$

and if the rows of D satisfy

$$\langle d^j, u \rangle < \langle C^T \lambda, u \rangle, \ j = 1, \cdots, k,$$

then $\bar{x}$ is not efficient for (P_{tD}) with $t > 0$.

Proof We note that a strictly positive vector λ stated in the corollary exists because $\bar{x}$ is efficient. By hypothesis we have

$$D^T \lambda \in \mathrm{cone}\{\mathrm{pos}\{a^i : i \in I(\bar{x})\} - C^T \lambda\}.$$

We find a positive number s such that

$$D^T \lambda \in s\big(\mathrm{pos}\{a^i : i \in I(\bar{x})\} - C^T \lambda\big).$$

If $s = 0$, then $D^T \lambda = 0$ and we are done. If $s > 0$, we deduce that

$$(C + \frac{1}{s}D)^T \lambda \in \mathrm{pos}\{a^i : i \in I(\bar{x})\}.$$

This combining with the fact that $C^T \lambda$ belongs to the cone $\mathrm{pos}\{a^i : i \in I(\bar{x})\}$ we deduce that

$$(C + tD)^T \lambda \in \mathrm{pos}\{a^i : i \in I(\bar{x})\}$$

which proves that $\bar{x}$ is efficient for (P_{tD}) for every $t \in [0, 1/s]$.

For the second part of the corollary we observe that in view of Theorem 6.4.11, $\bar{x}$ is not robust because the vector u separates the interior of the cone pos$\{a^i : i \in I(\bar{x})\}$ and the cone pos$\{c^j : j = 1, \cdots, k\}$. If the rows of D satisfies the hypothesis, then for every $t > 0$, the vector u strictly separates the cone pos$\{a^i : i \in I(\bar{x})\}$ and the relative interior of the cone pos$\{c^j + td^j : j = 1, \cdots, k\}$. This implies that $\bar{x}$ is not efficient for (P_{tD}) as requested. $\square$

We notice that if $\bar{x}$ is a vertex efficient solution which is not robust, then a vector u as stated in the corollary always exists, and the inequalities for c^j become equalities. This is because by Theorem 6.4.11, the interior of the cone pos$\{a^i : i \in I(\bar{x})\}$ does not meet the relative interior of the cone pos$\{c^j : j = 1, \cdots, k\}$, and so they can be separated. Moreover, since $C^T \lambda$ is a relative interior point of the latter cone and belongs also to the cone pos$\{a^i : i \in I(\bar{x})\}$, inequalities for c^j become equalities. Note further that the condition given in the first part of the corollary is sufficient, but not necessary for $\bar{x}$ to be robust in direction D, while the condition of the second part is sufficient but not necessary for $\bar{x}$ to not be robust in direction D. We turn next to robust solutions when adding objective functions and taking convex combinations of two objective matrices.

Corollary 6.4.17 *Assume that the row vectors of the matrices C and D are nonzero. If $\overline{x}$ is a robust efficient solution of (MOLP) and an efficient solution of $(MOLP)^D$, then it is a robust efficient solution of $(MOLP)^D$.*

Proof We wish to apply Theorem 6.4.11. Since the subspace span$\{c^i, d^j : i = 1, \cdots, k; j = 1, \cdots, k'\}$ contains the subspace span$\{c^i : i = 1, \cdots, k\}$ the second condition of that theorem holds true for $(MOLP)^D$. To check the first condition we pick any vector p from the intersection of the relative interior of the cone pos$\{a^i : i \in I(\overline{x})\}$ and the relative interior of the cone pos$\{c^j : j = 1, \cdots, k\}$ which is nonempty because $\overline{x}$ is a robust efficient solution of (MOLP). As $\overline{x}$ is an efficient solution of $(MOLP)^D$, there is some nonzero vector q from the cone pos$\{a^i : i \in I(\overline{x})\}$ and strictly positive numbers $\lambda_j, j = 1, \cdots, k$ and $\mu_l, l = 1, \cdots, k'$ such that

$$q = \sum_{j=1}^{k} \lambda_i c^i + \sum_{l=1}^{k'} \mu_l d^l.$$

In view of Lemma 6.4.10, the vector q belongs to the relative interior of the cone pos$\{c^j, d^l : j = 1, \cdots, k; l = 1, \cdots, k'\}$. It follows then that the vector $\frac{1}{2}(p + q)$ lies in both the relative interior of the cone pos$\{a^i : i \in I(\overline{x})\}$ and the relative interior of the cone pos$\{c^j, d^l : j = 1, \cdots, k; l = 1, \cdots, k'\}$. Thus condition (i) of Theorem 6.4.11 is satisfied and the proof is complete. $\square$

Corollary 6.4.18 *Assume that $\overline{x}$ is a robust efficient solution of (MOLP) and $(MOLP_0)$, and that there are a vector $d \in \mathbb{R}^n$ and real numbers of the same sign $p_1, \cdots, p_k$ such that the rows of C and D satisfy*

$$d^i = c^i + p_i d, \ i = 1, \cdots, k.$$

Then $\bar{x}$ is a robust efficient solution of (MOLP$_\alpha$) for all $\alpha \in [0, 1]$.

Proof The case when p_j, $j = 1, \cdots, k$ are all zero being trivial we assume that at least one of them is nonzero. Let $0 < \alpha < 1$ be given. We wish to prove the two conditions

(1) $\mathrm{ri}(\mathrm{pos}\{a^i : i \in I(\bar{x})\}) \cap \mathrm{ri}(\mathrm{pos}\{c^j + (1 - \alpha)p_j d : j = 1, \cdots, k\}) \neq \emptyset$, and
(2) $\mathrm{span}\{a^i : i \in I(\bar{x})\} + \mathrm{span}\{c^j + (1 - \alpha)p_j d : j = 1, \cdots, k\} = \mathbb{R}^n$,

which, according to Theorem 6.4.11, are sufficient for the solution $\bar{x}$ to be robust for the problem (MOLP$_\alpha$). We proceed by contradiction. If condition (2) is not true, then we find a nonzero vector v such that

$$\langle v, a^i \rangle = 0, \ i \in I(\bar{x})$$
$$\langle v, c^j \rangle = (\alpha - 1)p_j \langle v, d \rangle, \ j = 1, \cdots, k. \tag{6.23}$$

Since $\bar{x}$ is a robust efficient solution of (MOLP), the value $\langle v, d \rangle$ is nonzero in view of condition (ii) of Theorem 6.4.11. We note also that all the values on the right hand side of (6.23) are of the same sign and at least one of them is nonzero. Consequently, the relative interior of the cone $\mathrm{pos}\{a^i : i \in I(\bar{x})\}$ does not meet the relative interior of the cone $\mathrm{pos}\{c^j : j = 1, \cdots, k\}$. This is a contradiction with the necessary condition (i) of Theorem 6.4.11.

If condition (1) is not true, then by applying the separation theorem (Theorem 2.3.10) we find a nonzero vector v such that

$$\langle v, a^i \rangle \geq 0, \ i \in I(\bar{x})$$
$$\langle v, c^j \rangle \leqq (\alpha - 1)p_j \langle v, d \rangle, \ j = 1, \cdots, k. \tag{6.24}$$

As it has already been noticed the values on the right hand side of (6.24) are all nonpositive or nonnegative. If they are all nonpositive and at least one of them is strictly negative, we deduce

$$\langle v, a^i \rangle \geq 0, \ i \in I(\bar{x})$$
$$\langle v, c^j \rangle \leqq 0, \ j = 1, \cdots, k$$

in which at least one inequality is strict, and arrive at a contradiction that the relative interior of the cone $\mathrm{pos}\{a^i : i \in I(\bar{x})\}$ does not meet the relative interior of the cone $\mathrm{pos}\{c^j : j = 1, \cdots, k\}$. If the values on the right hand side of (6.24) are nonnegative and at least one of them is strictly positive, then

$$\alpha p_j \langle v, d \rangle \leqq 0, \ j = 1, \cdots, k.$$

We obtain a system

$$\langle v, a^i \rangle \geq 0, \ i \in I(\bar{x})$$
$$\langle v, c^j + p_j d \rangle \leqq 0, \ j = 1, \cdots, k.$$

If none of the above inequalities is strict, then v is orthogonal to the subspace span$\{a^i :$ $i \in I(\overline{x})\} + \text{span}\{c^1, \cdots, c^k\}$ which is a contradiction because $\overline{x}$ is a robust solution of (MOLP$_0$). If one of the inequalities is strict, then the two cones pos$\{a^i : i \in I(\overline{x})\}$ and pos$\{c^j + p_j d : j = 1, \cdots, k\}$ have no relative interior points in common. This again is a contradiction, and the proof is complete.   □

We saw in Chap. 4, Sect. 4.3 that the set of efficient solutions of a linear problem is arcwise connected. This property is no longer true for robust solutions as demonstrated by the next example.

Example 6.4.19 Let X be the unit cube of side length equal to one. Consider the bi-objective problem

$$\text{Maximize} \begin{pmatrix} 1 & 1 & 1 \\ 1 & -1 & -1 \end{pmatrix} \begin{pmatrix} x_1 \\ x_2 \\ x_3 \end{pmatrix}$$

subject to $x \in X$.

Then the efficient solution set is the two-dimensional face with $x_1 = 1$. A brute-force computation in combination with Theorem 6.4.11 shows that only two vertices $(1, 1, 1)^T$ and $(1, 0, 0)^T$ are robust.

Even being disconnected in some circumstances robust solutions of a problem form a set of nice structure.

Corollary 6.4.20 *The set of robust solutions consists of faces. Consequently if the feasible set has vertices and the problem has robust solutions, it has a robust vertex solution.*

Proof Let x be a robust solution and let F be a face that contains x in its relative interior. By definition there is a positive ε such that x remains an efficient solution of the problem (MOLP') with $\|C' - C\| < \varepsilon$. Hence every point of F is an efficient solution of (MOLP') and F is robust. The second part of the corollary follows from the fact that any feasible face has a vertex if the feasible set has vertices.   □

Radius of robustness

In this subsection we are interested in finding the largest number ϵ for which a given efficient solution of (MOLP) remains efficient with C' replacing C within the ϵ distance $\|C' - C\| < \epsilon$. Such a number (finite or infinite) is called *radius of robustness* of the problem at $\overline{x}$ and denoted $r(\overline{x})$. Given a vector a in $\mathbb{R}^k$, the negative vector a^- is obtained from a by substituting all positive components by zero. It is clear that $\|a^-\|$ is equal to the distance from a to the positive orthant $\mathbb{R}^k_+$.

Lemma 6.4.21 *Assume that u is a unit vector in $\mathbb{R}^n$ such that $Cu \notin \mathbb{R}^n_+$. Then*

$$\sup\left\{t : (C + tM)(u) \notin \mathbb{R}^k_+ \text{ for all matrix } M \text{ with } \|M\| \leq 1\right\} = \|(Cu)^-\|.$$

Proof Let us denote the supremum on the left hand side of the equality in the lemma by ρ. Let $0 \leq t < \|(Cu)^-\|$. Then for any unit norm matrix M, the distance from $(C + tM)(u)$ to the positive orthant is evaluated as follows

$$d((C + tM)(u), \mathbb{R}^k_+) \geq d(Cu, \mathbb{R}^k_+) - d(tM, \mathbb{R}^k_+)$$
$$\geq \|(Cu])^-\| - \|tM\|$$
$$> 0.$$

This implies that $\rho \geq \|(Cu)^-\|$. Now, let $t > \|(Cu)^-\| > 0$. Let v be a unit vector of $\mathbb{R}^n$ such that $\langle v, u \rangle = 1$. Define a linear operator M (which is represented by a matrix denoted M too) from $\mathbb{R}^n$ to $\mathbb{R}^k$ by

$$Mx = \langle v, x \rangle \left(-\frac{(Cu)^-}{\|(Cu)^-\|}\right).$$

Then $\|M\| = 1$ and

$$(C + tM)(u) = (Cu)^+ + (Cu)^- - \frac{t}{\|(Cu)^-\|}(Cu)^- \not\geq 0$$

which shows that $(C + tM)(u) \in \mathbb{R}^k_+$. By this $\rho \leq \|(Cu)^-\|$ and the lemma follows. $\square$

Here is a formula to compute the radius of robustness.

Theorem 6.4.22 *Assume that $\bar{x}$ is a robust solution and $u^1, \cdots, u^l$ are unit directions generating $cone(\Gamma - \bar{x})$. Then the radius of robustness at $\bar{x}$ is given by the formula*

$$r(\bar{x}) = \min\left\{\|[C(\alpha_1 u^1 + \cdots + \alpha_l u^l)]^-\| : \alpha_1, \cdots, \alpha_l \geq 0 \text{ and } \sum_{i=1}^{l} \alpha_i = 1\right\}.$$

Proof Let us denote the minimum on the right hand side by ρ. We show first that $\rho > 0$. Indeed, if $\|[C(\alpha_1 u^1 + \cdots + \alpha_l u^l)]^-\| = 0$ for some $\alpha_1, \cdots, \alpha_l \geq 0$ and $\sum_{i=1}^{l} \alpha_i = 1$, then $C(\alpha_1 u^1 + \cdots + \alpha_l u^l) = 0$ because $\bar{x}$ is efficient. For any $\epsilon > 0$, define M by

$$Mx = \frac{\epsilon}{2\sqrt{n}}\langle v, x \rangle e$$

where v is a unit vector such that $\langle v, \alpha_1 u^1 + \cdots + \alpha_l u^l \rangle = 1$ and e is the vector in $\mathbb{R}^k$ whose components are all equal to one. Then $\|M\| \leq \frac{\varepsilon}{2}$, and for $t > 0$ for which $\bar{x} + t(\alpha_1 u^1 + \cdots + \alpha_l u^l)$ is feasible, one has

$$(C + M)(\bar{x} + t(\alpha_1 u^1 + \cdots + \alpha_l u^l)) = (C + M)\bar{x} + tM(\alpha_1 u^1 + \cdots + \alpha_l u^l)$$
$$> (C + M)\bar{x}.$$

This proves that $\bar{x}$ is not efficient when C is perturbed to $C + M$, a contradiction.

Now, let M be a matrix with $\|M\| < \rho$. Then, in view of Lemma 6.4.21, $(C + M)(\alpha_1 u^1 + \cdots + \alpha_l u^l)$ does not meet $\mathbb{R}^k_+$ for all $\alpha_1, \cdots, \alpha_l \geq 0$ with $\sum_{i=1}^l \alpha_i = 1$. Consequently,

$$(C + M)(\Gamma - \bar{x}) \cap \mathbb{R}^k_+ = \{0\},$$

which shows that $\bar{x}$ is efficient solution when $C + M$ replacing C. Hence, $r(\bar{x}) \geq \rho$. Further, if $t > \rho$, then for a vector

$$u = \alpha_1 C u^1 + \cdots + \alpha_l C u^l$$

with $\alpha_1, \cdots, \alpha_l \geq 0$ and $\sum_{i=1}^l \alpha_i = 1$, at which ρ is realized, by Lemma 6.4.21, there is some matrix M of norm t such that $(C + M)u \in \mathbb{R}^k_+ \setminus \{0\}$. Choose $s > 0$ such that $\bar{x} + su$ is feasible. We deduce

$$(C + M)(\bar{x} + su) = (C + M)\bar{x} + s(C + M)(u)$$
$$\leq (C + M)\bar{x}$$

which proves that $\bar{x}$ is not efficient when $C + M$ replacing C. Thus, $r(\bar{x}) \leq \rho$ and equality follows. $\square$

There are some particular cases in which it is easy to find directions $u^i, i = 1, \cdots, l$ that generate the cone $\mathrm{cone}(\Gamma - \bar{x})$. For instance when x is a relative interior point of a proper face of Γ, that is a face of dimension $n - 1$, there is a column a^i of the matrix A^T such that the set $\mathrm{cone}(\Gamma - \bar{x})$ is the half-space determined by inequality $\langle a^i, x \rangle \leq 0$. It suffices then to choose any $n - 1$ unit norm vectors $u^1, \cdots, u^{n-1}$ forming a basis of the hyperplane $\langle a^i, x \rangle = 0$ and apply the formulae of the theorems to these u^i. When $\bar{x}$ is a vertex, the set $\mathrm{cone}(\Gamma - \bar{x})$ is generated by the directions of adjacent edges incident to $\bar{x}$ and the extreme rays of Γ. Any technique of linear programming for finding adjacent edges and extreme rays can be applied to obtain u^i.

6.5 Exercises

6.5.1 *Consider the following parametric problem with $\omega \in \mathbb{R}$:*

$$\text{Maximize} \quad \begin{pmatrix} \omega & 1 \\ 1 & \omega \\ 3 & 1 \end{pmatrix} \begin{pmatrix} x_1 \\ x_2 \end{pmatrix}$$
$$\text{subject to } x_1 + x_2 \leq 1$$
$$x_1, x_2 \geq 0.$$

Express the maximal solution set as a set-valued map of ω and find those points at which the map is not continuous.

6.5.2 *Assume that the matrix D in the problem $(MOLP)^D$ (Sect. 6.4) is a vector and Γ is the feasible set of (MOLP). Prove the following assertions:*

(i) *If a maximal solution of $(MOLP)^D$ minimizes the function $\langle D, . \rangle$ on Γ, then it is a maximal solution of (MOLP).*

(ii) *If a maximal solution of (MOLP) maximizes the function $\langle D, . \rangle$ on Γ, then it is a maximal solution of $(MOLP)^D$.*

6.5.3 *Let A be an $m \times n$-matrix, I the identity $n \times n$-matrix and b an m-vector. Consider two conditions:*

(i) $b \in \text{int}\left(A(\mathbb{R}_+^n)\right)$;

(ii) $\begin{pmatrix} b \\ -b \\ 0 \end{pmatrix} \in \text{int}\left[\begin{pmatrix} A \\ -A \\ -I \end{pmatrix} (\mathbb{R}^n) + \mathbb{R}^{m+m+n}\right].$

Show that (ii) implies (i), but the converse is not true.

Note that (i) expresses the regularity of the system $Ax = b, x \geq 0$, and (ii) expresses the regularity of the equivalent system $Ax \leq b, -Ax \leq -b, x \geq 0$. Consequently two systems having the same solution set are not necessarily regular simultaneously.

6.5.4 Best Hoffman constant *Assume that a point x is not a solution to the system (6.10) which means that $f(x) := \max\{\langle a^i, x \rangle - b_i : i = 1, \cdots, m\} > 0$. Define $J(x)$ to be the set of indices j such that $\langle a^j, x \rangle - b_j = f(x)$ and $J_0(x)$ the subset of $J(x)$ such that $a^j, j \in J_0(x)$ are extreme directions of the cone $\text{pos}\{a^j : j \in J(x)\}$. Denote by $\mathcal{J}$ the family of index sets $J \subseteq \{1, \cdots, m\}$ satisfying the condition (ii) of Theorem 6.3.3 and the following one: there is some $x \notin \Gamma$ such that $J \subseteq J_0(x)$. Define*

$$\bar{\sigma} = \min_{J \in \mathcal{J}, \lambda \in \mathbb{R}_+^{|J|}, \sum_{i \in J} \lambda_i = 1} \|A_J^T \lambda\|.$$

Prove that if for every $y \in \Gamma$, *the vectors* $a^i, i \in I(y)$ *are linearly independent, then* $\bar{\sigma} = \sigma$. *Actually* $\bar{\sigma}$ *is the best Hoffman constant in the sense that*

$$\bar{\sigma} = \inf_{x \notin \Gamma} \frac{\max_{i=1,\cdots,m}(\langle a^i, x \rangle - b_i)}{d(x, \Gamma)}.$$

6.5.5 Pareto eigenvalues and Hoffman constants *Let* $I \in \mathcal{I}$ *be given. Consider the following minimization problem*

$$\begin{array}{ll} \text{minimize} & y^T M_I y \\ \text{subject to } y \in \mathbb{R}^{|I|}_+, \|y\|^2 = 1 \end{array}$$

where $M_I = A_I A_I^T$. *The optimal value of this problem is denoted by* $\gamma(I)$. *Prove the following properties*

(i) $\beta := \min_{I \in \mathcal{I}} \sqrt{\gamma(I)} > 0$
(ii) $d(x, \Gamma) \leqq \frac{1}{\beta} d(Ax - b, -\mathbb{R}^m_+)$.

(Note: Any optimal solution y *of the above minimization problem satisfies the following optimality conditions*

$$\begin{aligned} M_I y - \theta y &\geqq 0 \\ \langle M_I y - \theta y, y \rangle &= 0 \\ \|y\| = 1, y &\geqq 0. \end{aligned} \tag{6.25}$$

Solutions y *of this system are called Pareto eigenvectors and the corresponding* θ *are called Pareto eigenvalues of the matrix* M_I. *The value* $\gamma(I)$ *is equal to the smallest Pareto eigenvalue of* M_I.)

6.5.6 Weak sharp maxima with respect to a function *Consider (MOLP) presented in Sect. 6.3. Assume that it has a weakly efficient solution. Define*

$$h(x) = \min_{\lambda \in \Delta} \max_{y \in \Gamma} \lambda^T C(y - x),$$

where Δ *is the standard simplex in* $\mathbb{R}^k$ *and* $\Gamma = \{x \in \mathbb{R}^n : Ax \leq b\}$ *is the feasible set of (MOLP). A set of weakly efficient solutions* E *of (MOLP) is called a set of* h-*weak sharp maxima if there is a strictly positive constant* r *such that*

$$d(x, E) \leqq r h(x) \text{ for all } x \in \Gamma.$$

(i) *Prove that for every* $x \in \Gamma$ *there are some* $\overline{\lambda} \in \Delta$ *and a weakly efficient solution* $\bar{x}$ *that maximizes* $\langle \overline{\lambda}, x \rangle$ *on* Γ *such that* $h(x) = \overline{\lambda}^T C(\bar{x} - x)$.
(ii) *Deduce that the weakly efficient solution set of (MOLP) is a set of* h-*weak sharp maxima.*

6.5.7 *For (MOLP) in the preceding exercise define*

$$g(x) = \max_{y \in \Gamma} \min\{c^i(y-x) : i = 1, \cdots, k\}$$

where c^i is the ith row of C.

(i) *Prove that $g(x) = \max_{y \in \Gamma} \min_{\lambda \in \Delta} \lambda^T C(y-x)$ (see also Exercice 4.4.20).*

(ii) *(Min-max theorem) Consider the linear problem*

$$\text{Maximize } (0\ 1) \begin{pmatrix} x \\ v \end{pmatrix}$$

$$\text{subject to } \begin{pmatrix} -C & e \\ A & 0 \end{pmatrix} \begin{pmatrix} x \\ v \end{pmatrix} \leqq \begin{pmatrix} 0 \\ b \end{pmatrix},$$

where e is the vector of ones, and its dual to prove existence of $\bar{x} \in \Gamma$ and $\bar{\lambda} \in \Delta$ such that $\min_{\lambda \in \Delta} \lambda^T C\bar{x} = \max_{x \in X} \bar{\lambda}^T Cx$.

(iii) *Show that if the weakly efficient solution set of (MOLP) is nonempty, then it is a set of g-weak sharp maxima.*

6.5.8 *Find robust efficient solutions of the problem*

$$\text{Maximize } \begin{pmatrix} 2 & 1 \\ 0 & 1 \\ 3 & 1 \end{pmatrix} \begin{pmatrix} x_1 \\ x_2 \end{pmatrix}$$

$$\text{subject to } \begin{pmatrix} 1 & 0 \\ 1 & 1 \\ 0 & 1 \end{pmatrix} \begin{pmatrix} x_1 \\ x_2 \end{pmatrix} \leqq \begin{pmatrix} 1 \\ 1 \\ 1 \end{pmatrix}.$$

6.5.9 *Let Γ be a polyhedron in $\mathbb{R}^2$, C a $k \times 2$-matrix and D a $k \times 2$-matrix. Prove that if $\bar{x}$ is an efficient solution of the problems*

$$\text{Maximize } Cx$$
$$\text{subject to } x \in \Gamma$$

and

$$\text{Maximize } Dx$$
$$\text{subject to } x \in \Gamma,$$

then it is also an efficient solution of the convex combination problem

$$\text{Maximize } [\alpha C + (1-\alpha)D]x$$
$$\text{subject to } x \in \Gamma,$$

for all $\alpha \in [0, 1]$. This shows that the conclusion of Theorem 6.4.3 is true without any condition on the objective functions C and D when the dimension of the decision variable x is two.

6.5.10 *Show that if the rank of the objective matrix C in the problem (MOLP) is equal to n, then every efficient solution of the problem (MOLP) is robust.*

6.5.11 *Let $p_1, p_2, q_1, q_2 \in \mathbb{R}^n$ with $n \geq 3$ be in general linear position (no plane contains all of them) and $v_1 \in [p_1, q_1]$, $v_2 \in [p_2, q_2]$. Prove that for every $p \in [p_1, p_2]$ there exist $v \in [v_1, v_2]$ and $q \in [q_1, q_2]$ such that $v \in [p, q]$.*
  Deduce that if $\bar{x}$ solves the bi-objective problems

$$\text{Maximize } \begin{pmatrix} c^1 \\ c^2 \end{pmatrix} x$$
$$\text{subject to } x \in \Gamma,$$

and

$$\text{Maximize } \begin{pmatrix} d^1 \\ d^2 \end{pmatrix} x$$
$$\text{subject to } x \in \Gamma,$$

where Γ is a nonempty polyhedron, then for every $\alpha \in [0, 1]$, there is $\beta \in [0, 1]$ such that $\bar{x}$ solves the problem

$$\text{Maximize } \begin{pmatrix} \alpha c^1 + (1 - \alpha)d^1 \\ \beta c^2 + (1 - \beta)d^2 \end{pmatrix} x$$
$$\text{subject to } x \in \Gamma.$$

Discuss the case of k-objective functions with $k > 2$.

6.5.12 *Let M be an $m \times n$-matrix. A Moore-Penrose pseudo-inverse of M, denoted M^+, is defined as an $n \times m$-matrix satisfying the following properties:*

  (i) $MM^+M = M^+$
  (ii) $M^+MM^+ = M$
  (iii) both M^+M and MM^+ are symmetric.

It is known that the Moore-Penrose pseudo-inverse exists and is unique. Moreover, if the system $Mz = a$ has any solutions, then $z = M^+a$ is a solution of minimum norm. Prove that if $\bar{x}$ is a robust solution and if $u^1, \cdots, u^l$ are unit directions generating $\text{cone}(\Gamma - \bar{x})$, then the radius of robustness at $\bar{x}$ is given by the formula

$$r(\bar{x}) = \min \ \{\|C^T (CC^T)^+ C(\alpha_1 u^1 + \cdots + \alpha_l u^l)^-\| : \alpha_1, \cdots, \alpha_l \geq 0$$

$$\text{and } \sum_{i=1}^{l} \alpha_i = 1\}.$$

Part III
Methods

Chapter 7
Multiobjective Simplex Method

The simplex method is the first method to solve linear programming problems and one of the most popular methods in computing mathematics. In this chapter we present an extension of it to multiobjective problems. The problem, denoted (MOLP), that we are going to study is given in standard form:

$$\text{Maximize } Cx$$
$$\text{subject to } Ax = b$$
$$x \geq 0,$$

where C is a $k \times n$-matrix, A is an $m \times n$-matrix and b is an m-vector.

7.1 Description of the Method

A general scheme for finding efficient vertices and efficient rays of (MOLP) consists of the following phases:

Phase 1. Find a feasible basis to start with. This phase is carried out by the simplex method described in Chap. 3. Let $\bar{x}$ be a basic feasible solution.

Phase 2. Check the efficiency of $\bar{x}$.

Phase 3. Move to adjacent vertices to find efficient adjacent vertices or efficient infinite rays.

Phase 4. Store all founded efficient vertices in one list and dominated vertices in another list, and stop when all vertices have been explored.

© Springer International Publishing Switzerland 2016
D.T. Luc, *Multiobjective Linear Programming*,
DOI 10.1007/978-3-319-21091-9_7

Checking efficiency of the current basis

In order to check whether a vertex $\bar{x}$ is efficient or not, we compute

$$\bar{b} = B^{-1}b$$
$$\overline{C}_N = C_N - C_B B^{-1}N,$$

where B is a feasible basis corresponding to $\bar{x}$ which is assumed to be non-degenerate and $A = (B\ N)$. The objective matrix C is decomposed into the basic and non-basic submatrices C_B and C_N correspondingly. The submatrix $\overline{C}_N$ is called *reduced cost matrix* at the basis B. As before the columns of the matrix C^T are denoted $c^1, \cdots, c^k$.

Theorem 7.1.1 *Let $\bar{x}$ be a basic feasible solution. The following conditions are equivalent.*

(i)   *The solution $\bar{x}$ is efficient.*
(ii)   *The solution $\bar{x}$ solves the following problem*

$$\text{maximize } \langle c^1 + \cdots + c^k, x \rangle$$
$$\text{subject to } Ax = b$$
$$Cx \geq C\bar{x}$$
$$x \geq 0.$$

If in addition $\bar{x}$ is non-degenerate, (i) is equivalent to each of the two conditions below.

(iii) *The following system has a solution*

$$-[\overline{C}_N]^T y \geq 0$$
$$y > 0, \ y \in \mathbb{R}^k.$$

(iv)   *The optimal value of the following problem is zero*

$$\text{maximize } y_1 + \cdots + y_k$$
$$\text{subject to } -\overline{C}_N z + y = 0,$$
$$y \geq 0, z \geq 0, \ y \in \mathbb{R}^k, \ z \in \mathbb{R}^{n-k}.$$

In particular the following assertions hold.

(a)   *If all entries of the reduced cost matrix $\overline{C}_N$ are negative or zero, then the current vertex $\bar{x}$ is an ideal efficient solution.*
(b)   *If the reduced cost matrix has a strictly negative row, then $\bar{x}$ is efficient.*
(c)   *If the reduced cost matrix has a nonzero positive column, then the current vertex $\bar{x}$ is not efficient.*

Proof We prove the equivalence between (i) and (ii) first. If $\bar{x}$ is an efficient solution, then for every feasible solution x of the problem in (ii), one has $Cx = C\bar{x}$, which shows that $\bar{x}$ is an optimal solution. Conversely, suppose that $\bar{x}$ is not efficient, then there is some $x \geq 0$ with $Ax = b$ and $Cx \geq C\bar{x}$. We deduce

$$\langle c^1 + \cdots + c^k, x \rangle > \langle c^1 + \cdots + c^k, \bar{x} \rangle$$

and $\bar{x}$ does not solve the problem of (ii).

Further, according to Theorem 4.3.1, $\bar{x}$ is efficient if and only if there is a strictly positive k-vector y such that $\bar{x}$ is an optimal solution of the scalarized problem

$$\begin{aligned}
\text{maximize} \quad & \langle y, Cx \rangle \\
\text{subject to} \quad & Ax = b \\
& x \geq 0.
\end{aligned}$$

Assume that $\bar{x}$ is non-degenerate. In view of Theorem 3.1.4 the latter problem has $\bar{x}$ as an optimal solution if and only if its reduced cost vector $(C^T y)_N$ has non-positive entries only. We have

$$\begin{aligned}
[(\overline{C^T y})_N]^T &= [(C^T y)_N]^T - [(C^T y)_B]^T B^{-1} N \\
&= y^T C_N - y^T C_B B^{-1} N \\
&= y^T \overline{C}_N
\end{aligned}$$

and obtain the system in (iii). Thus, (i) is equivalent to (iii). The equivalence between (iii) and (iv) is proved in the same manner as that of Proposition 3.3.2 in Chap. 3. Namely, the system of (iii) has a solution if and only if the problem

$$\begin{aligned}
\text{minimize} \quad & \langle 0, y \rangle \\
\text{subject to} \quad & -[\overline{C}_N]^T y \geq 0 \\
& y \geq e,
\end{aligned}$$

where e is the vector from $\mathbb{R}^k$, the components of which are all equal to one, has an optimal solution and its optimal value is zero. Its dual is exactly the maximization problem figured in (iv). Then the equivalence follows from the duality relation in Theorem 3.2.3 and Corollary 2.2.5.

Now, if all entries of $\overline{C}_N$ are non-positive, then for every strictly positive k-vector y the system in (iii) is true. Hence $\bar{x}$ is efficient. Moreover each feasible solution x of (MOLP) can be decomposed by the basic part x_B and the non-basic part x_N with $Bx_B + Nx_N = b$. We deduce that $x_B = B^{-1}b - B^{-1}Nx_N$ and

$$Cx = C_B x_B + C_N x_N = C_B B^{-1}b + (C_N - C_B B^{-1}N)x_N.$$

Since x_N is a positive vector and $C_B b = C\bar{x}$, we deduce that $Cx \geq C\bar{x}$ for every feasible solution x, thereby proving that $\bar{x}$ is an ideal solution of the problem.

Furthermore, under (b), a strictly positive k-vector y with all components equal to one except for the component corresponding to the negative row, which takes a sufficiently large positive value, will certainly satisfy the system in (iii) of the theorem. Hence $\bar{x}$ is efficient.

Finally, under condition (c) the system in (iii) has no solution, so $\bar{x}$ cannot be efficient. □

We note that the solution $\bar{x}$ is not necessarily efficient if the reduced cost matrix has a negative row. See the example below.

Example 7.1.2 Consider the problem

$$\text{Maximize} \begin{pmatrix} 1 & 0 & 0 & 0 \\ 0 & 1 & 0 & 0 \end{pmatrix} \begin{pmatrix} x_1 \\ x_2 \\ x_3 \\ x_4 \end{pmatrix}$$

$$\text{subject to} \begin{pmatrix} 1 & -1 & 1 & 0 \\ 0 & 1 & 0 & 1 \end{pmatrix} \begin{pmatrix} x_1 \\ x_2 \\ x_3 \\ x_4 \end{pmatrix} = \begin{pmatrix} 0 \\ 1 \end{pmatrix}$$

$$\text{and} \qquad x \geqq 0.$$

The basis corresponding to the variables x_2 and x_3 is given by $B = \begin{pmatrix} -1 & 1 \\ 1 & 0 \end{pmatrix}$. The associated basic solution is $\bar{x} = (0, 1, 1, 0)^T$ and the reduced cost matrix is

$$\bar{C}_{1,4} = C_{1,4} - C_{2,3} B^{-1} A_{1,4}$$

$$= \begin{pmatrix} 1 & 0 \\ 0 & 0 \end{pmatrix} - \begin{pmatrix} 0 & 0 \\ 1 & 0 \end{pmatrix} \begin{pmatrix} 0 & 1 \\ 1 & 1 \end{pmatrix} \begin{pmatrix} 1 & 0 \\ 0 & 1 \end{pmatrix}$$

$$= \begin{pmatrix} 1 & 0 \\ 0 & -1 \end{pmatrix}.$$

Here the matrix $C_{i,j}$ is constituted of the i-th and j-th columns of C. The first column of the reduced cost matrix $\bar{C}_{1,4}$ is nonzero positive, hence in view of Theorem 7.1.1 the basic solution $\bar{x}$ is not efficient. Observe, however, that its second row is negative, but not strictly negative.

Moving to adjacent vertices

Let s be a non-basic index. We know that if the column $\bar{a}_s = B^{-1} a_s$ is negative, then the ray $\bar{x} + tv^s$, $t \geq 0$ is feasible, where the basic and the non-basic components of v^s are given by

$$v_B^s = -\bar{a}_s$$
$$v_N^s = e^s \qquad\qquad (7.1)$$

in which e^s is the non-basic part of the s-th coordinate unit vector of $\mathbb{R}^n$. Otherwise, set

$$t_s := \frac{\overline{b}_\ell}{\overline{a}_{\ell s}} = \min \left\{ \frac{\overline{b}_i}{\overline{a}_{is}} : \overline{a}_{is} > 0 \right\} \tag{7.2}$$

and a new basis is obtained from B by introducing the column s in the place of the column ℓ. The new solution vertex is given by $\hat{x} = \overline{x} + t_s v^s$.

Theorem 7.1.3 *Assume that the column $\overline{a}_s$ is negative. The following assertions are true.*

(i)   *If $\overline{x}$ is non-efficient, then every element of the ray $\overline{x} + tv^s$, $t \geq 0$ is non-efficient.*

(ii) *If $\overline{x}$ is efficient and if the column $\overline{C}_s$ from the reduced cost matrix is zero, then the ray $\overline{x} + tv^s$, $t \geq 0$ is efficient.*

(iii) *If $\overline{x}$ is efficient and if the column $\overline{C}_s$ is negative, then every point of the ray $\overline{x} + tv^s$, $t \geq 0$ is dominated by $\overline{x}$.*

(iv) *If $\overline{x}$ is non-degenerate, efficient and if the column $\overline{C}_s$ has negative as well as positive components, then the ray $\overline{x} + tv^s$, $t \geq 0$, is efficient if and only if the following system has a solution:*

$$-(\overline{C}_N)^T y \geq 0$$
$$(\overline{C}_N e^s)^T y = 0$$
$$y \in \mathbb{R}^k, \, y > 0,$$

which is equivalent to the zero optimal value of the linear problem

$$\text{maximize } y_1 + \cdots + y_k$$
$$\text{subject to } \overline{C}_N(z + te^s) + y = 0$$
$$t \in \mathbb{R}, \, y \geq 0 \, z \geq 0.$$

(v)   *If $\overline{x}$ is efficient and if non-basic columns $\overline{C}_s$ and $\overline{C}_r$ of the reduced cost matrix $\overline{C}_N$ corresponding to negative columns $\overline{a}_s$ and $\overline{a}_r$, have mixed components with $\alpha \overline{C}_s \geq \overline{C}_r$ for some positive number α, then the ray corresponding to the column $\overline{a}_r$ is dominated.*

Proof We suppose, to the contrary, that $\overline{x}$ is not efficient, but $\overline{x} + t_0 v^s$ is efficient for some $t_0 > 0$. In view of Theorem 4.3.8, the solution $\overline{x} + t_0 v^s$ is contained in an efficient solution face of the feasible set. This face must contain also the ray $\overline{x} + tv^s$, $t \geq 0$, because the solution $\overline{x} + t_0 v^s$ is a relative interior point of the ray. Hence $\overline{x}$ is efficient, a contradiction.

To prove the remaining part of the theorem, we express the value $C(\overline{x} + tv^s)$ in terms of $C\overline{x}$ and $\overline{C}_s$ as

$$C(\overline{x} + tv^s) = (C_B \; C_N) \begin{pmatrix} \overline{x}_B - t\overline{a}_s \\ te^s \end{pmatrix}$$
$$= C_B \overline{x}_B + t(-C_B \overline{a}_s + C_N e^s)$$
$$= C\overline{x} + t\overline{C}_s.$$

We assume that $\overline{x}$ is an efficient solution. It is clear that if $\overline{C}_s = 0$, then $C(\overline{x}+tv^s) = C\overline{x}$ and hence $\overline{x} + tv$ is efficient for every $t \geq 0$ as well.

Moreover, if $\overline{C}_s \leq 0$, then $C(\overline{x} + tv^s) \leq C\overline{x}$, which means that the solution $\overline{x} + tv^s$ is dominated by $\overline{x}$.

Consider the case when $\overline{C}_s$ has both negative and positive components. In view of Theorem 3.3.5, the ray $\overline{x} + tv^s, t \geq 0$, is efficient if and only if there is a strictly positive k-vector y such that this ray is optimal for the scalarized problem

$$\text{maximize } \langle C^T y, x \rangle$$
$$\text{subject to } Ax = b$$
$$x \geq 0.$$

In other words, the ray $\overline{x} + tv^s, t \geq 0$, is efficient if and only if there is a strictly positive k-vector y such that $\overline{x}$ is optimal and the objective function $\langle C^T y, \cdot \rangle$ along the direction v^s is zero. In view of Theorem 7.1.1, the latter condition is equivalent to the consistence of the system given in (iii) of that theorem. To obtain equivalence between that system and the maximization problem mentioned in (iv), it suffices to write the dual problem and apply Theorem 3.2.3 as we did in the proof of Theorem 7.1.1.

Finally, under (v), for every element $\overline{x} + tv^r$ of the feasible ray corresponding to the negative column $\overline{a}_r$ one chooses the element $\overline{x} + \alpha t v^s$ of the ray corresponding to the negative column $\overline{a}_s$ and deduces

$$C(\overline{x} + \alpha t v^s) - C(\overline{x} + tv^r) = t(\alpha \overline{C}_s - \overline{C}_r) \geq 0$$

showing that the ray $\overline{x} + tv^r, t \geq 0$ is dominated.   □

When the column $\overline{a}_s$ is not negative, the ray $\overline{x} + tv^s, t \geq 0$ does not entirely lie in the feasible set. The feasible portion of this ray is the edge joining the vertex $\overline{x}$ and the new vertex $\hat{x} = \overline{x} + t_s v^s$. The variable x_s enters the basis while the variable x_ℓ leaves the basis because of (7.2). We shall explicitly write $\hat{x}^s$ instead of $\hat{x}$ if the vertex $\overline{x}$ has other feasible adjacent vertices. The next theorem tells us whether it is worthwhile to move to a new vertex if the current vertex is not efficient.

Theorem 7.1.4 *Assume that the column $\overline{a}_s$ has positive components. The following assertions are true.*

(i)   *If the column $\overline{C}_s$ is negative, then the new vertex $\hat{x}$ is dominated by the old vertex $\overline{x}$.*

(ii) *If for any non-basic indices s and r for which $\hat{x}^s$ and $\hat{x}^r$ are finite, one has $t_s \overline{C}_s - t_r \overline{C}_r \geq 0$, then the new vertex with x_r entering the basis is non-efficient.*

Proof We know that $C\hat{x} = C\overline{x} + t_s\overline{C}_s$. Therefore, when $\overline{C}_s$ is negative, one deduces that $C\hat{x} \leq C\overline{x}$.

Now, if $\hat{x}^s$ and $\hat{x}^r$ are two feasible adjacent vertices of $\overline{x}$ with corresponding $t_s > 0$ and $t_r > 0$, then

$$C\hat{x}^s - C\hat{x}^r = t_s\overline{C}_s - t_r\overline{C}_r.$$

Under the hypothesis of (ii), we have $C\hat{x}^s \geq C\hat{x}^r$, which means that $\hat{x}^r$ is not efficient. $\qquad\square$

Theorem 7.1.5 *If all feasible bases of the matrix A are non-degenerate, then the simplex algorithm terminates at a finite number of iterations.*

Proof Under the non-degeneracy condition, the simplex algorithm explores all vertices of the feasible set, and determines at each iteration whether the current vertex is efficient or not. Therefore the algorithm terminates after a finite number of iterations. $\qquad\square$

7.2 The Multiobjective Simplex Tableau

In order to solve the problem (MOLP), we assume that b is a positive vector and the matrix A is given by $(B\ N)$ where B is a non-degenerate feasible basis and N is the non-basic part of A. The initial simplex tableau, denoted T, is of the form

$$T = \begin{array}{|cc|c|} \hline C_B & C_N & 0_{k\times 1} \\ \hline B & N & b \\ \hline \end{array}$$

By pre-multiplying the tableau T by the extended inverse of B,

$$\begin{array}{|c|c|} \hline I_{k\times k} & -C_B B^{-1} \\ \hline 0_{m\times k} & B^{-1} \\ \hline \end{array}$$

we obtain the tableau T^* as follows

$$T^* = \begin{array}{|cc|c|c|} \hline 0_{k\times m} & \overline{C}_N = C_N - C_B B^{-1} N & -C_B B^{-1} b \\ \hline I_{m\times m} & \overline{N} = B^{-1} N & B^{-1} b \\ \hline \end{array}$$

The tableau T^* contains all information necessary for the simplex algorithm.

- The basic solution associated with B is found in the right bottom box: $x = \begin{pmatrix} x_B \\ x_N \end{pmatrix}$ with $x_B = B^{-1}b$ and $x_N = 0$.
- The value of the objective function at this basic solution is given in the right upper box $-Cx = -C_B B^{-1}b$.
- The reduced cost matrix $\overline{C}_N$ is given in the upper middle box.

With this tableau we move on finding a basic efficient solution.

Finding a basic efficient solution

The tableau T^* allows us to determine the efficiency or the non-efficiency of the basic feasible solution $\overline{x}$ in some evident situations listed below:

- $\overline{x}$ is not efficient if $\overline{C}_N$ has a nonzero positive column.
- $\overline{x}$ is ideal efficient if the entries of $\overline{C}_N$ are non-positive.
- $\overline{x}$ is efficient if at least one of the rows of $\overline{C}_N$ is strictly negative.

In all remaining situations checking the efficiency of $\overline{x}$ is necessary. This can be done by solving an auxiliary problem described in Theorem 7.1.1 (ii). By introducing m surplus variables $y_1, \cdots, y_m$ that problem, denoted (P), is written in the standard form

$$\text{maximize } \langle c^1 + \cdots + c^k, x \rangle$$
$$\text{subject to } \begin{pmatrix} -I & C \\ 0 & A \end{pmatrix} \begin{pmatrix} y \\ x \end{pmatrix} = \begin{pmatrix} C\overline{x} \\ b \end{pmatrix}$$
$$y \geq 0, \ x \geq 0.$$

The feasible solution $\begin{pmatrix} 0 \\ \overline{x} \end{pmatrix}$ is basic and associated with the basis $\hat{B} = \begin{pmatrix} -I_{k \times k} & C_B \\ 0_{m \times k} & B \end{pmatrix}$.
The simplex tableau at this basis is obtained from the tableau T^* as follows

$$\hat{T}^* = \begin{array}{|c|c|c|c|}
\hline
0_{1 \times k} & 0_{1 \times m} & c_N - c_B B^{-1} N & c_B B^{-1} b \\
\hline
I_{k \times k} & 0_{k \times m} & -C_N + C_B B^{-1} N & 0_{k \times 1} \\
0_{m \times k} & I_{m \times m} & B^{-1} N & B^{-1} b \\
\hline
\end{array}$$

In this tableau c stands for the sum of the rows of matrix C. From this tableau we deduce the following conclusion on the efficiency of $\overline{x}$:

- $\overline{x}$ is efficient if the reduced cost vector $\overline{c}_N$ is negative.

If the reduced cost vector $\bar{c}$ has all strictly positive components, nothing can be said about the efficiency of $\bar{x}$ because the basis $\hat{B}$ is degenerate. By introducing a non-basic variable x_j corresponding to a strictly positive component of $\bar{c}$ we may either remove a surplus variable or obtain some feasible solution $\begin{pmatrix} y \\ x \end{pmatrix}$ with $y \geq 0$. If no such feasible solutions are obtained, then $\bar{x}$ is efficient. Otherwise, $\bar{x}$ is non-efficient. Then apply usual pivots of the simplex algorithm (Chap. 3, Sect. 3.3) to (P). Its outcome provides either an optimal vertex solution (y, x) of (P) in which case x is a vertex efficient solution of (MOLP), or shows that (P) has unbounded optimal value. In the latter circumstance (MOLP) has no efficient solution because the value set consisting of Cx with $Cx \geq C\bar{x}$ and x feasible, is unbounded. Its nontrivial asymptotic cone lies in the positive orthant $\mathbb{R}_+^k$. Consequently, the value set of (MOLP) has nonzero positive asymptotic direction too. In view of Theorem 4.1.7, (MOLP) has no maximal element, hence it has no efficient solution.

When the matrix A is not given in the form $(B \ N)$, particular attention must be paid on the columns of B and N. For instance, in the next example, the matrix B of the second problem with three slack variables x_4, x_5 and x_6, consists of the last three columns of A, while the matrix N consists of the first three columns.

Example 7.2.1 Consider the problem

$$\text{Maximize } \begin{pmatrix} 5 & 4 & 3 \\ 3 & 1 & -1 \end{pmatrix} \begin{pmatrix} x_1 \\ x_2 \\ x_3 \end{pmatrix}$$

$$\text{subject to } \begin{pmatrix} 2 & 3 & 1 \\ 4 & 1 & 2 \\ 3 & 4 & 2 \end{pmatrix} \begin{pmatrix} x_1 \\ x_2 \\ x_3 \end{pmatrix} \leq \begin{pmatrix} 5 \\ 11 \\ 8 \end{pmatrix}$$

$$x \geq 0.$$

By introducing three slack variables x_4, x_5 and x_6 we transform the problem to standard form

$$\text{Maximize } \begin{pmatrix} 5 & 4 & 3 & 0 & 0 & 0 \\ 3 & 1 & -1 & 0 & 0 & 0 \end{pmatrix} \begin{pmatrix} x_1 \\ x_2 \\ x_3 \\ x_4 \\ x_5 \\ x_6 \end{pmatrix}$$

$$\text{subject to } \begin{pmatrix} 2 & 3 & 1 & 1 & 0 & 0 \\ 4 & 1 & 2 & 0 & 1 & 0 \\ 3 & 4 & 2 & 0 & 0 & 1 \end{pmatrix} \begin{pmatrix} x_1 \\ x_2 \\ x_3 \\ x_4 \\ x_5 \\ x_6 \end{pmatrix} = \begin{pmatrix} 5 \\ 11 \\ 8 \end{pmatrix}$$

$$x \geq 0.$$

An evident basic solution is $\bar{x} = (0, 0, 0, 5, 11, 8)^T$ and the associated basis is the identity matrix corresponding to the slack variables. The initial simplex tableau is given next

$$T = \begin{array}{|cccccc|c|}
5\,4 & 3\,0\,0\,0 & 0 \\
3\,1 & -1\,0\,0\,0 & 0 \\
2\,3 & 1\,1\,0\,0 & 5 \\
4\,1 & 2\,0\,1\,0 & 11 \\
3\,4 & 2\,0\,0\,1 & 8
\end{array}$$

Since the cost matrix C_B is null, the reduced cost matrix coincides with the non-basic part C_N of the objective matrix C. It has nonzero positive columns, and hence by Theorem 7.1.1 the basic solution $\bar{x}$ is not efficient. In order to find an efficient vertex we perform a pivot by choosing any non-basic index among columns of the reduced matrix with positive components, say the first one. The element $a_{11} = 2$ realizing the minimum min $\left\{ \frac{5}{2}; \frac{11}{4}; \frac{8}{3} \right\}$ is a pivot; the pivotal row $\ell = 1$ and the pivotal column $s = 1$. The variable x_1 enters the new basis and the variable x_4 leaves the basis. At this stage the basic variables are x_1, x_5 and x_6. By pre-multiplying T by the extended matrix for change of basis

$$S = \begin{array}{|ccc|}
1\,0 & -5/2\,0\,0 \\
0\,1 & -3/2\,0\,0 \\
0\,0 & 1/2\,0\,0 \\
0\,0 & -2\,1\,0 \\
0\,0 & -3/2\,0\,1
\end{array}$$

we obtain the new tableau

$$T^1 = \begin{array}{|ccccccc|c|}
0 & -7/2 & 1/2 & -5/2\,0\,0 & -25/2 \\
0 & -7/2 & -5/2 & -3/2\,0\,0 & -15/2 \\
1 & 3/2 & 1/2 & 1/2\,0\,0 & 5/2 \\
0 & -5 & 0 & -2\,1\,0 & 1 \\
0 & -1/2 & 1/2 & -3/2\,0\,1 & 1/2
\end{array}$$

From this tableau we observe that the second row of the reduced cost matrix is strictly negative. In view of Theorem 7.1.1 the new basic solution $x' = (2.5, 0, 0, 0, 1, 0.5)^T$ is efficient.

Moving from an efficient vertex to adjacent efficient vertices

After having found an efficient basic solution our task is to explore other efficient basic solutions. We knew that the set of efficient solutions is arcwise connected, and so if the solution value is not unique, there must exist basic solutions adjacent to the given basic solution. In this subsection we discuss how to find all these adjacent solutions using the simplex tableau T^*.

Assume that $\overline{x}$ is a basic, not ideal, efficient solution. In view of Theorem 7.1.1 the reduced cost matrix $\overline{C}_N$ has strictly positive entries among others and no nonzero positive columns. We classify the non-basic indices into two categories: those s with $\overline{C}_s$ negative and the remaining indices s with $\overline{C}_s$ not comparable with the zero vector. In view of Theorem 7.1.4 we deduce some rules for not introducing a non-basic variable x_s into a new basis:

- If $\overline{C}_s = 0$ and $\overline{a}_s \leq 0$, then the ray $\overline{x} + tv^s, t \geq 0$, where v^s is given by (7.1), is an efficient ray.
- If $\overline{C}_s \leq 0$, then x_s is not for introduction because x_s entering a basis leads to a dominated solution.
- If $\overline{C}_s$ and $\overline{C}_r$ have mixed components, where j is another non-basic index, with t_s and t_r strictly positive, and if $t_r\overline{C}_r - t_s\overline{C}_s \geq 0$, then x_s is not for introduction.

It is now clear that a non-basic variable x_s is eligible for introduction if $\overline{C}_s$ is not comparable with the zero vector and also not comparable with other non-basic reduced cost vectors. We assume that x_s is such a variable with $t_s > 0$. It enters the basis and x_ℓ leaves the basis. The pivot of the current simplex tableau is the element $\overline{a}_{\ell s}$. The new simplex tableau corresponding to the basis with x_s entering and x_ℓ leaving is obtained by pre-multiplying the tableau T^* by the extended matrix of change

$$
S = \begin{array}{|c|ccccc|}
\hline
I_{k\times k} & 0_{k\times 1} & \cdots & -\overline{C}_s/\overline{a}_{\ell s} & \cdots & 0_{k\times 1} \\
\hline
 & 1 & \cdots & -\overline{a}_{1s}/\overline{a}_{\ell s} & \cdots & 0 \\
 & & \cdots & & & \\
0_{m\times k} & & & 1/\overline{a}_{\ell s} & & \\
 & & & \cdots & & \\
 & 0 & \cdots & -\overline{a}_{ms}/\overline{a}_{\ell s} & \cdots & 1 \\
\hline
\end{array}
$$

The new tableau will allow us to determine whether the new basis produces an efficient adjacent vertex of $\overline{x}$ or not.

Example 7.2.2 We continue Example 7.2.1 by starting now with the efficient basic solution $x' = (2.5, 0, 0, 0, 1, 0.5)^T$ and the simplex tableau T^1. The basic variables are x_1, x_5 and x_6. Since the second and the fourth columns of the tableau are negative, the non-basic variables x_2 and x_4 are not for introduction into the basis. The variable x_3 is eligible for entering the basis. Thus, for $s = 3$ we have $t_3 = \min\left\{\frac{5/2}{1/2}, \frac{1/2}{1/2}\right\} = 1$. The variable x_3 enters the basis and x_6 leaves the basis. The pivot is $\overline{a}_{33} = 1/2$, the pivotal row is 3 and the pivotal column is 3. The extended matrix of change is given as

$$S = \begin{array}{|ccc|ccc|} \hline 1 & 0 & 0 & 0 & -1 \\ 0 & 1 & 0 & 0 & 5 \\ \hline 0 & 0 & 1 & 0 & -1 \\ 0 & 0 & 0 & 1 & 0 \\ 0 & 0 & 0 & 0 & 2 \\ \hline \end{array}$$

and the new tableau corresponding to the new basic variables x_1, x_3 and x_5 is given by

$$T^2 = ST^1 = \begin{array}{|ccccccc|c|} \hline 0 & -3 & 0 & -1 & 0 & -1 & -13 \\ 0 & -6 & 0 & -9 & 0 & 5 & -5 \\ 1 & 2 & 0 & 2 & 0 & -1 & 2 \\ 0 & -5 & 0 & -2 & 1 & 0 & 1 \\ 0 & -1 & 1 & -3 & 0 & 2 & 1 \\ \hline \end{array}$$

By reading this tableau we conclude that the new basic solution
$x'' = (2, 0, 1, 0, 1, 0)^T$ is an efficient adjacent vertex of x'.

Generating all efficient vertex solutions

As already discussed, due to the arcwise connectedness of the efficient solution set of
(MOLP), all efficient basic solutions can be generated by exploring adjacent vertices.
Here is a scheme among others to perform it. Given a basic feasible solution v, the
set of efficient adjacent vertices of v is denoted $G(v)$.

Method 1.

- Let v^0 be the initial basic efficient solution.
- Set $E_0 = \{v^0\}$ (the initial generation) and
  $$W_0 = G(v^0) \setminus E_0 \text{ (the first generation)}.$$
- Set $E_1 = E_0 \cup W_0$ and
  $$W_1 = \cup_{v \in W_0} G(v) \setminus E_1 \text{ (the second generation)}.$$
- For $s \geq 1$, if $W_s = \emptyset$, all efficient vertices are in E_s.
  Otherwise, set
  $$E_{s+1} = E_s \cup W_s \text{ and}$$
  $$W_{s+1} = \cup_{v \in W_s} G(v) \setminus E_s \text{ (it is the $(s+1)$-th generation)}.$$
Continue the procedure until no new generation is obtained, that is, until W_s is
empty.

Example 7.2.3 Let us assume that the feasible polytope of (MOLP) has 11 vertices
listed below together with efficient adjacent vertices and non-efficient adjacent vertices:

vertex	efficient adjacent vertices	non-efficient adjacent vertices
v^0	v^1, v^4, v^{10}	v^5
v^1	v^0, v^7, v^8, v^{10}	
v^2	v^7, v^8	
v^3	v^4	v_5
v^4	v^0, v^3, v^{10}	
v^5	v^0, v^3, v^7	v^6
v^6	v^7	v^5
v^7	v^1, v^8	v^2, v^6
v^8	v^1, v^7, v^9	v^2
v^9	v^1, v^8, v^{10}	
v^{10}	v^0, v^1, v^4, v^9	

By starting with v^0 we have

$$E_0 = \{v^0\} \qquad\qquad\qquad W_0 = \{v^1, v^4, v^{10}\}$$
$$E_1 = \{v^0, v^1, v^4, v^{10}\} \qquad W_1 = \{v^3, v^7, v^8, v^9\}$$
$$E_2 = \{v^0, v^1, v^3, v^4, v^7, v^8, v^9, v^{10}\} \quad W_2 = \emptyset$$

We stop the algorithm at the second generation. The set E_2 contains all efficient vertex solutions of the problem.

One should remark that in order to obtain the efficient vertices of the $(s + 1)$-th generation, all simplex tableaux of the s-th generation and their related information (inverse matrices, extended matrices of change etc.) are to be stored, which require a lot of computing memory. Another method consists of obtaining only one efficient vertex at each iteration, and so less memory is needed. Given two basic solutions u and v, the distance used below between them is measured by the number of columns that is needed to be introduced to the basis associated with u in order to obtain the basis associated with v. Thus, if $u \neq v$, the distance from u to v is at least one, which is the case when they are adjacent to each other, and at most m, the total number of columns of a basis.

Method 2.

- Let v^0 be the initial basic efficient solution.
- Set $E_0 = \{v^0\}$ and
    $$W_0 = G(v^0) \setminus E_0.$$
- Choose v^1 from W_0 that is closest to v^0 and set
    $$E_1 = E_0 \cup \{v^1\} \text{ and}$$
    $$W_1 = W_0 \cup G(v^1) \setminus E_1 .$$
- For $s \geq 1$, if $W_s = \emptyset$, all efficient vertices are in E_s. Otherwise choose v^s from W_s that is closest to v^{s-1} the last vertex adhered E_s and set

$$E_{s+1} = E_s \cup \{v^s\} \text{ and}$$
$$W_{s+1} = W_s \cup G(v^{s+1}) \setminus E_s.$$

Continue the procedure until W_s is empty.

Example 7.2.4 By starting with v^0 the method 2 applied to Example 7.2.3 is performed as follows

$$E_0 = \{v^0\} \qquad\qquad\qquad W_0 = \{v^1, v^4, v^{10}\}$$
$$E_1 = \{v^0, v^1\} \qquad\qquad\quad W_1 = \{v^4, v^{10}, v^7, v^8\}$$
$$E_2 = \{v^0, v^1, v^{10}\} \qquad\qquad W_2 = \{v^4, v^9, v^7, v^8\}$$
$$E_3 = \{v^0, v^1, v^{10}, v^4\} \qquad\quad W_3 = \{v^3, v^9, v^7, v^8\}$$
$$E_4 = \{v^0, v^1, v^{10}, v^4, v^3\} \qquad W_4 = \{v^9, v^7, v^8\}$$
$$E_5 = \{v^0, v^1, v^{10}, v^4, v^3, v^9\} \qquad W_5 = \{v^7, v^8\}$$
$$E_6 = \{v^0, v^1, v^{10}, v^4, v^3, v^9, v^8\} \quad W_6 = \{v^7\}$$
$$E_7 = \{v^0, v^1, v^{10}, v^4, v^3, v^9, v^8, v^7\} \quad W_7 = \emptyset$$

Note that after putting v^3 in E_4, there are no adjacent vertices to v^3 from the set W_4. At this stage we choose any from v^7, v^8 and v^9 because they are at equal distance to v^3. The set E_7 contains all efficient vertex solutions of the problem.

The simplex procedure

We summarize the simplex procedure to compute efficient vertices and store them in V as follows.

Step 1. Find the initial feasible vertex by solving the linear problem

$$\text{minimize } y_1 + \cdots + y_m$$
$$\text{subject to } Ax + y = b$$
$$x, y \geq 0.$$

- If no feasible vertex exists, stop. The problem is infeasible.
- Otherwise let x^0 be a feasible vertex. Set $i = 1$ and $v^i = x^0$.

Step 2. Compute the simplex tableau at v^i and go to Step 4 if $i \geq 1$.
Step 3. Check the rows of the reduced cost matrix $\overline{C}_N$.

- If $\overline{C}_N \leq 0$, stop. All efficient solutions x of (MOLP) satisfy equality $Cx = Cv^i$. They are ideal solutions.
- Otherwise go to Step 4.

Step 4. Check whether any row of $\overline{C}_N$ is strictly negative, or the sum of the rows of $\overline{C}_N$ is negative or not.

- If yes, this solution is efficient. Go to Step 6.
- Otherwise go to Step 5.

Step 5. Check the efficiency of the current solution or find an efficient vertex dominating the current solution by solving the linear problem

$$\begin{aligned}
\text{maximize } & \langle c^1 + \cdots + c^k, x \rangle \\
\text{subject to } & Ax = b \\
& Cx \geq Cv^i \\
& x \geq 0.
\end{aligned}$$

- If it has unbounded optimal value, stop. The problem (MOLP) has no efficient solutions.
- If the optimal value is finite and if v^i is an optimal solution, then it is efficient and go to Step 6.
- If $\bar{x}$ is an optimal vertex with $C\bar{x} \geq Cv^i$, then v^i is not efficient and $\bar{x}$ is efficient. Compute the simplex tableau at $\bar{x}$ and go to Step 6.

Step 6. Store this efficient vertex in V.
Step 7. Find all non-basic indices r of the last stored vertex satisfying

(i) $\overline{C}_r$ contains mixed components (negative and positive ones).
(ii) There is no other non-basic index s satisfying (i) such that $t_s \overline{C}_s \geq t_r \overline{C}_r$.

Step 8. Store all unexplored bases by introducing r from the preceding step.

- If no such bases exist, stop.
- Otherwise obtain v^{i+1} the basic solution with r entering the basis.

Step 9. Set $i = i + 1$ and return to Step 2.

Let us illustrate the simplex procedure by a hand computing example.

Example 7.2.5 Consider the problem

$$\text{Maximize } \begin{pmatrix} 6 & 4 & 5 \\ 0 & 0 & 1 \end{pmatrix} \begin{pmatrix} x_1 \\ x_2 \\ x_3 \end{pmatrix}$$

$$\text{subject to } \begin{pmatrix} 1 & 1 & 2 \\ 1 & 2 & 1 \\ 2 & 1 & 1 \end{pmatrix} \begin{pmatrix} x_1 \\ x_2 \\ x_3 \end{pmatrix} \leq \begin{pmatrix} 12 \\ 12 \\ 12 \end{pmatrix}$$

$$x_i \geq 0, \ i = 1, 2, 3.$$

We write this problem in standard form by introducing three slack variables x_4, x_5 and x_6:

$$\text{Maximize} \quad \begin{pmatrix} 6 & 4 & 5 & 0 & 0 & 0 \\ 0 & 0 & 1 & 0 & 0 & 0 \end{pmatrix} \begin{pmatrix} x_1 \\ x_2 \\ x_3 \\ x_4 \\ x_5 \\ x_6 \end{pmatrix}$$

$$\text{subject to} \quad \begin{pmatrix} 1 & 1 & 2 & 1 & 0 & 0 \\ 1 & 2 & 1 & 0 & 1 & 0 \\ 2 & 1 & 1 & 0 & 0 & 1 \end{pmatrix} \begin{pmatrix} x_1 \\ x_2 \\ x_3 \\ x_4 \\ x_5 \\ x_6 \end{pmatrix} = \begin{pmatrix} 12 \\ 12 \\ 12 \end{pmatrix}$$

$$x_i \geq 0, \ i = 1, \cdots, 6.$$

We will identify a basis B_J with its index set J. An evident basis to start with is $J^1 = [4, 5, 6]$, that is x_4, x_5 and x_6 are basic variables. The matrix B_{J^1} is the identity 3×3-matrix. The initial simplex tableau is below

$$T^1 = \begin{array}{|cccccc|c|}
\hline
6 & 4 & 5 & 0 & 0 & 0 & 0 \\
0 & 0 & 1 & 0 & 0 & 0 & 0 \\
\hline
1 & 1 & 2 & 1 & 0 & 0 & 12 \\
1 & 2 & 1 & 0 & 1 & 0 & 12 \\
2 & 1 & 1 & 0 & 0 & 1 & 12 \\
\hline
\end{array}$$

It is clear from this tableau that the basis J^1 is not efficient. We solve the linear problem of Step 5 to find an efficient vertex dominating the current solution $v^1 = (0, 0, 0, 12, 12, 12)^T$. As $Cv^1 = 0$, the problem takes the form

$$\text{maximize} \ 6x_1 + 4x_2 + 6x_3$$

$$\text{subject to} \quad \begin{pmatrix} 1 & 1 & 2 & 1 & 0 & 0 \\ 1 & 2 & 1 & 0 & 1 & 0 \\ 2 & 1 & 1 & 0 & 0 & 1 \end{pmatrix} \begin{pmatrix} x_1 \\ x_2 \\ x_3 \\ x_4 \\ x_5 \\ x_6 \end{pmatrix} = \begin{pmatrix} 12 \\ 12 \\ 12 \end{pmatrix}$$

$$6x_1 + 4x_2 + 5x_3 \geq 0$$
$$x_i \geq 0, \ i = 1, \cdots, 6.$$

Note that inequality $Cx \geq Cv^1$ consists of two inequalities in which the second one is exactly $x_3 \geq 0$. Note further that the constraint $6x_1 + 4x_2 + 5x_3 \geq 0$ is consequence of the positivity of the variables, and so it is superfluous. We deal again with the linear problem whose simplex tableau at the basis $J^1 = [4, 5, 6]$ is obtained from the tableau T by substituting the two rows of the objective function by their sum. We perform a pivot with the pivot $\bar{a}_{13} = 2$ (the pivotal row $\ell = 1$ and the pivotal column $s = 3$). The new basis is $J^2 = [3, 5, 6]$, that is the basic variables are x_3, x_5, x_6. We write the new simplex tableau for both the linear problem and the (MOLP):

$$
T^2 = \begin{array}{|ccccccc|}
\hline
3 & 1 & 0 & -3 & 0 & 0 & -36 \\
\hline
7/2 & 3/2 & 0 & -5/2 & 0 & 0 & -30 \\
-1/2 & -1/2 & 0 & -1/2 & 0 & 0 & -6 \\
\hline
1/2 & 1/2 & 1 & 1/2 & 0 & 0 & 6 \\
1/2 & 3/2 & 0 & -1/2 & 1 & 0 & 6 \\
3/2 & 1/2 & 0 & -1/2 & 0 & 1 & 6 \\
\hline
\end{array}
$$

The basic solution associated with the basis J^2 is $v^2 = (0, 0, 6, 0, 6, 6)^T$. Although it is not optimal for the linear problem, it is efficient for the (MOLP) because the second row of the reduced cost matrix is strictly negative. From the tableau we obtain two columns eligible for introduction into the basis: $\overline{C}_1$ and $\overline{C}_2$. We compute $t_1 = \min\left\{\frac{6}{1/2}, \frac{6}{1/2}, \frac{6}{3/2}\right\} = 4$ and $t_2 = \min\left\{\frac{6}{1/2}, \frac{6}{3/2}, \frac{6}{1/2}\right\} = 4$ yielding respectively the pivots $\overline{a}_{31} = 3/2$ and $\overline{a}_{22} = 3/2$. Notice, however, that $t_1\overline{C}_1 \geq t_2\overline{C}_2$, which in view of Theorem 7.1.3 means that the basic solution obtained from v^2 by introducing x_2 into the basis is dominated by the solution obtained from v^2 by introducing x_1 into the basis. The pivot $\overline{a}_{31} = 3/2$ gives the following tableau including the sum of the reduced cost rows

$$
T^3 = \begin{array}{|ccccccc|}
\hline
0 & 0 & 0 & -2 & 0 & -2 & -48 \\
\hline
0 & 1/3 & 0 & -4/3 & 0 & -7/3 & -44 \\
0 & -1/3 & 0 & -2/3 & 0 & 1/3 & -4 \\
\hline
0 & 1/3 & 1 & 2/3 & 0 & -/3 & 4 \\
0 & 4/3 & 0 & -1/3 & 1 & -1/3 & 4 \\
1 & 1/3 & 0 & -1/3 & 0 & 2/3 & 4 \\
\hline
\end{array}
$$

Although the reduced cost matrix does not allow us to conclude about the efficiency of this basis $J^3 = [1, 3, 5]$, but reading the reduced cost vector sum of the reduced cost rows proves that this basis is efficient. Its associated solution is $v^3 = (4, 0, 4, 0, 4, 0)^T$.

From the tableau we have two non-basic variables that are eligible for introduction: x_2 and x_6. The introduction of x_6 leads to the basis $[3, 5, 6]$ that was already explored. We introduce x_2 by performing the pivot $\overline{a}_{32}$. The basic variable x_5 leaves the basis. The new basis is $J^4 = [1, 2, 3]$ and its associated tableau is below

$$
T^4 = \begin{array}{|ccccccc|}
\hline
0 & 0 & 0 & -2 & 0 & -2 & -48 \\
\hline
0 & 0 & 0 & -5/4 & -1/4 & -9/4 & -45 \\
0 & 0 & 0 & -3/4 & 1/4 & 1/4 & -3 \\
\hline
1 & 0 & 0 & -1/4 & -1/4 & 3/4 & 3 \\
0 & 1 & 0 & -1/4 & 3/4 & -1/4 & 3 \\
0 & 0 & 1 & 3/4 & -1/4 & -1/4 & 3 \\
\hline
\end{array}
$$

It is clear that this basis is optimal. Its associated solution is the vector $v^4 =$ $(3, 3, 3, 0, 0, 0)^T$. The fourth column of $\overline{C}$ being strictly negative, introduction of x_4 leads to a dominated solution. For the fifth and the sixth columns we compute $t_5 = t_6 = 4$. Since $t_5\overline{C}_5 \geq t_6\overline{C}_6$ the introduction of x_6 leads to a dominated solution. Thus, this time the pivot is $\overline{a}_{52}$ with x_2 leaving the basis. The basis obtained from this pivot is $[1, 3, 5]$ which is J^3 already explored. The algorithm terminates. The initial problem has three efficient vertices $(0, 0, 6)^T$, $(4, 0, 4)^T$ and $(3, 3, 3)^T$.

7.3 Exercises

7.3.1 Let B be a basis of (MOLP) considered in Sect. 7.1 and $\overline{x}$ the feasible basic solution associated with B. One defines $\Lambda(B) = \{\lambda \in \mathbb{R}^k : \lambda^T\overline{C}_N \leq 0\}$ where $\overline{C}_N$ is the reduced cost matrix at $\overline{x}$.

(a) Prove that $\overline{x}$ maximizes the function $\lambda^T Cx$ over the feasible set of (MOLP) for every $\lambda \in \Lambda(B)$.
(b) Show that if $\Lambda(B) \cap \mathrm{int}(\mathbb{R}^k_+) \neq \emptyset$, then $\overline{x}$ is efficient. Conversely, if $\overline{x}$ is efficient and if B is non-degenerate, then $\Lambda(B) \cap \mathrm{int}(\mathbb{R}^k_+) \neq \emptyset$.
(c) If B' is the new feasible basis obtained from B by introducing a non-basic variable x_r into the basis, then $\Lambda(B) = \Lambda(B')$ if the reduced cost column $\overline{C}_r$ is zero, and $\mathrm{ri}(\Lambda(B)) \cap \mathrm{ri}(\Lambda(B')) = \emptyset$ if $\overline{C}_r$ is nonzero.

7.3.2 Maximize two functions $f_1(x, y) = 3x + y$ and $f_2(x, y) = x - y$ on the polyhedral set defined by the system

$$
\begin{cases}
x + y \leq 4 \\
2x + y \leq 5 \\
x, y \geq 0.
\end{cases}
$$

7.3.3 Maximize three functions $f_1(x, y) = x+y$, $f_2(x, y) = x+2y$ and $f_3(x, y) = x - y$ on the polyhedral set defined by the system

$$
\begin{cases}
x - y \leq 4 \\
-x + y \leq 4 \\
x + y \leq 8 \\
x, y \geq 0.
\end{cases}
$$

7.3.4 Consider the problem

$$\text{Maximize } \begin{pmatrix} x \\ 2x - y \end{pmatrix}$$
$$\text{subject to } \quad y - 2x \geq 1$$
$$x - 3y \leq 1$$
$$y - x \leq 1.5$$
$$x \geq 0, \ y \geq 0$$

Find asymptotic directions of the feasible set and show that the problem is unbounded.

7.3.5 Find a basic feasible solution of the problem

$$\text{Maximize } \begin{pmatrix} x_1 + x_2 + x_3 \\ -x_1 - x_2 + x_3 \end{pmatrix}$$
$$\text{subject to } \quad 2x_1 + x_2 - 2x_3 = 4$$
$$3x_1 - 3x_2 + x_3 = 3$$
$$x_1, x_2, x_3 \geq 0$$

and prove that the problem is unbounded.

7.3.6 Let C be a 2×3-matrix. Consider the problem of maximizing Cx under the constraints as defined in the preceding exercise. Find conditions on C such that the problem has efficient solutions.

7.3.7 What can be said about the solution set of a multiobjective problem in which the objective matrix has rank one?

7.3.8 Solve the following problem using the simplex method

$$\text{Maximize } \begin{pmatrix} x_1 + x_2 + x_3 \\ -3x_1 + x_2 + 3x_3 - x_4 \end{pmatrix}$$
$$\text{subject to } \quad x_1 - x_2 + 2x_3 - x_4 = 6$$
$$2x_1 - 2x_2 + 3x_3 + 3x_4 = 9$$
$$x_1, x_2, x_3, x_4 \geq 0$$

7.3.9 Solve the following problem by simplex method

$$\text{Maximize } \begin{pmatrix} x_1 + 2x_2 - x_3 + 3x_4 + 2x_5 + x_7 \\ x_2 + x_3 + 2x_4 + 3x_5 + x_6 \\ x_1 + x_3 - x_4 - x_6 - x_7 \end{pmatrix}$$
$$\text{subject to } \quad x_1 + 2x_2 + x_3 + x_4 + 2x_5 + x_6 + 2x_7 \leq 16$$
$$-2x_1 - x_2 + x_4 + 2x_5 + x_7 \leq 16$$
$$x_2 + 2x_3 - x_4 + x_5 - 2x_6 - x_7 \leq 16$$
$$x_i \geq 0, \ i = 1, \cdots, 7.$$

(The efficient vertices are:

$$(0, 0, 0, 0, 8, 0, 0)^T \quad (0, 0, 0, 16, 0, 0, 0)^T \quad (16, 0, 0, 0, 0, 0, 0)^T$$
$$(8, 0, 8, 0, 0, 0, 0)^T \quad (0, 0, \tfrac{32}{3}, \tfrac{16}{3}, 0, 0, 0)^T \quad (0, 0, \tfrac{16}{3}, 0, \tfrac{16}{3}, 0, 0)^T)$$

Chapter 8
Normal Cone Method

An important characterization of efficient faces of a polyhedral set is the fact that their normal cones contain strictly positive vectors. This will be utilized to develop an algorithm to find the efficient solution set and the efficient value set of a multiobjective linear problem.

8.1 Normal Index Sets

We consider a finite system of linear inequalities

$$\langle a^i, x \rangle \leqq b_i, \quad i = 1, \cdots, m, \tag{8.1}$$

where $a^1, \cdots, a^m$ are n-dimensional column vectors and $b_1, \cdots, b_m$ are real numbers. The solution set of this system is denoted X. Throughout this chapter we assume the following hypothesis:

(A1) The system (8.1) is non-redundant and consistent.

Recall that the set of active indices at a point $x^0 \in X$ is denoted $I(x^0)$, which consists of indices $i \in \{1, \cdots, m\}$ such that $\langle a^i, x^0 \rangle = b_i$ for $i \in I(x^0)$ and $\langle a^i, x^0 \rangle < b_i$ for $i \notin I(x^0)$. We also recall that given a face F of X, the index set $I(F)$ (sometimes denoted I_F) of F is the active index set at a relative interior point of F, and $\mathrm{pos}(X)$ is the positive hull of X.

Definition 8.1.1 A nonempty index set $I \subseteq \{1, \cdots, m\}$ is said to be normal if there is some point $x \in X$ such that

$$N_X(x) = \mathrm{pos}\{a^i : i \in I\}.$$

It is clear that when X has a boundary point, that is, at least one vector among $a^i, i = 1, \cdots, m$ is nonzero, then normal index sets exist. Moreover, not every subset of the index set $\{1, \cdots, m\}$ is normal.

© Springer International Publishing Switzerland 2016
D.T. Luc, *Multiobjective Linear Programming*,
DOI 10.1007/978-3-319-21091-9_8

Example 8.1.2 Consider a system of four inequalities in $\mathbb{R}^2$:

$$\begin{aligned}
x_1 &\leqq 2 \\
-x_1 &\leqq -1 \\
x_1 + x_2 &\leqq 3 \\
-x_1 - x_2 &\leqq -2.
\end{aligned}$$

This system consists of four inequalities that we enumerate from one to four. It is clear that the index sets $\{1\}$, $\{2\}$, $\{3\}$, $\{4\}$, $\{1, 3\}$, $\{1, 4\}$, $\{2, 3\}$ and $\{2, 4\}$ are all normal, while the remaining subsets of the index set $\{1, 2, 3, 4\}$ are not. For instance $I = \{1, 2\}$ is not normal because

$$\text{pos}\{a^1, a^2\} = \text{pos}\left\{\begin{pmatrix} 1 \\ 0 \end{pmatrix}, \begin{pmatrix} -1 \\ 0 \end{pmatrix}\right\} = \text{pos}\left\{\begin{pmatrix} t \\ 0 \end{pmatrix}, t \in \mathbb{R}\right\}$$

is a normal cone to X at no point.

Lemma 8.1.3 *An index set $I \subseteq \{1, \cdots, m\}$ is normal if and only if the following system has a solution*

$$\begin{aligned}
\langle a^i, x \rangle &= b_i, i \in I \\
\langle a^j, x \rangle &< b_j, j \in \{1, \cdots, m\} \backslash I.
\end{aligned} \tag{8.2}$$

Proof Assume that the system (8.2) has a solution, denoted x. Then x is a boundary point of X and the active index set at x is I. In view of Theorem 2.3.24, we have $N_X(x) = \text{pos}\{a^i : i \in I\}$. By definition I is a normal index set.

Conversely, let I be a normal index set and let $x \in X$ be a point such that $N_X(x) = \text{pos}\{a^i : i \in I\}$. Since x is an element of X, it satisfies the system

$$\begin{aligned}
\langle a^i, x \rangle &= b_i, i \in I(x) \\
\langle a^j, x \rangle &< b_j, j \in \{1, \cdots, m\} \backslash I(x).
\end{aligned}$$

In view of Theorem 2.3.24, we have $N_X(x) = \text{pos}\{a^i : i \in I(x)\}$ and deduce

$$\text{pos}\{a^i : i \in I(x)\} = \text{pos}\{a^i : i \in I\}.$$

We claim that every vector a^i for $i \in I(x)$ is an extreme ray of the cone $\text{pos}\{a^i : i \in I(x)\}$. Indeed, assume to the contrary, that for some index $i_0 \in I(x)$ one finds $t_i \geqq 0, i \in I(x) \backslash \{i_0\}$ such that $a^{i_0} = \sum_{i \in I(x) \backslash \{i_0\}} t_i a^i$. Then

$$b_{i_0} = \langle a^{i_0}, x \rangle = \sum_{i \in I(x) \backslash \{i_0\}} t_i \langle a^i, x \rangle = \sum_{i \in I(x) \backslash \{i_0\}} t_i b_i.$$

Let $y \in \mathbb{R}^n$ satisfy

$$\langle a^i, y \rangle \leqq b_i \text{ for } i \in I(x) \setminus \{i_0\}.$$

We deduce

$$\langle a^{i_0}, y \rangle = \sum_{i \in I(x) \setminus \{i_0\}} t_i \langle a^i, y \rangle \leqq \sum_{i \in I(x) \setminus \{i_0\}} t_i b_i = b_{i_0}.$$

This shows that inequality $\langle a^{i_0}, y \rangle \leqq b_{i_0}$ is redundant in (8.1), a contradiction to (A1). By this, $I(x) \subseteq I$ and each of vectors a^j, $j \in I$ can be expressed as a positive combination of the vectors a^i, $i \in I(x)$. Again, by a similar argument as above, one proves that $I = I(x)$ due to the non-redundancy hypothesis (A1). Consequently the system (8.2) is consistent. $\qquad\qquad\qquad\qquad\qquad\qquad\qquad\qquad\qquad\qquad\qquad\square$

There is a close relation between normal index sets and faces of X. We remember that an index set $I \subseteq \{1, \cdots, m\}$ is said to determine a face F of X if F is the solution set to the system

$$\begin{aligned} \langle a^i, x \rangle &= b_i, i \in I \\ \langle a^j, x \rangle &\leqq b_j, j \in \{1, \cdots, m\} \setminus I \end{aligned} \qquad (8.3)$$

and if no inequality can be replaced by equality. We know from Corollary 2.3.5 that I, denoted also $I(F)$, coincides with the active index set of a relative interior point of F. Moreover, if F is the solution set of another system corresponding to another index set $I' \subseteq \{1, \cdots, m\}$, then $I' \subseteq I(F)$.

Theorem 8.1.4 *A nonempty index set $I \subseteq \{1, \cdots, m\}$ is normal if and only if the solution set to the system (8.3) is determined by I. Moreover, a nonempty subset F of X is a face if and only if it is determined by a normal index set.*

Proof Let $I \subseteq \{1, \cdots, m\}$ be a normal index set and let F be the solution set to the system (8.3). By definition there is some point $\bar{x} \in X$ such that $N_X(\bar{x}) = \text{pos}\{a^i : i \in I\}$. We wish to prove that I determines F. Towards this end, we first show that F is nonempty, namely it contains $\bar{x}$, that is

$$\langle a^i, \bar{x} \rangle = b_i \text{ for } i \in I.$$

Suppose, to the contrary, that there is some index $i_0 \in I$ such that $\langle a^{i_0}, \bar{x} \rangle < b_{i_0}$. It follows from the definition of the normal cone that

$$\langle a^{i_0}, y \rangle \leqq \langle a^{i_0}, \bar{x} \rangle < b_{i_0} \qquad (8.4)$$

for every $y \in X$. Consider the system

$$\langle a^i, y \rangle \leqq b_i \text{ for } i \in \{1, \cdots, m\} \setminus \{i_0\}.$$

We claim that every solution y of this system lies in X. This is because if for some solution y one has $\langle a^{i_0}, y \rangle > b_{i_0}$, then there is a real number $t \in [0, 1]$ such that $\langle a^{i_0}, t\bar{x} + (1-t)y \rangle = b_{i_0}$. As the point $t\bar{x} + (1-t)y$ belongs to X, the latter equality contradicts (8.4). Thus, inequality $\langle a^{i_0}, y \rangle \leq b_{i_0}$ is redundant, which contradicts (A1). Next, we prove $I = I(\bar{x})$. Inclusion $I \subseteq I(\bar{x})$ is evident. If there is some $j \in I(\bar{x}) \setminus I$, then in view of Theorem 2.3.24, a^j is a normal vector at $\bar{x}$, and hence there are $t_i \geq 0, i \in I$ such that $a^j = \sum_{i \in I} t_i a^i$. This expression of a^j leads to a contradiction that inequality $\langle a^j, y \rangle \leq b_j$ is redundant. Hence we conclude that I determines the face F.

To show the second part of the theorem, let F be a face of X. Pick any relative interior point $\bar{x}$ of F. Then $\langle a^j, \bar{x} \rangle < b_j$ for $j \in \{1, \cdots, m\} \setminus I(\bar{x})$ and $\langle a^i, \bar{x} \rangle = b_i$ for $i \in I(\bar{x})$. In view of Lemma 8.1.3, $I(\bar{x})$ is a normal index set that determines F. The converse statement is clear because if F is nonempty and given by the system (8.3), then it is a face of X by Theorem 2.3.3. $\square$

Let $\bar{x}$ be a vertex of X. It is a zero-dimensional face, hence the index set $I(\bar{x})$ has at least n elements. We recall that a point $\bar{x}$ is a non-degenerate vertex of X if there are exactly n linearly independent inequalities in (8.1) that are satisfied as equalities at $\bar{x}$. It follows that the active index set at a non-degenerate vertex has n elements. The next result tells us when a subset $I \subseteq I(\bar{x})$ is a normal set.

Corollary 8.1.5 *Let $\bar{x}$ be a non-degenerate vertex of X. Then every nonempty subset $I \subseteq I(\bar{x})$ is normal.*

Proof Without loss of generality we may assume that $I(\bar{x}) = \{1, \cdots, n\}$. Let I be a nonempty subset of $I(\bar{x})$. Consider the system

$$\langle a^i, x \rangle = b_i, \ i \in I$$
$$\langle a^j, x \rangle \leq b_j, j \in \{1, \cdots, m\} \setminus I.$$

This system is consistent, for instance $\bar{x}$ is a solution. Hence the solution set, denoted F, is a face of X. Since the vectors $a^i, i = 1, \cdots, n$ are linearly independent, the dimension of F is equal to $n - |I|$. Let I_F be the index set that determines the face F and is the active index set at a relative interior point of F. Then $I \subseteq I_F$ and $\dim F = n - \text{rank}\{a^i : i \in I_F\} = n - |I|$. We conclude $I = I_F$. By Theorem 8.1.4, I is a normal index set. $\square$

Corollary 8.1.6 *Let I^1 and I^2 be two normal index sets. Then the intersection $I^1 \cap I^2$ is normal if it is nonempty.*

Proof Since I^1 and I^2 are normal, when $I = I^1$ (respectively $I = I^2$) the system (8.2) has at least one solution, say x (respectively y). Set $z = (x + y)/2$. Then for $i \in I^1 \cap I^2$ we have

$$\langle a^i, z \rangle = \frac{1}{2}\left(\langle a^i, x \rangle + \langle a^i, y \rangle\right) = \frac{1}{2}(b_i + b_i) = b_i \ .$$

For $j \in \{1, \cdots, m\} \setminus (I^1 \cap I^2)$ we have either $j \notin I^1$ which implies $\langle a^j, x \rangle < b_j$, or $j \notin I^2$ which implies $\langle a^j, y \rangle < b_j$. This implies

$$\langle a^j, z \rangle = \frac{1}{2}\left(\langle a^j, x \rangle + \langle a^j, y \rangle\right) < \frac{1}{2}(b_j + b_j) = b_j$$

for all $j \in \{1, \cdots, m\} \setminus (I^1 \cap I^2)$. Consequently, z is a solution to the system (8.2), in which $I = I^1 \cap I^2$. In view of Lemma 8.1.3, the index set $I^1 \cap I^2$ is normal. $\square$

Assume that there exist ℓ edges $F_1, \cdots, F_\ell$ emanating from a vertex $\bar{x}$. Then, each of $I(F_1), \cdots, I(F_\ell)$ has at least $(n-1)$ elements (remember that $I(F_i)$ denotes the active index set of a relative interior point of F_i). Let $J \subseteq \{1, \cdots, \ell\}$ with $|J| = r \leq \min\{\ell, n-1\}$. Take $x^i \in F_i \setminus \{\bar{x}\}$, $i = 1, \cdots, \ell$ and set

$$x^J = \frac{\bar{x}}{r+1} + \sum_{j \in J} \frac{x^j}{r+1}.$$

The next result allows us to determine the largest face that contains x^J as a relative interior point.

Proposition 8.1.7 *Assume that the active index set $I(x^J)$ is nonempty. Then it is a normal set and the face F determined by the system*

$$\langle a^i, x \rangle = b_i, \ i \in I(x^J)$$
$$\langle a^j, x \rangle \leqq b_j, \ j \in \{1, \cdots, m\} \setminus I(x^J),$$

contains the convex hull of all edges F_j, $j \in \{1, \cdots, m\}$ that satisfy the containment $I(F_j) \supseteq I(x^J)$, including $j \in J$.

Proof Since x^J belongs to X, it is a solution of the system described in the proposition. Hence F is a face of X. It follows from the definition of active index sets, x^J is a solution to the system

$$\langle a^i, x^i \rangle = b_i, \ i \in I(x^J)$$
$$\langle a^j, x^i \rangle < b_j, \ j \in \{1, \cdots, m\} \setminus I(x^J).$$

By Lemma 8.1.3, $I(x^J)$ is a normal index set. Furthermore, if for some index $j \in \{1, \cdots, m\}$ one has $I(F_j) \supseteq I(x^J)$, then $N_X(F_j) \supseteq N_X(F)$, and hence $F_j \subseteq F$. Being convex, the face F contains the convex hull of all such edges. Finally, for $i \in I(\bar{x})$, equality

$$b_i = \langle a^i, x^J \rangle = \langle a^i, \frac{\bar{x}}{r+1} + \sum_{j \in J} \frac{x^j}{r+1} \rangle$$

holds if and only if $\langle a^i, x^j \rangle = b_i$ for all $i \in I(\bar{x})$ and $j \in J$. We conclude that $F_j \subseteq F$ for all $j \in J$. $\square$

8.2 Positive Index Sets

Let C be a $k \times n$-matrix and let the columns of C^T be denoted by $c^1, \cdots, c^k$.

Definition 8.2.1 A vector $v \in \mathbb{R}^n$ is called C-positive if there exist strictly positive numbers $\lambda_1, \cdots, \lambda_k$ such that $v = \sum_{i=1}^{k} \lambda_i c^i$, and it is C-negative if $-v$ is C-positive.

In the matrix form, a column vector v is C-positive if and only if $v = C^T \lambda$ for some strictly positive vector $\lambda \in \mathbb{R}^k$. Throughout this chapter we also assume the following

(A2) the cone pos$\{c^1, \cdots, c^k\}$ is not a linear subspace.

This assumption is clearly equivalent to the fact that the origin of the space is not C-positive. When the zero vector of $\mathbb{R}^n$ is a strictly positive combination of $c^1, \cdots, c^k$, the problem of vector maximizing Cx over a set $X \subseteq \mathbb{R}^n$ becomes trivial because every feasible solution is maximal. Some more properties of C-positive vectors are given next.

Lemma 8.2.2 *The following properties hold true.*

(i) *If C is the identity matrix, then a vector $v \in \mathbb{R}^n$ is C-positive if and only if it is strictly positive.*

(ii) *The set of C-positive vectors coincides with the relative interior of the cone pos$\{c^1, \cdots, c^k\}$.*

(iii) *If there is a vector simultaneously C-positive and C-negative, then the rows of C are linearly dependent.*

(iv) *For $x \in \mathbb{R}^n$, one has $Cx \geqq 0$ (respectively $Cx > 0$) in $\mathbb{R}^k$ if and only if $\langle v, x \rangle \geqq 0$ (respectively $\langle v, x \rangle > 0$) for every C-positive vector v of $\mathbb{R}^n$.*

Proof The first property is immediate from the definition. The second one follows from Lemma 6.4.10. For the third property, we notice that when a vector is simultaneously C-positive and C-negative, then the zero vector is a linear combination of the rows of C. Hence the rows of C are linearly dependant. Let us prove the last property. We have $Cx \geqq 0$ if and only if $0 \leqq \langle Cx, \lambda \rangle = \langle x, C^T \lambda \rangle$ for every $\lambda \in \mathbb{R}^k, \lambda > 0$, or equivalently $\langle x, v \rangle \geqq 0$ for every C-positive vector $v \in \mathbb{R}^n$. The strict inequality $Cx > 0$ is proven in a similar way. $\square$

Definition 8.2.3 Let $a^1, \cdots, a^m$ be (column) vectors in $\mathbb{R}^n$. An index set $I \subseteq \{1, \cdots, m\}$ is said to be positive if pos$\{a^i : i \in I\}$ contains a C-positive vector.

It is clear that if an index set is positive, any index set that contains it is also positive, while a smaller subset is not necessarily positive.

Example 8.2.4 Consider a matrix $C = \begin{pmatrix} 2 & -3 & -1 \\ 3 & 1 & 0 \end{pmatrix}$ and a system of inequalities given by

$$\begin{pmatrix} 1 & 1 & 1 \\ -2 & -2 & -1 \\ -1 & 0 & 0 \\ 0 & -1 & 0 \\ 0 & 0 & -1 \end{pmatrix} \begin{pmatrix} x_1 \\ x_2 \\ x_3 \end{pmatrix} \leqq \begin{pmatrix} 1 \\ -1 \\ 0 \\ 0 \\ 0 \end{pmatrix}.$$

The transposes of the row vectors of this system are denoted by $a^1, \cdots, a^5$. Direct calculation shows that the index set $\{2, 4\}$ is not positive. The set $\{1, 4, 5\}$ is positive because the cone $\text{pos}\{a^1, a^4, a^5\}$, where $a^1 = (1, 1, 1)^T, a^4 = (0, -1, 0)^T$ and $a^5 = (0, 0, -1)^T$, contains a C-positive vector $v = (7/2, -5/2, -1)^T = (2, -3, -1)^T + \frac{1}{2}(3, 1, 0)^T$.

Proposition 8.2.5 *An index set $I \subseteq \{1, \cdots, m\}$ is positive if and only if the following system is consistent*

$$\sum_{i \in I} \mu_i a^i - \sum_{j=1}^{k} \lambda_j c^j = 0 \tag{8.5}$$
$$\mu_i \geq 0, i \in I \text{ and } \lambda_j \geq 1, \ j = 1, \cdots, k.$$

Proof It is clear that if the system has a solution, then $\text{pos}\{a^i : i \in I\}$ contains the C-positive vector $\sum_{j=1}^{k} \lambda_j c^j$. Conversely, if there are strictly positive numbers $\lambda_1, \cdots, \lambda_k$ such that $\sum_{j=1}^{k} \lambda_j c^j$ belongs to the cone $\text{pos}\{a^i : i \in I\}$, then the vector $\sum_{j=1}^{k} \frac{\lambda_j}{\min\{\lambda_1, \cdots, \lambda_k\}} c^j$ belongs to that cone too, by which the system given in the proposition has a solution. $\square$

Given a family of vectors $a^1, \cdots, a^m$ we denote

$$I^1 = \{i \in \{1, \cdots, m\} : a^i \text{ is } C-\text{positive}\}$$
$$I^3 = \{i \in \{1, \cdots, m\} : a^i \text{ is } C-\text{negative}\}$$
$$I^2 = \{1, \cdots, m\} \setminus (I^1 \cup I^3).$$

Under (A2) we have a partition of the index set $\{1, \cdots, k\} = I^1 \cup I^2 \cup I^3$ by disjoint subsets. The next result shows how to find positive normal sets outside I^1.

Theorem 8.2.6 *Assume that $I \subseteq \{1, \cdots, m\}$ is a positive and normal index set such that the cone $\text{pos}\{a^i : i \in I\}$ is not a linear subspace. Then there exists a positive and normal set $I_0 \subseteq I \cap (I^1 \cup I^2)$.*

Proof Let $I = \{i_1, \cdots, i_l\}$ be a positive and normal index set. We prove the theorem by induction on l. If $l = 1$, then $t a^{i_1}$ is C-positive for some $t \geq 0$ since $\text{pos}\{a^{i_1}\}$ is a positive normal cone. Actually $t > 0$ because otherwise the zero vector would belong to the relative interior of the cone $\mathrm{pos}\{c^1, \cdots, c^k\}$, which contradicts the assumption that $\mathrm{pos}\{c^1, \cdots, c^k\}$ is not a linear subspace. It follows that a^{i_1} is C-positive and $I_0 = I$ satisfies the requirements of the theorem.

Now, let $l > 1$. We claim that $I \cap (I^1 \cup I^2) \neq \emptyset$. Indeed, if not, then one has $I \subseteq I^3$. Let v be a C-positive vector that belongs to the cone $\mathrm{pos}\{a^i : i \in I\}$. As all $a^i, i \in I$ are C-negative, the vector v is C-negative too. We deduce that $0 = v - v$ is a C-positive vector and arrive at the same contradiction as above. Consider two possible cases: $I \cap I^3 = \emptyset$ and $I \cap I^3 \neq \emptyset$, say i_l is a common element of I and I^3. In the first case, the index set $I_0 = I$ will be suitable to achieve the proof. In the second case we claim that the cone $\mathrm{pos}\{a^i : i \in I\}$ does not contain all C-positive vectors in its relative interior. In fact, if not, this relative interior should contain the vector $-a^{i_l}$ because a^{i_l} is C-negative, and hence $\mathrm{pos}\{a^i : i \in I\}$ contains the zero vector $0 = a^{i_l} - a^{i_l}$ in its relative interior, which contradicts the hypothesis. Let u be a C-positive vector outside the relative interior of the cone $\mathrm{pos}\{a^i : i \in I\}$. Joining v and u we find a C-positive vector w on a proper face of the cone $\mathrm{pos}\{a^i : i \in I\}$. Let I' be a proper subset of I such that the face is the cone $\mathrm{pos}\{a^i : i \in I'\}$. In view of Theorem 2.3.26 the index set I' determines a face of P, which contains the face determined by I. In other words, I' is a normal index set. It is positive because it contains the C-positive vector w. By induction, there is a positive normal index set $I_0 \subseteq I' \cap (I^1 \cup I^2) \subseteq I \cap (I^1 \cup I^2)$ as requested.   □

Efficient solution faces

Let us consider the following multiobjective linear programming problem (MOLP)

$$\text{Maximize} \quad Cx$$
$$\text{subject to } Ax \leqq b,$$

where C is a real $k \times n$-matrix, A is a real $m \times n$-matrix and b is a column m-vector. The feasible solution set of (MOLP) is denoted X and its efficient (maximal) solution set is denoted $S(MOLP)$. We will assume throughout that X is nonempty. Moreover, if the zero vector of $\mathbb{R}^n$ is C-positive, then every feasible solution is efficient because there is a strictly positive vector $\lambda \in \mathbb{R}^k$ such that $0 = C^T \lambda$ which implies that every element of X is an optimal solution of the scalarized problem (P_λ)

$$\text{maximize} \quad \langle \lambda, Cx \rangle$$
$$\text{subject to} \quad Ax \leqq b,$$

and hence, in view of Theorem 4.3.1, it is an efficient solution of (MOLP). For this reason, we will assume henceforth (A2) as before, that is, the cone $\mathrm{pos}\{c^1, \cdots, c^k\}$ is not a linear subspace.

Theorem 8.2.7 *A feasible solution of (MOLP) is an efficient solution if and only if the active index set at this solution is positive.*

Proof Let $x^0 \in X$ be an efficient solution. In view of Theorem 4.3.1 there is a strictly positive vector λ such that x^0 solves the problem (P_λ). In particular,

$$\langle C^T \lambda, x^0 - x \rangle \geq 0 \text{ for all } x \in X \qquad (8.6)$$

which proves that the vector $C^T \lambda$ is a normal vector to X at x^0. If $I(x^0)$ is empty, then x^0 is an interior point of X. Hence $C^T \lambda$ is the zero vector and contradicts the assumption. In view of Theorem 2.3.24 the normal cone to X at x^0 is the cone $\mathrm{pos}\{a^i : i \in I(x^0)\}$. Hence $I(x^0)$ is a positive and normal index set.

Conversely, assume that $I(x^0)$ is positive. There is a strictly positive vector λ such that $C^T \lambda$ belongs to the cone $\mathrm{pos}\{a^i : i \in I(x^0)\}$. In particular $C^T \lambda$ is normal to the set X at x^0. Consequently, (8.6) is true, and therefore x^0 solves (P_λ). We deduce from Theorem 4.3.1 that x^0 is an efficient solution of (MOLP). □

Corollary 8.2.8 *(MOLP) has an efficient solution if and only if the index set $\{1, \cdots, m\}$ is positive, or equivalently, the following system is consistent*

$$\sum_{i=1}^{m} \mu_i a^i - \sum_{j=1}^{k} \lambda_j c^j = 0$$

$$\mu_i \geq 0, i = 1, \cdots, m \text{ and } \lambda_j \geq 1, \ j = 1, \cdots, k.$$

Proof Let x^0 be an efficient solution of (MOLP). In view of Theorem 8.2.7 the index set $I(x^0)$ is positive. As the set $\{1, \cdots, m\}$ contains $I(x^0)$, it is positive too. Conversely, if the set $\{1, \cdots, m\}$ is positive, the cone $\mathrm{pos}\{a^1, \cdots, a^m\}$ which is exactly the normal cone of X contains a C-positive vector. Hence there is some point $x^0 \in X$ such that the normal cone to X at x^0 contains that C-positive vector. Again, by Theorem 8.2.7, the point x^0 is efficient, and therefore (MOLP) has efficient solutions. The equivalence between the consistency of the linear system mentioned in the corollary and the positivity of the index set $\{1, \cdots, m\}$ is immediate from the definition. □

Corollary 8.2.9 *Let F be a face of X and $I(F)$ the index set of F. Then F is efficient if and only if $I(F)$ is positive, in which case the dimension of F is equal to $n - \mathrm{rank}\{a^i : i \in I_F\}$. In particular, when X is of full dimension, (MOLP) admits an $(n-1)$-dimensional efficient face if and only if there is an index $i_0 \in \{1, \cdots, m\}$ such that a^{i_0} is C-positive, in which case the $(n-1)$-dimensional face determined by the linear system*

$$\langle a^{i_0}, x \rangle = b_{i_0}$$
$$\langle a^j, x \rangle \leq b_j, j \in \{1, \cdots, m\} \backslash \{i_0\},$$

is an efficient face.

Proof Let x be a relative interior point of the face F. Then $I(x) = I(F)$. By Theorem 4.3.8 the face F is efficient if and only if x is efficient. It remains to apply Theorem 8.2.7 to conclude the first part of the corollary. If the dimension of X is equal to n, then a face F is of dimension $n - 1$ if and only if it is given by the (non-redundant) system determining X in which only one inequality is equality. In other words the index set of an $(n-1)$-dimensional face consists of one element. Therefore, this face is efficient if and only if that unique index is positive. $\qquad\square$

Let $\bar{x}$ be an efficient vertex of (MOLP) and $F_1, \cdots, F_\ell$ the efficient edges emanating from $\bar{x}$. The active index set of each F_i is denoted $I(F_i)$. Below is a condition for an efficient face adjacent to $\bar{x}$ to be maximal, that is, it is not a proper face of any other efficient face of the problem.

Corollary 8.2.10 *Let $J \subseteq \{1, \ldots, \ell\}$ and $I_J = \bigcap_{i \in J} I(F_i)$. Then the face F adjacent to $\bar{x}$ determined by the system*

$$\langle a^i, x \rangle = b_i, i \in I_J$$
$$\langle a^j, x \rangle \leqq b_j, j \in \{1, \cdots, m\} \setminus I_J,$$

is a maximal efficient face if the following conditions hold:

(i) I_J is positive;
(ii) For every $i \notin I_J$ such that $I(F_i) \not\supseteq I_J$, the index set $I_J \cap I(F_i)$ is either empty or not positive.

Proof It is clear that under (i), F is an efficient face. If it is not maximal, then it is contained in a bigger efficient face, say F'. We may find an edge F_j of F' emanating from $\bar{x}$ which does not belong to F. Then

$$I_J \cap I(F_j) \neq \emptyset.$$

Since this index set contains the positive index set $I(F')$ we conclude that $I_J \cap I(F_j)$ is positive, which contradicts the hypothesis. $\qquad\square$

The support of a vector $\mu \in \mathbb{R}^m_+$ is denoted by $\text{supp}(\mu)$ and defined as

$$\text{supp}(\mu) = \big\{i \in \{1, \cdots, m\} : \mu_i > 0\big\}.$$

We shall use also the following notations: Γ is the solution set to the system formulated in Corollary 8.2.8 and

$$\mathcal{I}_0 = \big\{I \subseteq \{1, \cdots, m\} : I = \text{supp}(\mu) \text{ for some } (\mu, \lambda) \in \Gamma\big\}$$
$$\mathcal{I}_1 = \big\{I \in \mathcal{I}_0 : I = \text{supp}(\mu) \text{ for some vertex } (\mu, \lambda) \in \Gamma\big\}$$

and $\mathcal{I}$ denotes the set of all minimal elements of $\mathcal{I}_1$ with respect to inclusion. We recall also that for a subset $I \subseteq \{1, \cdots, m\}$, the set $F(I)$ consists of feasible solutions

x of (MOLP) such that $\langle a^i, x \rangle = b_i, i \in I$. In particular when I is the empty set, $F(I) = X$.

Corollary 8.2.11 *The following statements hold.*

(i) *Let $(\mu^i, \lambda^i), i = 1, \cdots, l$ be vertices of Γ. Then for all $t_i \in (0, 1)$ with $\sum\limits_{i=1}^{l} t_i = 1$ one has*

$$F\Big(\mathrm{supp}\big(\sum_{i=1}^{l} t_i \mu^i\big)\Big) = \bigcap_{i=1}^{l} F\big(\mathrm{supp}(\mu^i)\big).$$

(ii) $S(MOLP) = \bigcup\limits_{I \in \mathcal{I}_0} F(I) = \bigcup\limits_{I \in \mathcal{I}_1} F(I) = \bigcup\limits_{I \in \mathcal{I}} F(I)$

(iii) *Given an index set $I \subseteq \{1, \cdots, m\}$, the set $F(I)$ is a maximal efficient face if and only if it is nonempty and $I \in \mathcal{I}$.*

Proof It follows from the definition that

$$\mathrm{supp}\big(\sum_{i=1}^{l} t_i \mu^i\big) = \bigcup_{i=1}^{l} \mathrm{supp}(\mu^i).$$

This implies the equality in the first statement.

For the second statement, it is clear that for every $I \in \mathcal{I}_0$, if nonempty, the set $F(I)$ is a face of X. Hence the index set of $F(I)$ that is included in I is positive. By Corollary 8.2.9, $F(I)$ is efficient. Thus, $\bigcup_{I \in \mathcal{I}_0} F(I) \subseteq S(MOLP)$ is true. Conversely, let $\overline{x}$ be an efficient solution of (MOLP). In view of Theorem 8.2.7, the active index set $I(\overline{x})$ is positive, which means that the system

$$\sum_{i \in I(\overline{x})} \mu_i a^i - \sum_{j=1}^{k} \lambda_j c^j = 0$$
$$\mu_i \geq 0, i \in I(\overline{x}) \text{ and } \lambda_j \geq 1, \ j = 1, \cdots, k$$

is solvable. Let (μ, λ) be a solution. Define $\overline{\mu}$ to be the vector the coordinates of which are given by

$$\overline{\mu}_i = \begin{cases} \mu_i & \text{for } i \in I \\ 0 & \text{else.} \end{cases}$$

It is then clear that

$$\overline{x} \in F\big(I(\overline{x})\big) \subseteq F\big(\mathrm{supp}(\overline{\mu})\big)$$

with $\mathrm{supp}(\overline{\mu}) \in \mathcal{I}_0$ and the first equality in (ii) follows. Furthermore, since $\mathcal{I} \subseteq \mathcal{I}_1 \subseteq \mathcal{I}_0$, we deduce inclusions

$$\bigcup_{I\in\mathcal{I}_0} F(I) \supseteq \bigcup_{I\in\mathcal{I}_1} F(I) \supseteq \bigcup_{I\in\mathcal{I}} F(I).$$

For the converse inclusions we notice that for each element $I^1 \in \mathcal{I}_1$, we find an element $I \in \mathcal{I}$ such that $I \subseteq I^1$. Then $F(I^1) \subseteq F(I)$, which proves

$$\bigcup_{I\in\mathcal{I}_1} F(I) \subseteq \bigcup_{I\in\mathcal{I}} F(I).$$

Moreover, for every element $I^0 \in \mathcal{I}_0$, say $I = \mathrm{supp}(\mu)$ for some $(\mu, \lambda) \in \Gamma$, there exist vertices $(\mu^i, \lambda^i) \in \Gamma$ and positive numbers $t_i, i = 1, \cdots, l$ such that $\sum_{i=1}^{l} t_i = 1$ and $(\mu, \lambda) = \sum_{i=1}^{l} t_i (\mu^i, \lambda^i)$. It follows from the first part that

$$F(I^0) \subseteq F\big(\mathrm{supp}(\mu^i)\big) \quad \text{for every } i = 1, \cdots, l.$$

We conclude that

$$\bigcup_{I\in\mathcal{I}_0} F(I) \subseteq \bigcup_{I\in\mathcal{I}_1} F(I),$$

by which equalities in the second statement hold.

To prove the last statement we assume that $F(I)$ is a maximal efficient face of (MOLP). By Corollary 8.2.9 the index set I belongs to $\mathcal{I}_0$. According to (ii), there is a minimal index set $I' \in \mathcal{I}$ such that $F(I) \subseteq F(I')$. Since $F(I)$ is maximal, we deduce $F(I) = F(I')$. Under the non-redundancy hypothesis (A1), we obtain $I = I'$. The converse statement is clear because if the efficient face $F(I)$ were not maximal for $I \in \mathcal{I}$, then one would find an efficient face F' that contains $F(I)$ as a proper face. Then the index set of F' is strictly smaller than the index set of $F(I)$, which is a contradiction because the index set of $F(I)$ is equal to I. $\qquad \square$

Notice that the family $\mathcal{I}$ as well as the families $\mathcal{I}_0$ and $\mathcal{I}_1$ gathers positive index sets which uniquely depend on the objective matrix C and the constraint matrix A of (MOLP) and do not depend on the second term b of the constraints. This latter term intervenes in the normality of the index sets, that is the nonemptiness of the faces determined by these index sets. Therefore, for a given b, some of subsets in the unions described in (ii) of Corollary 8.2.11 may be empty, which are precisely the case when the corresponding index sets are not normal.

8.3 The Normal Cone Method

In this section we shall give a method for numerically solving the problem (MOLP). The study of normal cones and their relationship with efficient faces that we have developed in the previous sections allow us to construct simple algorithms to determine efficient faces of any dimension. Throughout this section we will make the following assumption.

(A3) The feasible set X is an n-dimensional polyhedral set and contains no lines and there is no redundant inequality in the constraint system of (MOLP).

The method we are going to describe consists of three main procedures:

1—Determine whether (MOLP) has efficient solutions and if it has, find an initial efficient solution.

2—Starting from an efficient vertex, find all efficient edges and efficient rays emanating from it. Since the efficient solution set of (MOLP) is arcwise connected, this procedure allows us to find all efficient vertices and all efficient edges of the problem.

3—Find all efficient faces adjacent to (i.e. containing) a given efficient vertex when all the efficient edges adjacent to this vertex are already known.

Existence of efficient solutions and finding an initial efficient solution for (MOLP)

According to Corollary 8.2.8, (MOLP) has an efficient solution if and only if the system

$$A^T \mu - C^T \lambda = 0$$
$$\mu \in \mathbb{R}^m, \mu \geqq 0 \qquad\qquad (8.7)$$
$$\lambda \in \mathbb{R}^k, \lambda \geqq e,$$

where e is the vector of ones, has a solution. Remember that the columns of A^T are $a^1, \cdots, a^m$ and the columns of C^T are $c^1, \cdots, c^k$.

Procedure 1.

- *Step 1.* Solve the system (8.7).
  (a) If the system has no solution, then stop. (MOLP) has no efficient solution.
  (b) Otherwise, go to Step 2.
- *Step 2.* Let $\lambda > 0$ be a solution and $v = C^T \lambda$. If $v = 0$, then every feasible solution of (MOLP) is an efficient solution. Otherwise, solve the linear programming problem

$$\text{maximize } \langle v, x \rangle$$
$$\text{subject to } Ax \leqq b.$$

This problem has a solution, say $\bar{x}$. Then $\bar{x}$ is an initial efficient solution of (MOLP).

Example 8.3.1 We consider the following (MOLP)

$$\text{Maximize} \quad \begin{pmatrix} 2 & -3 & 1 \\ 3 & 1 & 0 \end{pmatrix} \begin{pmatrix} x_1 \\ x_2 \\ x_3 \end{pmatrix}$$

$$\text{subject to} \quad \begin{pmatrix} 1 & 1 & 1 \\ -2 & -2 & -1 \\ -1 & 0 & 0 \\ 0 & -1 & 0 \\ 0 & 0 & -1 \end{pmatrix} \begin{pmatrix} x_1 \\ x_2 \\ x_3 \end{pmatrix} \leq \begin{pmatrix} 1 \\ -1 \\ 0 \\ 0 \\ 0 \end{pmatrix}.$$

By solving the system (8.7) we find a solution $\lambda = (2, 1)^T$. The vector v in Step 2 is $v = (7, -5, 2)^T$ and the problem to solve in this step is to maximize the function $7x_1 - 5x_2 + 2x_3$ over the feasible set of (MOLP). The simplex method of Chap. 3 yields a vertex solution $x^0 = (1, 0, 0)^T$, which is also an efficient solution of (MOLP).

Determination of efficient vertices and efficient edges

When (MOLP) has an efficient solution, by solving the linear problem in Step 2 of Procedure 1 one may obtain an efficient vertex. Let $\bar{x}$ be such a vertex. Then the active index set $I(\bar{x})$ has at least n elements. Any edge emanating from x is determined by $n - 1$ linearly independent equations among the m inequality constraints, and of course its index set is a subset of $I(\bar{x})$. An index set $I \subseteq I(\bar{x})$ the cardinality of which is equal to $n - 1$ determines a one-dimensional space that may give rise to an edge of X by the system

$$\langle a^i, v \rangle = 0, \ i \in I$$

provided that the vectors $a^i, i \in I$ are linearly independent. Otherwise the solution set of this system would be of higher dimension. Moreover, if I is normal, then there is some real number $t \neq 0$ such that $\bar{x} + tv$ is a solution to the system

$$\langle a^i, x \rangle = b_i, \ i \in I$$
$$\langle a^j, x \rangle \leq b_j, \ j \in \{1, \cdots, m\} \setminus I.$$

The edge emanating from $\bar{x}$ in direction v above is efficient if, in addition, I is positive. We are now able to describe the second procedure to solve (MOLP).

Procedure 2.

- *Step 0 (Initialization).* Determine the active index set

$$I(\bar{x}) = \left\{ i \in \{1, \cdots, m\} : \langle a^i, \bar{x} \rangle = b_i \right\}.$$

Choose $I \subset I(\bar{x})$ with $|I| = n - 1$.

- *Step 1.* Check the linear independence of the family $\{a^i : i \in I\}$.
  If not, choose another $I \subseteq I(\bar{x})$ with $|I| = n - 1$.
  If yes, go further.
- *Step 2. (I is positive?)* Solve

$$\sum_{i \in I} \mu_i a^i - \sum_{j=1}^{k} \lambda_j c^j = 0$$
$$\mu_i \geqq 0, i \in I \text{ and } \lambda_j \geqq 1, \ j = 1, \cdots, k.$$

(a) If it has no solution, pick another $I \subseteq I(\bar{x})$ with $|I| = n - 1$ and return to Step 1.
(b) Otherwise, I is a positive set, go further.
- *Step 3. (I is normal? If yes, find the corresponding efficient edge)*

  - *Step 3.1.* Find a direction $v \neq 0$ of a possible edge emanating from $\bar{x}$ by solving

$$\langle a^i, v \rangle = 0, i \in I.$$

  - *Step 3.2.* Solve the following system

$$\langle a^i, \bar{x} + tv \rangle \leqq b_i, i = 1, \cdots, m.$$

Let the solution set be $[t_0, 0]$ or $[0, t_0]$ according to $t_0 < 0$ or $t_0 > 0$. The values $t_0 = -\infty$ and $t_0 = \infty$ are possible.

(a) If $t_0 = 0$, no edge of X emanating from $\bar{x}$ along v. I is not normal. Pick another $I \subseteq I(\bar{x})$ and go to Step 1.
(b) If $t_0 \neq 0$ and is finite, then $\bar{x} + t_0 v$ is an efficient vertex and $[\bar{x}, \bar{x} + t_0 v]$ is an efficient edge. Store them if they have not been stored before. Pick another $I \subseteq I(\bar{x})$ and go to Step 1.
(c) If t_0 is infinite, say $t_0 = \infty$, then the ray $\{\bar{x} + tv : t \geq 0\}$ is efficient. Store the result. Pick another $I \subseteq I(\bar{x})$ and go to Step 1.

We notice that if $\bar{x}$ is a non-degenerate vertex, that is $|I(\bar{x})| = n$, then Step 1 can be skipped because the family of vectors $a^i, i \in I(\bar{x})$ is already linearly independent, and any subset of $I(\bar{x})$ is normal (see Corollary 8.1.5).

Moreover, by solving the system of Step 3.2 we mean finding the solution set of type $[t_0, 0]$ or $[0, t_0]$ with t_0 negative or positive respectively. The infinite values $+\infty$ and $-\infty$ are possible. If $t_0 \neq 0$, then this solution set is a 1-dimensional face of X determined by the system

$$\langle a^i, x \rangle = b_i, i \in I$$
$$\langle a^j, x \rangle \leqq b_j, j \in \{1, \cdots, m\} \setminus I.$$

The active index set of this face contains I, but does not necessarily coincide with I, unless $\bar{x}$ is non-degenerate. It can be found by checking the number of equalities $\langle a^i, \bar{x} + tv \rangle = b_i$ with $i \in I(\bar{x})$ for some $0 < t < t_0$ if $t_0 > 0$, or $t_0 < t < 0$ if $t_0 < 0$.

Example 8.3.2 The aim of this example is to apply Procedure 2 to find all efficient edges adjacent to an initial efficient vertex. We consider the following (MOLP)

$$\text{Maximize} \qquad \begin{pmatrix} 2 & 3 & 4 \\ 3 & 2 & 0 \end{pmatrix} \begin{pmatrix} x_1 \\ x_2 \\ x_3 \end{pmatrix}$$

$$\text{subject to} \qquad \begin{pmatrix} 1 & 1 & 1 \\ -2 & -2 & -1 \\ -1 & 0 & 0 \\ 0 & -1 & 0 \\ 0 & 0 & -1 \end{pmatrix} \begin{pmatrix} x_1 \\ x_2 \\ x_3 \end{pmatrix} \leqq \begin{pmatrix} 1 \\ -1 \\ 0 \\ 0 \\ 0 \end{pmatrix}.$$

To find an initial efficient vertex we solve the system (8.7) that takes the form

$$\mu_1 \begin{pmatrix} 1 \\ 1 \\ 1 \end{pmatrix} + \mu_2 \begin{pmatrix} -2 \\ -2 \\ -1 \end{pmatrix} + \mu_3 \begin{pmatrix} -1 \\ 0 \\ 0 \end{pmatrix} + \mu_4 \begin{pmatrix} 0 \\ -1 \\ 0 \end{pmatrix} + \mu_5 \begin{pmatrix} 0 \\ 0 \\ -1 \end{pmatrix} \qquad (8.8)$$

$$= \lambda_1 \begin{pmatrix} 2 \\ 3 \\ 4 \end{pmatrix} + \lambda_2 \begin{pmatrix} 3 \\ 2 \\ 0 \end{pmatrix}$$

$$\mu_i \geqq 0, i = 1, \cdots, 5, \ \lambda_1 \geqq 1, \lambda_2 \geqq 1.$$

A solution can be given as $\lambda_1 = 3, \lambda_2 = 1, \mu_1 = 12, \mu_2 = 0, \mu_3 = 3, \mu_4 = 1$, and $\mu_5 = 0$. The normal cone of the feasible set contains a C-positive vector $v = \lambda_1 c^{1T} + \lambda_2 c^{2T} = (9, 11, 12)^T$, which generates a denegerate efficient vertex $x^0 = (0, 0, 1)^T$. The active index set of this solution is $I(x^0) = \{1, 2, 3, 4\}$. In order to find efficient edges emanating from x^0 we check the normality and the positivity of each of the 2-element index subsets of $I(x_0)$ and also the linear independence of the corresponding vectors.

(1) For $I_1 = \{1, 2\}$ we have the vectors $a^1 = (1, 1, 1)^T$ and $a^2 = (-2, -2, -1)^T$ linearly independent. To check its positivity we solve (8.8) by setting $\mu_3 = \mu_4 = \mu_5 = 0$, which leads to

$$\mu_1 - 2\mu_2 = 2\lambda_1 + 3\lambda_2 = 3\lambda_1 + 2\lambda_2$$
$$\mu_1 - \mu_2 = 4\lambda_1.$$

In particular $\mu_2 = -\lambda_1$, which contradicts the constraints $\mu_2 \geqq 0$ and $\lambda_1 \geqq 1$. Hence I_1 is not positive.
(2) For $I_2 = \{1, 3\}$, we notice that the vectors $a^1 = (1, 1, 1)^T$ and $a^3 = (-1, 0, 0)^T$

are linearly independent. To check its positivity we solve (8.8) by setting $\mu_2 = \mu_4 = \mu_5 = 0$. A solution of it can be given as $\lambda_1 = 2$, $\lambda_2 = 1$, $\mu_1 = 8$, $\mu_3 = 1$. As consequence, I_2 is positive. To see whether it is normal we solve the constraint inequalities of (MOLP) in which the first and the third inequations are equations. It is easy to check that the solution set is the segment connecting the vertex x^0 and the vertex $x^2 = (0, 1, 0)^T$. Thus, I_2 is positive and normal. By this, the segment $[x^0, x^2]$ is an efficient edge.

(3) For $I_3 = \{1, 4\}$, we notice again that the vectors $a^1 = (1, 1, 1)^T$ and $a^4 = (0, -1, 0)^T$ are linearly independent. We set $\mu_2 = \mu_3 = \mu_5 = 0$ in (8.8) for checking the positivity of I_3. Similarly to the case of I_1, the system yields $\mu_4 = -\lambda_2/2$, which is a contradiction. Hence I_3 is not positive.

(4) The index sets $\{2, 3\}$, $\{2, 4\}$ and $\{3, 4\}$ are evidently not positive because in the corresponding systems obtained from (8.8) by setting at least $\mu_1 = 0$, the vector on the left hand side is negative, while the vector on the right hand side is strictly positive.

We conclude that there is only one efficient edge emanating from the vertex x^0 and ending at the vertex x^2.

Determination of higher dimensional efficient solution faces

Assume that $\bar{x}$ is an efficient vertex of problem (MOLP) and $[\bar{x}, \bar{x} + t_i v_i]$, $i = 1, \cdots, r$ are efficient edges emanating from $\bar{x}$ with $t_i > 0$. Here, for the convenience we use $t_i = \infty$ if the ray edge $\{\bar{x} + t v_i : t \geq 0\}$ is efficient and $[\bar{x}, \bar{x} + t_i v_i]$ denotes this ray. Let $I_i \subseteq I(\bar{x})$, $i = 1, \cdots, r$ be the positive index sets determining these edges.

Observe that except for the pathological case when the entire set X is efficient, the largest dimension that an efficient face adjacent to $\bar{x}$ may have is $\min\{r, n - 1\}$. For $1 < l \leq \min\{r, n - 1\}$, we have the following procedure to find l-dimensional efficient faces adjacent to the given efficient vertex $\bar{x}$.

Procedure 3.

- *Step 0 (Initialization).* Pick $J \subseteq \{1, \cdots, r\}$ with $|J| = l$ and determine $I = \bigcap_{j \in J} I_j$.

  Find rank$\{a^i : i \in I\}$.
  If rank$\{a^i : i \in I\} \neq n - l$, choose another J and repeat this step.
  Otherwise go to the next step.
- *Step 1. (I is positive?)* Solve the system (8.5).
  (a) If it has no solution, return to Step 0 by choosing another J.
  (b) Otherwise, I is positive, go to Step 2.
- *Step 2.* Determine $J_0 = \{j \in \{1, \cdots, r\} : I_j \supseteq I\}$. (It is evident that $J \subseteq J_0$.) The convex hull of the edges $[\bar{x}, \bar{x} + t_j v_j]$, $j \in J_0$ forms an l-dimensional efficient face. Return to Step 0 by picking another not yet explored J that is not contained in J_0 with $|J| = l$ until no such J left.

The index set I obtained in the initialization step is normal whenever it is nonempty because it is the intersection of normal sets I_j, $j \in J$ (Corollary 8.1.6). If in addition the rank of the family $\{a^i : i \in I\}$ is equal to $n - l$, then the face determined by the system

$$\langle a^i, x \rangle = b_i, i \in I$$
$$\langle a^j, x \rangle \leqq b_j, j \in \{1, \cdots, m\} \backslash I,$$

has its dimension equal to l, and its active index set contains I. When the efficient vertex $\bar{x}$ is non-degenerate, the condition rank$\{a^i : i \in I\} = n - l$ is equivalent to the fact that I has l elements.

Example 8.3.3 The aim of this example is to apply Procedure 3 to find all efficient faces adjacent to an initial efficient vertex. We consider the following (MOLP)

$$\text{Maximize} \quad \begin{pmatrix} 1 & 0 & 0 \\ 0 & 1 & 0 \\ 0 & 0 & 1 \end{pmatrix} \begin{pmatrix} x_1 \\ x_2 \\ x_3 \end{pmatrix}$$

$$\text{subject to} \quad \begin{pmatrix} -2 & -2 & -1 \\ 2 & 1 & 2 \\ 1 & 2 & 2 \\ -1 & 0 & 0 \\ 0 & -1 & 0 \\ 0 & 0 & -1 \end{pmatrix} \begin{pmatrix} x_1 \\ x_2 \\ x_3 \end{pmatrix} \leqq \begin{pmatrix} -1 \\ 2 \\ 2 \\ 0 \\ 0 \\ 0 \end{pmatrix}.$$

Applying Procedure 1 and Procedure 2 one finds an initial vertex $x^0 = (0, 0, 1)^T$ to start and the following efficient edges adjacent to x^0: $F_i = [x^0, x^i]$, $i = 1, 2, 3$, where $x^1 = (1, 0, 0)^T$, $x^2 = (0, 1, 0)^T$ and $x^3 = (2/3, 2/3, 0)^T$. The index sets of x^0 and F_i are respectively given by

$$I(x^0) = \{1, 2, 3, 4, 5\}$$
$$I_1 = \{2, 5\}$$
$$I_2 = \{3, 4\}$$
$$I_3 = \{2, 3\}.$$

We use Procedure 3 to determine a two-dimensional efficient face adjacent to x^0. At Step 0, we choose for instance $J = \{1, 3\}$ and consider $I = I_1 \cap I_3 = \{2\}$. The rank of $a^2 = (2, 1, 2)^T$ is equal to 1, hence we may go further. In Step 2, we check the positivity of I by solving the system (8.5) applied to our example:

$$\mu \begin{pmatrix} 2 \\ 1 \\ 2 \end{pmatrix} = \lambda_1 \begin{pmatrix} 1 \\ 0 \\ 0 \end{pmatrix} + \lambda_2 \begin{pmatrix} 0 \\ 1 \\ 0 \end{pmatrix} + \lambda_3 \begin{pmatrix} 0 \\ 0 \\ 1 \end{pmatrix}.$$

A solution is given by $\lambda_1 = 2, \lambda_2 = 1, \lambda_3 = 2$ and $\mu = 1$. We conclude that the two-dimensional face determined by the system of constraints in which the second inequality is set to equation is efficient. Since I is contained in I_1 and I_3, this face contains the edges F_1 and F_3. The edge F_2 is not included in it because I_2 does not contain I.

Determination of all maximal efficient faces adjacent to an efficient vertex

Let $\bar{x}$ be an efficient vertex of (MOLP) and let $\{p_1, \cdots, p_r\}$ be the collection of all efficient edges (possibly rays) emanating from $\bar{x}$ which have been obtained by Procedure 2. The positive index sets determining these edges are denoted $I_1, \cdots, I_r$. Thus, each edge p_i is the solution to the system

$$\langle a^j, x \rangle = b_j \, , \; j \in I_i$$
$$\langle a^j, x \rangle \leqq b_j \, , \; j \in \{1, \cdots, m\} \setminus I_i \, .$$

The next algorithm determines all maximal efficient faces adjacent to $\bar{x}$. The biggest dimension of these faces does not exceed $\min\{r, n-1\}$ as we have already discussed.

Procedure 4.

- *Step 0.* For $l = 2, \cdots, r$ pick $J \subseteq \{1, \cdots, r\}$ with $|J| = l$ and compute

$$I = \bigcap_{j \in J} I_j.$$

  If either $I = \emptyset$, or I is not positive, then choose another J.
  If I is nonempty positive, go to the next step.
- *Step 1.* For each $j \in \{1, \cdots, r\} \setminus J$, compute $I' = I \cap I_j$.

  (a) If either $I' = \emptyset$ or I' is not positive, proceed for other j. If this is the case for all $j \in \{1, \cdots, r\} \setminus I$, the face F_J generated by the edges p_j with $I \subseteq I_j$ including $j \in J$, is a maximal efficient face to be stocked together with the index set $J = \{i : I \subseteq I_i\}$. Return to Step 0 for other J.
  (b) If I' is nonempty positive, set $J = J \cup \{j\}$ (then $|J| \geqq l + 1$) and repeat Step 1 until no J left.

- *Step 2.* Set $l := l + 1$ and return to Step 0 by choosing J not yet exploited or not contained in any index subset already stocked in Step 1.

  The positivity of I in Step 0 and of I' in Step 1 is checked by solving the system (8.5). The maximality of the efficient faces stocked in Step 1 is due to Corollary 8.2.10.

Example 8.3.4 We continue Example 8.3.3 by choosing $x^0 = (0, 0, 1)^T$ as an initial vertex. Its efficient edges $F_i = [x^0, x^i], i = 1, 2, 3$ where $x^1 = (1, 0, 0)^T, x^2 = (0, 1, 0)^T, x^3 = (2/3, 2/3, 0)^T$. The index set of x^0 is $I(x^0) = \{1, 2, 3, 4, 5\}$ and the index sets of $F_i, i = 1, 2, 3$ are respectively

$$I(F_1) = \{2, 5\}, I(F_2) = \{3, 4\}, I(F_3) = \{2, 3\}.$$

We apply Procedure 4 to determine all maximal efficient faces adjacent to x^0. In Step 0, for $l = 2$, we choose $J \subseteq \{1, 2, 3\}$ with $|J| = 2$.

- For $J = \{1, 3\}$ one has $I = I(F_1) \cap I(F_3) = \{2, 5\} \cap \{2, 3\} = \{2\}$. This index set I is positive (see Example 8.3.3), we go to Step 1. Let $j \in \{1, 2, 3\} \setminus J = \{2\}$. Compute $I' = I \cap I(F_2) = \{2\} \cap \{3, 4\} = \emptyset$. Thus, the face generated by the index set $\{2\}$ is a maximal efficient face that contains the edges $F_j, j \in J$.
- For $J = \{1, 2\}$ we have $I = I(F_1) \cap I(F_2) = \emptyset$. Return to Step 0.
- For $J = \{2, 3\}$ we have $I = I(F_2) \cap I(F_3) = \{3\}$. It is positive because the system

$$\mu \begin{pmatrix} 1 \\ 2 \\ 2 \end{pmatrix} = \lambda_1 \begin{pmatrix} 1 \\ 0 \\ 0 \end{pmatrix} + \lambda_2 \begin{pmatrix} 0 \\ 1 \\ 0 \end{pmatrix} + \lambda_3 \begin{pmatrix} 0 \\ 0 \\ 1 \end{pmatrix}, \quad \lambda_1, \lambda_2, \lambda_3 \geq 1, \mu \geq 0$$

admits a solution $\lambda_1 = 1, \lambda_2 = 2, \lambda_3 = 2$ and $\mu = 1$. We go to Step 1. Let $j \in \{1, 2, 3\} \setminus J = \{1\}$. Compute $I' = I \cap I(F_1) = \{3\} \cap \{2, 5\} = \emptyset$. The face generated by the index set $I = \{1\}$ is maximal efficient and contains the edges $F_j, j \in J = \{2, 3\}$.

The algorithm yields two maximal efficient faces that contain respectively the edges F_1, F_3 and F_2, F_3.

Determination of the entire efficient solution set of (MOLP)

Since every efficient solution of (MOLP) is contained in a maximal efficient face, the efficient solution set of the problem will be completely found if we can identify all maximal efficient faces. The next algorithm for generating all maximal efficient faces is based on Procedures 1, 2 and 4 and on the fact that the solution set of a multiobjective linear problem is arcwise connected, that is any two efficient vertices can be joined by a finite number of efficient edges.

The Algorithm

- *Step 1.* Determine whether (MOLP) has maximal solutions. If yes, find an efficient vertex to start by using Procedure 1.
- *Step 2.* Find all efficient edges adjacent to this efficient vertex by Procedure 2.
- *Step 3.* Determine all maximal efficient faces adjacent to the given vertex by Procedure 4 and stock them together with the active index set of each such a face.

- *Step 4*. Choose a new efficient vertex adjacent to the given vertex and return to Step 2 with this vertex to start unless no such efficient vertex left.

So far the analysis made in this chapter requires three assumptions (A1–A3), under which the algorithm terminates after a finite number of iterations because this is so for the three procedures we apply. There are some simple situations when one or some of the above mentioned assumptions do not hold and there is no need to solve the problem. For instance when (1) the feasible set is empty; or (2) the cone $\mathrm{pos}\{c^1, \cdots c^k\}$ is a linear subspace (the zero vector is C-positive, and hence every feasible solution is maximal).

Particular case 1: Efficient sets in $\mathbb{R}^2$

Sometimes we wish to compute the efficient set of a polyhedron (a bounded polyhedral convex set) which corresponds to the efficient solution set of the problem (MOLP) with C being the identity matrix. Below we provide an effective and direct algorithm to do this in the case $X \subseteq \mathbb{R}^2$.

By renumbering the indices if necessary, we may assume

$$0 < \theta_1 < \theta_2 < \cdots < \theta_l < \frac{1}{2}\pi \leq \theta_{l+1} < \cdots < \theta_m \leq 2\pi,$$

where $\theta_1, \cdots, \theta_m$ are the angular coordinates of $a^1, \cdots, a^m$ in the polar coordinate system of $\mathbb{R}^2$ (Fig. 8.1).

Fig. 8.1 Angular coordinates

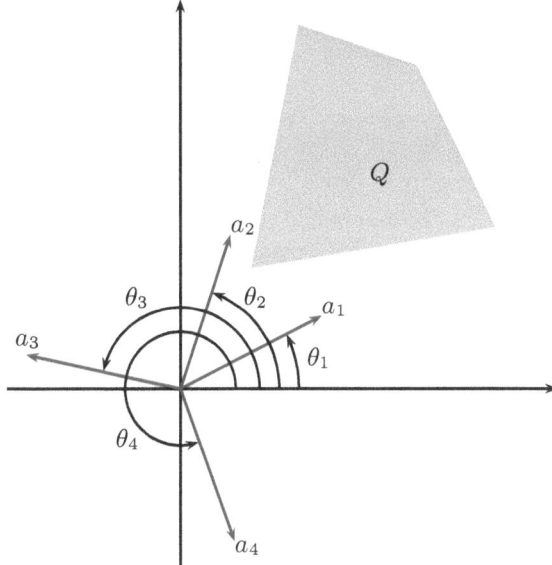

It is evident that $a^1, \cdots, a^l$ are positive vectors and hence each of them determines an efficient edge. Moreover, each of the pairs of indices $\{m, 1\}, \{1, 2\}, \cdots, \{l, l+1\}$ is positive and normal. So, by Corollary 8.2.9 they determine all 0-dimensional efficient faces (vertices) of X. Denote by x^i the intersection point of the lines

$$\langle a^i, x \rangle = b_i$$
$$\langle a^{i+1}, x \rangle = b_{i+1},$$

$i = 0, \cdots, l$, where $a^0 = a^m$, $b_0 = b_m$. Then the efficient set of X is given by

$$\bigcup_{i=0}^{l} [x^i, x^{i+1}].$$

Particular case 2: Efficient sets in $\mathbb{R}^3$

If the dimension of X is three, then it may have efficient faces of dimension 0 or 1 or 2. We recall that a point $\bar{x} \in X$ is said to be an ideal efficient point if $\bar{x} \geq x$ for all $x \in X$. It is easy to see that X does not possess ideal efficient points if and only if it has efficient faces of dimension 1 or 2, or it has no efficient point at all. Now we describe an algorithm to determine the set of all efficient points of $X \subseteq \mathbb{R}^3$. With one exceptional case when $\text{Max}(X)$ consists of only one point, the efficient set $\text{Max}(X)$ can be completely determined if we know all efficient edges.

- *Step 1 (Determine whether X possesses an ideal efficient point).*
  Solve the linear problem

$$\text{maximize} \quad \langle e^i, x \rangle$$

$$\text{subject to} \quad x \in X.$$

for $i = 1, 2, 3$, where $e^1 = (1, 0, 0)^T$, $e^2 = (0, 1, 0)^T$, $e^3 = (0, 0, 1)^T$. Let x_1^*, x_2^*, x_3^* be the optimal values of these problems. If $x^* = (x_1^*, x_2^*, x_3^*) \in X$, then x^* is an ideal efficient point of X and $\text{Max}(X) = \{x^*\}$;
Otherwise, go to Step 2.
- *Step 2.* Decompose the index set $\{1, \cdots, m\}$ into I^1, I^2, I^3, where $I^1 = \{i : a^i > 0\}$, $I^3 = \{i : a^i < 0\}$, $I^2 = \{1, \cdots, m\} \setminus (I^1 \cup I^3)$.
If $I^1 = \emptyset$, then there are no efficient faces of dimension 2. Go to Step 3 to find efficient faces of smaller dimension.
Otherwise, each $a^i, i \in I^1$ determines an efficient face of dimension 2 by the system

$$\langle a^i, x \rangle = b_i$$
$$\langle a^j, x \rangle \leq b_j, j \in \{1, \cdots, m\} \setminus \{i\}.$$

Go to Step 3 to find efficient faces of smaller dimension, not included in the above 2-dimensional efficient faces.

- *Step 3.* Choose $i, j \in I^2$.

  - *Step 3.1. (Is $\{i, j\}$ positive ?)*
    Solve the system

  $$ta^i + (1 - t)a^j > 0$$
  $$1 \geqq t \geqq 0.$$

  If it has a solution, then $\{i, j\}$ is positive. Go to Step 3.2.
  Otherwise, $\{i, j\}$ is not positive. Pick other pair $i, j \in I^2$ and return to Step 3.1.
  - *Step 3.2. (Is $\{i, j\}$ normal?)*
    Determine the set $\Delta_{ij} := \{x \in X : \langle a^i, x \rangle = b_i, \langle a^j, x \rangle = b_j\}$. If $\Delta_{ij} = \emptyset$ or Δ_{ij} is a point , then either $\{i, j\}$ is not normal or $\dim\Delta_{ij} = 0$. Pick other pair $i, j \in I^2$ and Return to Step 3.1.
    Otherwise Δ_{ij} is a segment. This segment is an efficient edge. Store it. Pick another $i, j \in I^2$ and return Step 3.1.

Remark According to Corollary 8.2.9, Step 2 and Step 3 allow us to generate the entire efficient set of X because other efficient faces are included in those that were found in these steps.

Computing weakly efficient solutions

The normal cone method we presented above to compute efficient solutions of (MOLP) is also suitable to find weakly efficient solutions. The only difference is that the C-positivity must be substituted by the weak C-positivity in all procedures. Namely, we say that an index set I is weakly C-positive if there are nonnegative numbers $\lambda_1, \cdots, \lambda_k$, not all zero, such that $v = \sum_{i=1}^{k} \lambda_i c^i$. Here are some results that provide theoretical basis of the normal method for weakly efficient solutions:

- A face F of X is weakly efficient if and only if its active index set $I(F)$ is weakly C-positive.
- (MOLP) has weakly efficient solutions if and only if the following system is consistent

$$\sum_{i=1}^{m} \mu_i a^i - \sum_{j=1}^{k} \lambda_j c^j = 0$$
$$\mu_i \geqq 0, \quad \lambda_j \geqq 0, \sum_{j=1}^{k} \lambda_j = 1.$$

• If the above system has a solution $(\mu_1, \cdots, \mu_m, \lambda_1 \cdots, \lambda_k)$, then the function $x \mapsto \langle \sum_{j=1}^{k} \lambda_j c^j, x \rangle$ attains its maximum on X and every maximum point is a weakly efficient solution of (MOLP).

8.4 Exercises

8.4.1 Find a counter-example to show that without the non-redundancy hypothesis (A1), an index set I may be normal while the system

$$\langle a^i, x \rangle = b_i, \ i \in I$$
$$\langle a^j, x \rangle < b_j, \ j \in \{1, \cdots, m\} \backslash I$$

is inconsistent.

8.4.2 Consider (MOLP) described in Sect. 8.2 and assume that $I \subseteq \{1, \cdots, m\}$ is a positive and normal index set. Prove that the efficient face determined by the system

$$\langle a^i, x \rangle = b_i, \ i \in I$$
$$\langle a^j, x \rangle \leqq b_j, \ j \in \{1, \cdots, m\} \backslash I,$$

is not a maximal efficient face if the system

$$\sum_{i \in I} \mu_i a^i - \sum_{j=1}^{k} \lambda_j c^j = 0$$
$$\mu_i \geqq 1, i \in I \text{ and } \lambda_j \geqq 1, \ j = 1, \cdots, k,$$

is inconsistent. Show that the converse statement is not always true.

8.4.3 Find all normal index sets of the following systems:

$$(1) \begin{cases} x_1 - x_2 + 2x_3 \leqq 3 \\ x_1 - 2x_2 \quad\quad \leqq -2 \\ \quad\quad - x_2 \quad\quad \leqq 0 \end{cases} \quad\quad (2) \begin{cases} x_1 + x_2 + x_3 \leqq 6 \\ 5x_1 + 3x_2 + 6x_3 \leqq 15 \\ -x_1 - x_2 - x_3 \leqq 0. \end{cases}$$

8.4.4 Consider (MOLP) described in Sect. 8.2. Let X be the solution set of a linear system $Ax \leqq b$.

(1) Prove that if the polar cone of the asymptotic cone of X contains a C-positive vector, then positive and normal index sets exist. In particular, when X is bounded, there always exists a positive and normal index set.

(2) Assume that the system $\langle a^i, x \rangle \leqq b_i, i = 1 \cdots, m$ is non-redundant. Prove that it has an $(n-1)$-dimensional efficient face if and only if there is some $i_0 \in$

$\{1, \cdots, m\}$ such that a^{i_0} is C-positive, and that such an index is unique for each $(n-1)$-dimensional efficient face.

8.4.5 Solve the following problem by the normal cone method

$$
\text{Minimize} \quad \begin{pmatrix} -x_1 - x_2 - 0.25x_3 \\ x_1 + x_2 + 1.5x_3 \end{pmatrix}
$$

$$
\text{subject to} \quad \begin{pmatrix} 2 & 1 & 2 \\ 1 & 2 & 1 \\ -1 & -1 & -1 \end{pmatrix} \begin{pmatrix} x_1 \\ x_2 \\ x_3 \end{pmatrix} \geq \begin{pmatrix} 2 \\ 2 \\ -6 \end{pmatrix}
$$

$$
x_1, x_2, x_3 \geq 0.
$$

and prove that the index set $I = \{6\}$ determines a 2-dimensional maximal efficient face whose vertices are $x_1 = (0.67, 0.67, 0)^T$, $x_2 = (2, 0, 0)^T$, $x_3 = (0, 2, 0)^T$, $x_4 = (6, 0, 0)^T$, $x_5 = (0, 6, 0)^T$.

8.4.6 Solve the problem

$$
\text{Minimize} \quad \begin{pmatrix} -x_1 + 100x_2 + 0x_3 \\ -x_1 - 100x_2 + 0x_3 \\ 0x_1 + 0x_2 - 1x_3 \end{pmatrix}
$$

$$
\text{subject to} \quad \begin{pmatrix} 1 & 2 & 2 \\ 2 & 1 & 2 \\ 5 & 5 & 6 \end{pmatrix} \begin{pmatrix} x_1 \\ x_2 \\ x_3 \end{pmatrix} \leq \begin{pmatrix} 10 \\ 10 \\ 30 \end{pmatrix}
$$

$$
x_1, x_2, x_3 \geq 0.
$$

We have obtained the following list of efficient vertices and faces.

(a) 3 two dimensional efficient faces F_1, F_2, F_3 which are determined by $I(F_1) = \{1\}$, $I(F_2) = \{2\}$, $I(F_3) = \{3\}$;
(b) Face F_1 has 3 efficient vertices : $x_2 = (2, 4, 0)$, $x_4 = (0, 0, 5)$, $x_5 = (0, 5, 0)$ and 3 efficient edges : $:[x_2, x_4]$, $[x_2, x_5]$, $[x_4, x_5]$.
(c) Face F_2 has 3 efficient vertices : $x_1 = (4, 2, 0)$, $x_3 = (5, 0, 0)$, $x_4 = (0, 0, 5)$ and 3 efficient edges : $:[x_1, x_3]$, $[x_1, x_4]$, $[x_3, x_4]$.
(d) Face F_3 has 3 efficient vertices : $x_1 = (4, 2, 0)$, $x_2 = (2, 4, 0)$, $x_4 = (0, 0, 5)$ and 3 efficient edges : $:[x_1, x_2]$, $[x_1, x_4]$, $[x_2, x_4]$.

Write problems to determine

• an initial weakly efficient solution;
• all weakly efficient edges emanating from a given weakly efficient vertex;
• all maximal weakly efficient faces adjacent to a given weakly efficient vertex;
• the weakly efficient solution set of (MOLP).

8.4.7 Consider the following problem

$$\text{Minimize} \begin{pmatrix} -1 & -1 & -0.25 \\ 1 & 1 & 1.5 \\ 0 & 1 & 1 \end{pmatrix} \begin{pmatrix} x_1 \\ x_2 \\ x_3 \end{pmatrix}$$

$$\text{subject to} \begin{pmatrix} -1 & -1 & 1 \\ -1 & -1 & -1 \\ -2 & -2 & -1 \end{pmatrix} \begin{pmatrix} x_1 \\ x_2 \\ x_3 \end{pmatrix} \geqq \begin{pmatrix} -3 \\ -5 \\ -8 \end{pmatrix}$$

$$x_1, \ x_2, \ x_3 \geqq 0.$$

Using the normal method establish the following result

(a) 5 weakly efficient vertices:

$$v_1 = (3, 0, 0), \quad v_2 = (0, 0, 0), \quad v_3 = (0, 3, 0),$$
$$v_4 = (3.667, 0, 0.667), \quad v_5 = (0, 3.667, 0.667);$$

(b) 6 weakly efficient edges:

$$[v_1, v_2], \ [v_1, v_3], \ [v_1, v_4],$$
$$[v_2, v_3], \ [v_3, v_5], \ [v_4, v_5];$$

(c) 2 maximal weakly efficient faces of dimension 2:

$$F_1 = \text{co}\{v_1, v_3, v_4, v_5\}$$
$$F_2 = \text{co}\{v_1, v_2, v_3\}.$$

8.4.8 Consider the following problem

$$\text{Minimize} \begin{pmatrix} -1 & 100 & 0 \\ -1 & -100 & 0 \\ 0 & 0 & -1 \end{pmatrix} \begin{pmatrix} x_1 \\ x_2 \\ x_3 \end{pmatrix}$$

$$\text{subject to} \begin{pmatrix} -1 & -2 & -2 \\ -2 & -1 & -2 \\ -5 & -5 & -6 \end{pmatrix} \begin{pmatrix} x_1 \\ x_2 \\ x_3 \end{pmatrix} \geqq \begin{pmatrix} -10 \\ -10 \\ -30 \end{pmatrix}$$

$$x_1, \ x_2, \ x_3 \geqq 0.$$

Using the normal cone method to obtain the following efficient vertices and faces.

(a) 6 weakly efficient vertices:

$$v_1 = (5, 0, 0); \ v_2 = (4, 2, 0); \ v_3 = (0, 0, 5)$$
$$v_4 = (2, 4, 0); \ v_5 = (0, 0, 0); \ v_6 = (0, 5, 0).$$

(b) 8 weakly efficient edges:

$$[v_1, v_2], [v_1, v_3], [v_2, v_4], [v_2, v_3],$$
$$[v_3, v_5], \ [v_4, v_6], [v_4, v_3], \ [v_6, v_3],$$

(c) 3 maximal weakly efficient faces of dimension 2:

$$F_1 = \text{co}\{v_1, v_2, v_3\},$$
$$F_2 = \text{co}\{v_2, v_3, v_4\},$$
$$F_3 = \text{co}\{v_3, v_4, v_6\}.$$

Chapter 9
Outcome Space Method

In many applications the number of decision variables is large. The feasible decision set is defined by many constraints in a high dimensional space, and therefore it has a lot of vertices and faces. Its descriptive analysis becomes costly and time consuming. On the other hand, the number of objective functions is limited, frequently does not exceed four or five. This leads to the outcome or value set having far fewer vertices and faces and much simpler structure. Because of this advantage outcome space methods are aimed at developing algorithms to compute efficient vertices and efficient faces of the value set in the outcome space.

9.1 Analysis of the Efficient Set in the Outcome Space

We consider the problem (MOLP) in a general form

$$\text{Maximize} \quad Cx$$
$$\text{subject to} \ x \in X,$$

where C is a $k \times n$-matrix and $X \subseteq \mathbb{R}^n$ is the feasible set or the decision set, which is defined by a system of linear equations. The set $Q = \{Cx : x \in X\} \subseteq \mathbb{R}^k$ is the value set or the outcome set of the problem. A face of X that consists of efficient solutions is called an efficient face of X. We know that the value sets $\text{Max}(Q)$ and $\text{WMax}(Q)$ as well as the solution sets $S(MOLP)$ and $WS(MOLP)$ are arcwise connected and consist of faces. We wish to know how efficient faces of Q and efficient faces of X are linked. This will help us to understand the structure of the set of efficient solutions once properties of the set of efficient outcomes are established.

Theorem 9.1.1 *Assume X has a vertex. For every vertex $y^0 \in \text{Max}(Q)$ (respectively, $y^0 \in \text{WMax}(Q)$) there exists a vertex $x^0 \in S(MOLP)$ (respectively, $x^0 \in WS(MOLP)$) such that $Cx^0 = y^0$.*

© Springer International Publishing Switzerland 2016
D.T. Luc, *Multiobjective Linear Programming*,
DOI 10.1007/978-3-319-21091-9_9

Proof Let $y^0 \in \text{Max}(Q)$ be a vertex. By definition, there is $x \in S(MOLP)$ such that $Cx = y^0$. If x is a vertex of X, we are done. Otherwise, there exist vertices $x^0, \cdots, x^p$ of X and strictly positive numbers $\lambda_0, \cdots, \lambda_p$ with $\sum_{i=0}^{p} \lambda_i = 1$ such that $\sum_{i=0}^{p} \lambda_i x^i = x$. Then $y^0 = \sum_{i=0}^{p} \lambda_i Cx^i$, by which y^0 is a relative interior point of $Cx^0, \cdots, Cx^p$. Since y^0 is a vertex of Q we deduce that $Cx^0 = \cdots = Cx^p$ and hence $y^0 = Cx^0$. It is clear that $x^0 \in S(MOLP)$ because $y^0 \in \text{Max}(Q)$. The proof for a weakly efficient vertex of Q follows on the same line.   □

It is quite evident that for a given efficient vertex $y \in \text{Max}(Q)$ there may exist a number of vertices $x \in S(MOLP)$ such that $Cx = y$. In a single objective problem the optimal value is a real number, which is also a vertex of Q. Hence every optimal solution vertex is mapped by C into the optimal outcome vertex. In a multiple objective problem, an efficient solution vertex of X may map into a non-vertex point of Q.

Example 9.1.2 Let $X = \{x \in \mathbb{R}^2 : -1 \leq x_i \leq 1, i = 1, 2\}$ and let $C = \begin{pmatrix} 1 & -1 \\ -1 & 1 \end{pmatrix}$.

Then Q is a line segment in $\mathbb{R}^2$ connecting two points $\begin{pmatrix} 2 \\ -2 \end{pmatrix}$ and $\begin{pmatrix} -2 \\ 2 \end{pmatrix}$. With these data every feasible solution of (MOLP) is efficient. The vertices $\begin{pmatrix} -1 \\ -1 \end{pmatrix}$ and $\begin{pmatrix} 1 \\ 1 \end{pmatrix}$ of X map into a relative interior point of Q.

According to Theorem 9.1.1 and the remarks above, the number of vertices of $\text{Max}(Q)$ when Q is bounded, is generally less than the number of vertices of $S(MOLP)$. If Q is unbounded, then the efficient outcome set may contain more vertices than the efficient solution set.

Example 9.1.3 Let $X = \{x \in \mathbb{R}^3 : x_1 \leq 0, x_2 = 0\}$ and $C = \begin{pmatrix} 1 & 0 & 0 \\ 0 & 1 & 0 \end{pmatrix}$. The set $Q \subseteq \mathbb{R}^2$ is the negative ray of the x_1-axis. We have $\text{Max}(Q) = \left\{ \begin{pmatrix} 0 \\ 0 \end{pmatrix} \right\}$, which consists of the unique vertex of Q, while the set X has no vertex.

Theorem 9.1.4 *Let G be an arbitrary face of Q. Then the set $F = \{x \in X : Cx \in G\}$ is a face of X, and $\dim F \geq \dim G$.*

Proof Choose y^0 from the relative interior of G and λ from the relative interior of the normal cone $N_Q(y_0)$. We have

$$G = \{y \in Q : \langle \lambda, y \rangle = \langle \lambda, y^0 \rangle \geq \langle \lambda, y' \rangle \text{ for all } y' \in Q\}. \tag{9.1}$$

Define

$$F_0 = \{x \in X : \langle C^T \lambda, x \rangle = \langle \lambda, y^0 \rangle\}.$$

We claim that F_0 is a face of X and $F = F_0$. In fact, F_0 is nonempty because it contains all $x^0 \in X$ such that $Cx^0 = y^0$. Moreover, for every $x \in X$ one has

$$\langle C^T \lambda, x \rangle = \langle \lambda, Cx \rangle \leq \langle \lambda, y^0 \rangle = \langle C^T \lambda, x^0 \rangle.$$

Hence the set $H = \{x \in \mathbb{R}^n : \langle C^T \lambda, x \rangle = \langle \lambda, y^0 \rangle\}$ is a supporting hyperplane of X at x^0, by which $F_0 = H \cap X$ is a face of X. Furthermore, if $x \in F$, then it follows from (9.1) that

$$\langle C^T \lambda, x \rangle = \langle \lambda, Cx \rangle = \langle \lambda, y^0 \rangle,$$

which implies $x \in F_0$. Conversely, $x \in F_0$ implies $\langle C^T \lambda, x \rangle = \langle \lambda, y^0 \rangle$. Hence $Cx \in G$, by which $x \in F$ as well.

Finally, let $\{y^1, \cdots, y^p\} \subseteq G$ be an affinely independent family in G. There are some $x \in F$ such that $Cx^i = y^i, i = 1, \cdots, p$. Then $\{x^1, \cdots, x^p\}$ is an affinely independent family in F. Consequently, $dim F \geq dim G$. □

Similar to the case of vertices, a face of X needs not be mapped by C onto a face of Q, that is, if F is a face of X, its image $C(F)$ is not necessarily a face of Q. This fact, of course, is also true for efficient faces and weakly efficient faces of X. Here is an exception when a face of X is maximal efficient or maximal weakly efficient.

Theorem 9.1.5 *Let F be a maximal efficient (respectively maximal weakly efficient) face of X. Then $C(F) = \{Cx : x \in F\}$ is a maximal efficient (respectively maximal weakly efficient) face of Q.*

Proof Let F be a maximal efficient face. Choose $x^0 \in ri F$. In view of Theorem 4.3.1, there is a strictly positive k-vector λ such that

$$\langle \lambda, Cx^0 \rangle \geqq \langle \lambda, Cx \rangle \text{ for all } x \in X.$$

Claim 1. $F = \{x \in X : \langle \lambda, Cx \rangle = \langle \lambda, Cx^0 \rangle\}$.
Indeed, let F be the convex hull of its vertices $x^1, \cdots, x^p$. By Corollary 6.4.8 there exist strictly positive numbers $t_1, \cdots, t_p$ with $\sum_{i=1}^{p} t_i = 1$ such that $x^0 = \sum_{i=1}^{p} t_i x^i$. We have

$$\langle \lambda, Cx^0 \rangle = \sum_{i=1}^{p} t_i \langle \lambda, Cx^i \rangle$$

Since $\langle \lambda, Cx^0 \rangle \geqq \langle \lambda, Cx^i \rangle$ for $i = 1, \cdots, p$, we deduce

$$\langle \lambda, Cx^0 \rangle = \langle \lambda, Cx^i \rangle \text{ for } i = 1, \cdots, p.$$

In other words, all x^i belong to the set on the right hand side of Claim 1. Consequently, $F \subseteq \{x \in X : \langle \lambda, Cx \rangle = \langle \lambda, Cx^0 \rangle\}$. Conversely, let $x \in X$ be such that $\langle \lambda, Cx \rangle = \langle \lambda, Cx^0 \rangle$. If $x \notin F$, then $\frac{1}{2}(x + x^0) \notin F$. Let F' be the face that contains $\frac{1}{2}(x + x^0)$ in its relative interior. We have

$$\langle \lambda, C(\frac{1}{2}(x + x^0)) \rangle = \frac{1}{2}(\langle \lambda, Cx \rangle + \langle \lambda, Cx^0 \rangle) = \langle \lambda, Cx^0 \rangle.$$

It follows that $\frac{1}{2}(x + x^0)$ is an efficient point, and therefore, F' is an efficient face. Because F' contains the relative point x^0 of F and also the point x outside of F, this face strictly contains F, which contradicts the maximality hypothesis. Claim 1 is established.

Claim 2. $C(F)$ is an efficient face.
Consider the linear function $\langle \lambda, . \rangle$. If $y \in C(F)$ and $y' \in Q$, then there are some $x \in F$ and $x' \in X$ such that $Cx = y$ and $Cx' = y'$. Then $\langle \lambda, y \rangle \geq \langle \lambda, y' \rangle$ and $C(F) = \{y \in Q : \langle \lambda, y \rangle = \langle \lambda, Cx^0 \rangle \geq \langle \lambda, y' \rangle \text{ for all } y' \in Q\}$. Consequently, $C(F)$ is an efficient face of Q.

Claim 3. $C(F)$ is a maximal efficient face.
Indeed, if this is not true, there is an efficient face G of Q that contains $C(F)$ and satisfies $C(F) \neq G$. Let $y^0 \in riG$. Then $y^0 \notin C(F)$ and there is some strictly positive k-vector λ^0 such that

$$G = \{y \in Q : \langle \lambda^0, y \rangle = \langle \lambda^0, Cx^0 \rangle \geq \langle \lambda^0, y' \rangle \quad \text{for all } y' \in Q\}.$$

Set $F' = \{x \in X : Cx \in G\}$. By Theorem 9.1.4, F' is a face of X and efficient. It strictly contains F because $C(F) \subseteq G$. For any $x \in X$ such that $Cx = y^0$, we have $x \in F' \setminus F$, which contradicts the maximality of F.
For weakly efficient faces apply the same argument. □

Theorem 9.1.6 *Let G be a maximal efficient (respectively maximal weakly efficient) face of Q. Then the set $F = \{x \in X : Cx \in G\}$ is a maximal efficient (respectively maximal weakly efficient) face of X.*

Proof In view of Theorem 9.1.4, F is a face and by definition it is efficient. If it were not maximal, one would find another efficient face F' of X that strictly contains F. We may assume F' is maximal. By Theorem 9.1.5, $C(F')$ is an efficient face of Q. Since $F \subseteq F'$ and $F \neq F'$, $C(F') \supseteq G$ and $C(F') \neq G$. This contradicts the maximality of G. □

Corollary 9.1.7 *The linear operator C induces a one-to-one correspondence between maximal efficient faces of the solution set X and the value set Q of (MOLP).*

Proof Let F be a maximal efficient face of X. By Theorem 9.1.5, $C(F)$ is a maximal efficient face of Q. If F' is another maximal efficient face of X such that $C(F') = C(F)$, then in view of Theorem 9.1.6,

$$F'' = \{x \in X : Cx \in C(F)\}$$

is a maximal efficient face of X. But $F \subseteq F''$, $F' \subseteq F''$, we conclude $F = F' = F''$ and so the correspondence $F \mapsto C(F)$ is injective. The surjectivity of this correspondence follows from Theorem 9.1.6. □

9.2 Free Disposal Hull

Definition 9.2.1 Let P be a nonempty set in $\mathbb{R}_+^k$. We say that P is free disposal if it contains all positive vectors y smaller than some element of P.

Equivalently, P is free disposal if it satisfies $P = (P - \mathbb{R}_+^k) \cap \mathbb{R}_+^k$ (Figs. 9.1 and 9.2).

Definition 9.2.2 Given an arbitrary set $Q \subseteq \mathbb{R}_+^k$, the set

$$\hat{Q} := (Q - \mathbb{R}_+^k) \cap \mathbb{R}_+^k$$

is called the free disposal hull of Q.

Throughout this section we assume that Q is a polytope included in the set $e + \mathbb{R}_+^k$, where e is the k-vector of ones. We will present some relationships between efficient elements of Q and efficient elements of its free disposal hull $\hat{Q}$ (Fig. 9.3).

Theorem 9.2.3 *The following statements hold.*

Fig. 9.1 A free disposal set

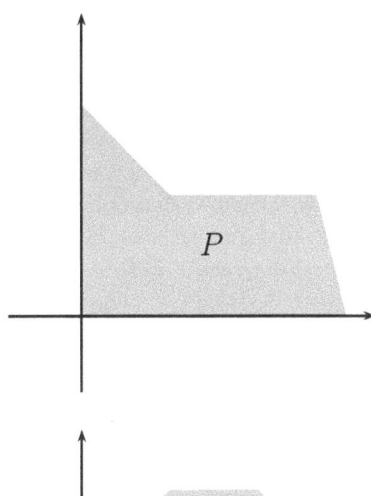

Fig. 9.2 A non-free disposal set

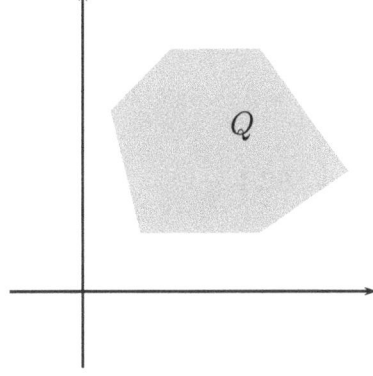

Fig. 9.3 Free disposal hull

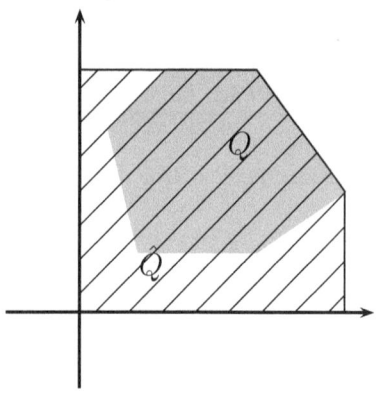

(i) $\hat{Q}$ is a bounded, free disposal polyhedron of full dimension.
(ii) $\mathrm{Max}(Q) = \mathrm{Max}(\hat{Q})$.
(iii) $\mathrm{WMax}(Q) = \mathrm{WMax}(\hat{Q}) \cap Q$.

Proof Since Q is a polytope set, the set $Q - \mathbb{R}_+^k$ is a polyhedral set and there is some $t > 0$ such that $Q \subseteq te - \mathbb{R}_+^k$. Being an intersection of two polyhedral sets, the set $\hat{Q}$ is a polyhedron satisfying $0 \leq y \leq te$ for every $y \in \hat{Q}$. In other words, $\hat{Q}$ is a polytope. Further, let y be a positive vector such that $y \leq y'$ for some $y' \in \hat{Q}$. There are some $y'' \in Q$ and $v \geq 0$ such that $y' = y'' - v$. Setting $u = y' - y \geq 0$, we deduce $y = y'' - (u + v) \in Q - \mathbb{R}_+^k$ proving that $y \in \hat{Q}$. By this, $\hat{Q}$ is free disposal. We also notice that every point y satisfying $0 < y < e$ is an interior point of $\hat{Q}$, which shows that $\hat{Q}$ is of full dimension.

To prove (ii), let $y \in \mathrm{Max}(Q)$. Let $z \in \mathrm{Max}(\hat{Q})$ such that $z \geq y$. There are some $y' \in Q$ and $v \geq 0$ such that $z = y' - v$. We deduce $y' \geq y + v$. Since y is efficient, we have $v = 0$ and $y = y'$, which shows that $y \in \mathrm{Max}(\hat{Q})$. Conversely, let $z \in \mathrm{Max}(\hat{Q})$ with $z = y' - v$ for some $y' \in Q \subseteq \hat{Q}$ and some $v \geq 0$. It follows that $v = 0$ and $z = y'$ because otherwise $y' \geq z$, which contradicts the efficiency of z. Hence $z \in Q$ and $z \in \mathrm{Max}(Q)$.

For the last statement, let $y \in \mathrm{WMax}(Q)$. Then $y \in Q \subseteq \hat{Q}$. If $z > y$ for some $z \in \hat{Q}$, where $z = y' - v$ with $y' \in Q$ and $v \geq 0$, then $y' = z + v > y$, which contradicts the efficiency of y. By this, $y \in \mathrm{WMax}(\hat{Q}) \cap Q$. Conversely, let $z \in \mathrm{WMax}(\hat{Q}) \cap Q$. Since $Q \subseteq \hat{Q}$, we deduce $z \in \mathrm{WMax}(Q)$ too. The proof is complete.   □

Lemma 9.2.4 *A point $w \in \hat{Q}$ is a weakly efficient point of $\hat{Q}$ if and only if the optimal value of the following problem, denoted* (P_w), *is zero,*

$$\begin{aligned} \text{maximize} \quad & t \\ \text{subject to} \quad & y - te \geq w \\ & y \in Q, t \in \mathbb{R}. \end{aligned}$$

Moreover, w is an efficient point of Q if and only if $(w, 0)$ is the unique optimal solution of (P_w).

Proof Assume that w is a weakly efficient element of $\hat{Q}$. Then the point (y, t), where y is a vector from Q such that $y \geq w$ and $t = 0$, is a feasible solution of (P_w). Hence the optimal value (P_w) is not less than 0. On the other hand, if for some feasible solution (y', t) one has $t > 0$, then $y' > y' - te \geq w$ implying that w is not weakly efficient. Thus, the optimal value of (P_w) is equal to zero. Conversely, if w is not weakly efficient, then there is some $y \in Q$ such that $y > w$. For $t > 0$ sufficiently small one has $y - te > w$, by which the optimal value of (P_w) is strictly positive.

To prove the second assertion, we assume that w is an efficient element of Q. By Theorem 9.2.3 it is also an efficient element of $\hat{Q}$. It follows that if $(y', 0)$ is a feasible solution of (P_w), then $y' = w$. Hence $(w, 0)$ is a unique optimal solution of (P_w). The converse is clear because the uniqueness of the optimal solution $(w, 0)$ shows that w is an efficient element of $\hat{Q}$, and hence an efficient element of Q by Theorem 9.2.3.   □

Corollary 9.2.5 *Every vertex of $\hat{Q}$ having all strictly positive components is an efficient element of Q.*

Proof Let w be a vertex of $\hat{Q}$ such that $w > 0$. Consider the problem (P_w) introduced in the preceding lemma. If there is some feasible solution $(y, 0)$ satisfying $y \geq w$, then w is an interior point of the segment $[w + \epsilon(w - y), y] \subseteq \hat{Q}$ when $\epsilon > 0$ is so small that $w + \epsilon(w - y) > 0$ (such an ϵ exists because $w > 0$). This contradicts the hypothesis that w is a vertex. Hence $(w, 0)$ is a unique optimal solution of (P_w). By Lemma 9.2.4, w is an efficient element of Q.   □

In the rest of this section we consider a particular case of Q. Namely, we consider the problem (MOLP) in standard form

$$\text{Maximize} \quad Cx$$
$$\text{subject to} \quad Ax = b$$
$$x \geq 0,$$

where C is a $k \times n$-matrix, A is an $m \times n$-matrix and b is an m-vector. The value set of (MOLP), denoted Q, consists of the values Cx with $x \in X$, where X is the feasible set. We assume that Q is bounded. Then the infimum of Q, denoted y^0, is the vector the components of which are given by

$$y_i^0 = \min\{y_i : y = (y_1, \cdots, y_k)^T \in Q\}, \ i = 1 \cdots, k$$

is finite. By shifting Q to $Q' = Q - y^0 + e$ we obtain $Q' \subseteq e + \mathbb{R}_+^k$. Moreover, it is clear that $\text{Max}(Q) = \text{Max}(Q') + y^0 - e$ and $\text{WMax}(Q) = \text{WMax}(Q') + y^0 - e$, so when solving (MOLP) we may assume without loss of generality that Q is contained in $e + \mathbb{R}_+^k$.

By applying Lemma 9.2.4 we deduce that w a weakly efficient point of $\hat{Q}$ if and only if the optimal value of the following problem, denoted also (P$_w$), is zero,

$$
\begin{aligned}
\text{maximize} \quad & t \\
\text{subject to} \quad & Cx - te \geqq w \\
& Ax = b \\
& x \geqq 0, t \geqq 0, t \in \mathbb{R}.
\end{aligned}
$$

We consider the dual program of (P$_w$), denoted (D$_w$) given by

$$
\begin{aligned}
\text{minimize} \quad & -\langle w, \lambda \rangle + \langle b, \gamma \rangle \\
\text{subject to} \quad & -\lambda^T C + \gamma^T A \geqq 0 \\
& \langle e, \lambda \rangle \geqq 1, \lambda \geqq 0.
\end{aligned}
$$

Theorem 9.2.6 *Assume that the feasible set X of (MOLP) is bounded and that w is a weakly efficient element of $\hat{Q}$. Then w is contained in a weakly efficient face of $\hat{Q}$ determined by*

$$
\langle \lambda, y \rangle \leqq \langle \lambda, w \rangle = \langle b, \gamma \rangle, \ y \in \hat{Q}
$$

where (λ, γ) is an optimal solution of (D$_w$).

Proof Under the hypothesis of the theorem and in view of Lemma 9.2.4, the program (P$_w$) has the zero optimal value, hence so does its dual program (D$_w$). Let (λ, γ) be an optimal solution. Then $\lambda \geq 0$ and satisfies

$$
\begin{aligned}
& -\langle w, \lambda \rangle + \langle b, \gamma \rangle = 0, \\
& -\lambda^T C + \gamma^T A \geqq 0.
\end{aligned}
$$

Let $y \in \hat{Q}$, say $y = Cx - v$ for some $x \in X$ and $v \geqq 0$. Then

$$
\begin{aligned}
-\langle \lambda, y \rangle + \langle \gamma, b \rangle &= -\lambda^T (Cx - v) + \gamma^T Ax \\
&\geqq -\lambda^T Cx + \gamma^T Ax \geqq 0.
\end{aligned}
$$

In other words,

$$
\langle w, \lambda \rangle = \langle \gamma, b \rangle \geqq \langle \lambda, y \rangle
$$

for every $y \in \hat{Q}$. Since $\lambda \geq 0$, the solution set to the latter inequality is a weakly efficient face of $\hat{Q}$ and contains w. $\qquad\square$

9.3 Outer Approximation

We consider a polytope $Q \subseteq e + \mathbb{R}^k_+$ and its free disposal hull $\hat{Q}$. When Q is not explicitly given by a system of linear inequalities or by a family of vertices, it is difficult to describe the efficient set of Q. One of the methods to find the efficient set of Q is to construct a finite sequence of free disposal polytopes $Q^0, \cdots, Q^\ell$ that have a simple structure and provide more and more efficient elements of Q as the construction progresses until the whole efficient set is produced.

Let us begin with the construction of a first outer approximation for $\hat{Q}$. Consider the linear function $\langle e, y \rangle = \sum_{i=1}^k y_i$ on Q. Let α be its maximum, which is finite because Q is bounded. Set (Fig. 9.4)

$$Q^0 = \mathbb{R}^k_+ \cap \{y \in \mathbb{R}^k : \langle e, y \rangle \leqq \alpha\}.$$

Theorem 9.3.1 *The set Q^0 is a k-dimensional simplex containing $\hat{Q}$. Its vertex set consists of the origin 0 and the vectors αe^i, $i = 1, \cdots, k$. Moreover, the set $\mathrm{Max}(Q^0)$ contains an efficient face of $\hat{Q}$.*

Proof It is clear that $\alpha > 0$ because $y \geq e$ for every $y \in Q$, and that Q is contained in Q^0. We deduce that the vectors αe^i, $i = 1, \cdots, k$ are linearly independent. Consequently, the vectors αe^i, $i = 1, \cdots, k$ and 0 are in general position (affinely independent). Moreover, they are all satisfy the system defining Q^0, therefore the convex hull of these vectors which is a k-dimensional simplex, is contained in Q^0. Let $y \in Q^0$. Since $y \geq 0$ there are positive numbers $t_1, \cdots, t_k$ such that $y = \sum_{i=1}^k t_i e^i$. Moreover, we have

$$\langle e, y \rangle = \sum_{i=1}^k t_i \langle e, e^i \rangle = \sum_{i=1}^k t_i \leqq \alpha,$$

Fig. 9.4 The simplex Q^0

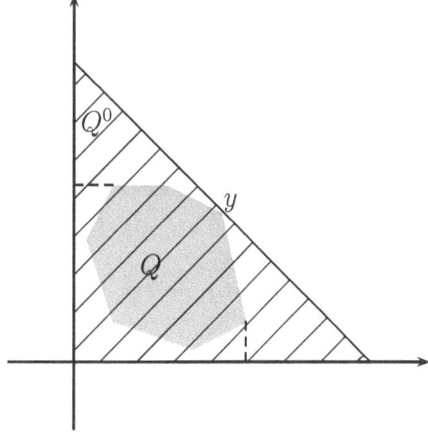

which implies $t_i/\alpha \in [0, 1], i = 1, \cdots, k$ and $1 - \frac{1}{\alpha} \sum_{i=1}^{k} t_i \in [0, 1]$. We express y as

$$y = (1 - \frac{1}{\alpha} \sum_{i=1}^{k} t_i)0 + \sum_{i=1}^{k} \frac{t_i}{\alpha}(\alpha e^i).$$

In other words every element y of Q^0 is a convex combination of the vertices αe^i, $i = 1, \cdots, k$ and 0. Hence Q^0 is a simplex. It is plain that $\hat{Q}$ is contained in Q^0.

The second part of the theorem is clear because the vector e is strictly positive and so by Corollary 4.1.12 the face $\{y \in Q^0 : \langle e, y \rangle = \alpha\}$ is an efficient face of Q^0 and its intersection with $\hat{Q}$ is nonempty and an efficient face of $\hat{Q}$. $\qquad\square$

To proceed further, let us fix an interior point $q \in \hat{Q}$, for instance $q = e/2$, and show that if a free disposal polytope Q^0 is an outer approximation of $\hat{Q}$ and does not coincide with it, then one can find a suitable cut in order to obtain a smaller outer approximation that is a free disposal polytope, the weakly efficient set of which contains a weakly efficient face of $\hat{Q}$ not previously found in the weakly efficient set of Q^0.

Lemma 9.3.2 *Let $y \geq 0$ be a point outside $\hat{Q}$. Then the unique point $w = q + t_0(y - q)$ on the segment connecting q and y, where $t_0 \in (0, 1)$ is the optimal solution of the problem*

$$\begin{aligned} \text{maximize} \qquad & t \\ \text{subject to} \quad q + t(y - q) & \in \hat{Q}, \end{aligned}$$

is a weakly efficient element on the boundary of $\hat{Q}$.

Proof Since $\hat{Q}$ is bounded, the problem stated in this lemma has an optimal solution, denoted t_0. As q is an interior point of $\hat{Q}$, we have $t_0 > 0$, and as y is outside of $\hat{Q}$, $t_0 < 1$. The uniqueness of the optimal solution is clear. We show that w (Fig. 9.5) is on the boundary and a weakly efficient point of $\hat{Q}$. Indeed, if w were an interior point, one would find some $t > t_0$ such that $q + t(y - q) \in \hat{Q}$, which contradicts the optimality of t_0. Further, if w were not a weakly efficient point of $\hat{Q}$, there would exist some $w' \in \hat{Q}$ such that $w' > w$. Then w belongs to the interior of $w' - \mathbb{R}_+^k \subseteq \hat{Q} - \mathbb{R}_+^k$. It is also clear that for $\epsilon > 0$ sufficiently small, $q + (t_0 + \epsilon)(y - q) \in w' - \mathbb{R}_+^k$, hence $q + (t_0 + \epsilon)(y - q) \in \hat{Q}$ as well. This contradicts the definition of t_0 and the proof is complete. $\qquad\square$

Theorem 9.3.3 *Let P be a free disposal polytope that contains $\hat{Q}$. If there is some $y^0 \in P$ such that $y^0 \notin \hat{Q}$, then there exist a nonzero positive vector λ and a weakly efficient element w of $\hat{Q}$ that is not weakly efficient element of P such that the polytope*

$$P' = \{y \in P : \langle \lambda, y \rangle \leq \langle \lambda, w \rangle\}$$

Fig. 9.5 The point
$w = q + t_0(y - q)$

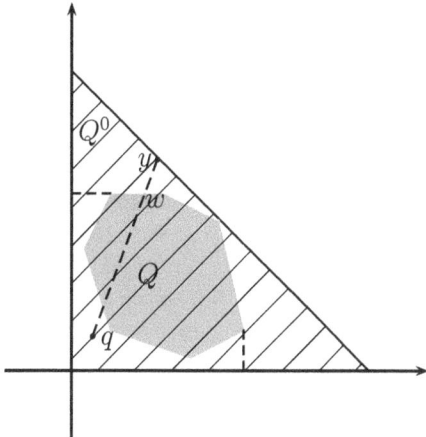

is a free disposal polytope containing $\hat{Q}$ and its weakly efficient set contains a weakly efficient face of $\hat{Q}$ that contains w.

Proof Since $y^0 \in P$ we have $y^0 \geq 0$. Apply Lemma 9.3.2 to find a weakly efficient point w of $\hat{Q}$ on the segment joining q and y_0. It is clear that w is not a weakly efficient element of P because it is an interior point of P being strictly inside of the segment $[q, y_0]$. According to Theorem 9.2.6, there exists a vector $\lambda \geq 0$ such that

$$\langle \lambda, y \rangle \leq \langle \lambda, w \rangle \text{ for all } y \in \hat{Q}.$$

It is clear that the polytope P' as defined in the theorem is free disposal and satisfies the inclusions $\hat{Q} \subseteq P' \subset P$. Of course P' is smaller than P because y^0 belongs to P, but does not belong to P'. As the vector λ is positive, in view of Corollary 4.1.12, the set $\{y \in \hat{Q} : \langle \lambda, y \rangle = \langle \lambda, w \rangle\}$ is a weakly efficient face of $\hat{Q}$ and contains w. It is contained in the face $\{y \in P' : \langle \lambda, y \rangle = \langle \lambda, w \rangle\}$ of P' that is also a weakly efficient face because $\lambda \geq 0$. $\qquad \square$

9.4 The Outcome Space Algorithm

The outcome space method, called also *outer approximation method*, is aimed at computing the efficient vertex set of the value set Q for problem (MOLP) described in Sect. 9.2. Normally, numerical methods in mathematical programming are developed for finding an optimal solution of the problem and thereby the optimal value is obtained as a by-product. However, in multiple objective programming it is sometimes preferable to compute efficient values and then efficient solutions are deduced as a by-product. There are some reasons for this. First, in most practical situations the number of objective functions are much less than the number of variables. It is often the case when k is two or three, while n may be hundreds, thousands or more. Then under the objective matrix C several feasible solutions may be mapped onto a single value in the outcome space $\mathbb{R}^k$. As a result, the value set Q may have a structure much simpler than the structure of X, for its number of vertices is less than the number of vertices of X, and so computing the set $\text{Max}(Q)$ is less demanding than computing the efficient solution set S(MOLP). Another reason is with post-optimal analysis. In fact, without a decision maker's interaction during the computing process, after the problem has been solved the entire set S(MOLP) or at least a good portion thereof must be presented to the decision maker for a final choice. The very high dimensionality of the decision space and the complicated nature of the set S(MOLP) will certainly overwhelm him or her. On the contrary, the efficient value set in the outcome space of low dimension, especially when k is equal to 2 or 3, is easily visualized and makes the choice much more manageable. Below is a description of the algorithm. The vertex set of Q is denoted $V(Q)$.

Step 0 (Initialization). Solve

$$\begin{aligned} \text{maximize} \quad & \langle e, Cx \rangle \\ \text{subject to} \quad & Ax = b \\ & x \geqq 0. \end{aligned}$$

Let α be the optimal value of the problem. Set Q^0 to be the simplex given in Theorem 9.3.1. Store $V(Q^0)$ and the inequality representation of Q^0.
Set $i = 0$ and go to the next step.

Step 1. For $i \geqq 0$ check $V(Q^i) \subseteq \hat{Q}$.

If this is true, stop.
Otherwise, choose $y \in V(Q^i) \setminus \hat{Q}$ and find $w^i \in \text{WMax}(\hat{Q})$ (Lemma 9.3.2) by solving

$$\begin{aligned} \text{maximize} \quad & t \\ \text{subject to} \quad & q + t(y - q) \leqq Cx \\ & Ax = b \\ & x \geqq 0, \ t \geqq 0. \end{aligned}$$

Step 2. Find $\lambda^i \geq 0$ (Theorem 9.2.6) by solving

$$\begin{aligned} \text{minimize} \quad & -\langle w^i, \lambda \rangle + \langle b, \gamma \rangle \\ \text{subject to} \quad & -\lambda^T C + \gamma^T A \geq 0 \\ & \langle e, \lambda \rangle \geqq 1, \lambda \geqq 0. \end{aligned}$$

Set $Q^{i+1} = Q^i \cap \{y \in \mathbb{R}^k : \langle \lambda^i, y \rangle \leqq \langle \lambda^i, w^i \rangle\}$.

Step 3. Store $V(Q^{i+1})$. Set $i := i + 1$ and return to Step 1.

Let us make some comments on implementations of the algorithm.
(1) In Step 1, in order to test a vertex $y \in V(Q^i)$ for membership in $\hat{Q}$ it suffices to find a feasible solution of the linear program

$$
\begin{aligned}
\text{maximize} \quad & 0 \\
\text{subject to} \quad & Cx \geqq y \\
& Ax = b, x \geqq 0.
\end{aligned}
$$

(2) In Step 0, the vertex set of Q^0 was already given in Theorem 9.3.1. In Step 3, the vertex set of Q^{i+1} is computed by using the vertices of Q^i and the cutting hyperplane that defines Q^{i+1} from Q^i. Namely, those vertices that satisfy strict inequality $\langle \lambda^i, v \rangle < \langle \lambda^i, w^i \rangle$ remain in the collection $V(Q^{i+1})$ and those satisfy strict opposite inequality $\langle \lambda^i, v \rangle > \langle \lambda^i, w^i \rangle$ do not enter $V(Q^{i+1})$. New vertices will appear in $V(Q^{i+1})$, which are determined by equality $\langle \lambda^i, v \rangle = \langle \lambda^i, w^i \rangle, v \in Q^i$.

Example 9.4.1 We solve the following problem by the outcome space method.

$$
\begin{aligned}
\text{Maximize} \quad & \begin{pmatrix} x_1 + 3x_2 + 2x_3 \\ x_1 + x_2 + 2x_3 \end{pmatrix} \\
\text{subject to} \quad & x_1 + x_2 + x_3 = 1 \\
& x_1, x_2, x_3 \geqq 0.
\end{aligned}
$$

At the initialization step we solve the problem

$$
\begin{aligned}
\text{maximize} \quad & (x_1 + 3x_2 + 2x_3) + (x_1 + x_2 + 2x_3) \\
\text{subject to} \quad & x_1 + x_2 + x_3 = 1 \\
& x_1, x_2, x_3 \geqq 0.
\end{aligned}
$$

The optimal value α is equal to 4. The simplex Q^0 (Fig. 9.6) is given by the system

$$
\begin{aligned}
y_1 + y_2 &\leqq 4 \\
y_1, y_1 &\geqq 0.
\end{aligned}
$$

Its vertex set is given by

$$
V(Q^0) = \left\{ \begin{pmatrix} 0 \\ 0 \end{pmatrix}, \begin{pmatrix} 4 \\ 0 \end{pmatrix}, \begin{pmatrix} 0 \\ 4 \end{pmatrix} \right\}.
$$

Iteration $i = 0$. We have $V(Q^0) \not\subseteq \hat{Q}$. Choose $v^0 = \begin{pmatrix} 0 \\ 4 \end{pmatrix} \in V(Q^0) \setminus \hat{Q}$ and solve the problem formulated in Step 1 to find

$$
w^0 = \begin{pmatrix} 0 \\ 2 \end{pmatrix} \in \text{WMax}(\hat{Q}).
$$

By solving the problem formulated in Step 2 we find $\lambda^0 = \begin{pmatrix} 0 \\ 1 \end{pmatrix}$. The second approximation set Q^1 (Fig. 9.7) of $\hat{Q}$ is determined by the system

$$y_1 + y_2 \leq 4$$
$$y_1, y_1 \geq 0$$
$$\left\langle \begin{pmatrix} 0 \\ 1 \end{pmatrix}, \begin{pmatrix} y_1 \\ y_2 \end{pmatrix} \right\rangle \leq \left\langle \begin{pmatrix} 0 \\ 1 \end{pmatrix}, \begin{pmatrix} 0 \\ 2 \end{pmatrix} \right\rangle.$$

Then we store the vertex set of Q^1:

$$V(Q^1) = \left\{ \begin{pmatrix} 0 \\ 0 \end{pmatrix}, \begin{pmatrix} 0 \\ 2 \end{pmatrix}, \begin{pmatrix} 2 \\ 2 \end{pmatrix}, \begin{pmatrix} 4 \\ 0 \end{pmatrix} \right\}.$$

Iteration $i = 1$. We still have $V(Q^1) \not\subseteq \hat{Q}$. Choose $v^1 = \begin{pmatrix} 4 \\ 0 \end{pmatrix} \in V(Q^1) \setminus \hat{Q}$ and solve the problem formulated in Step 1 to find

$$w^1 = \begin{pmatrix} 3 \\ 0 \end{pmatrix} \in \mathrm{WMax}(\hat{Q}).$$

By solving the problem formulated in Step 2 we find $\lambda^1 = \begin{pmatrix} 1 \\ 0 \end{pmatrix}$. The third approximation set Q^2 (Fig. 9.8) of $\hat{Q}$ is determined by the system

$$y_1 + y_2 \leq 4$$
$$y_1, y_1 \geq 0$$
$$y_2 \leq 2$$
$$\left\langle \begin{pmatrix} 1 \\ 0 \end{pmatrix}, \begin{pmatrix} y_1 \\ y_2 \end{pmatrix} \right\rangle \leq \left\langle \begin{pmatrix} 1 \\ 0 \end{pmatrix}, \begin{pmatrix} 3 \\ 0 \end{pmatrix} \right\rangle.$$

Then we store the vertex set of Q^2:

$$V(Q^2) = \left\{ \begin{pmatrix} 0 \\ 0 \end{pmatrix}, \begin{pmatrix} 0 \\ 2 \end{pmatrix}, \begin{pmatrix} 2 \\ 2 \end{pmatrix}, \begin{pmatrix} 3 \\ 0 \end{pmatrix}, \begin{pmatrix} 3 \\ 1 \end{pmatrix} \right\}$$

and return to Step 1.

Iteration $i = 2$. At this iteration we have $V(Q^2) \subseteq \hat{Q}$. The algorithm terminates. The efficient vertex set of Q consists of two strictly positive vertices of $V(Q^2)$: $\begin{pmatrix} 2 \\ 2 \end{pmatrix}$ and $\begin{pmatrix} 3 \\ 1 \end{pmatrix}$.

Fig. 9.6 The first
approximation of $\hat{Q}$

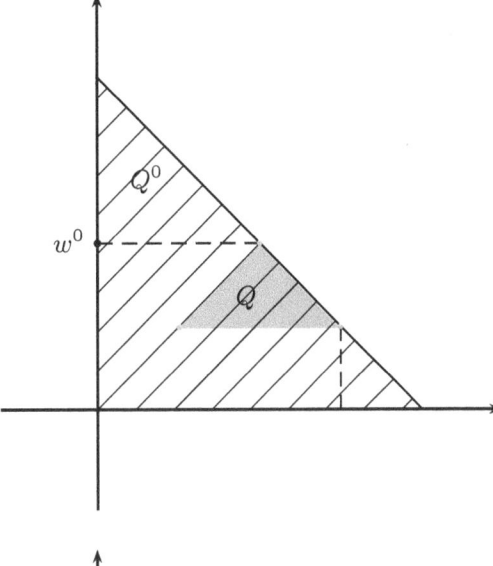

Fig. 9.7 The second
approximation of $\hat{Q}$

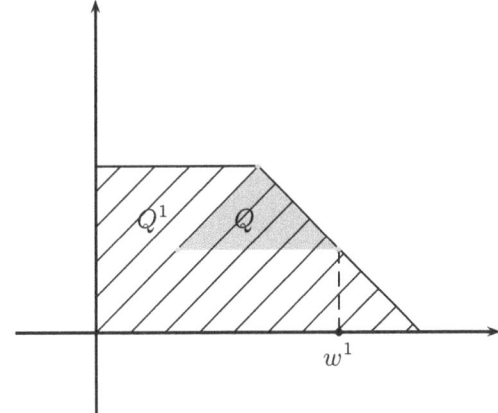

Theorem 9.4.2 *The outcome space algorithm is finite. At the final iteration $\ell \geq 0$,
one has $Q^\ell = \hat{Q}$ and the efficient vertex set of Q consists of all vertices of Q^ℓ that
are strictly positive.*

Proof By construction one has $\hat{Q} \subseteq Q^i$ for all iteration number $i \geq 0$. If $Q^i \neq \hat{Q}$,
then there is some vertex y^i of Q^i and $y^i \notin \hat{Q}$. By Theorem 9.3.3, Step 2 generates a
new polytope Q^{i+1} that includes a weakly efficient face of $\hat{Q}$ in its weakly efficient
set, not previously encountered in the weakly efficient set of Q^i. Since the number of
faces of $\hat{Q}$ is finite, the algorithm must terminate after a finite number of iterations,
say at the iteration $\ell \geq 0$. We then have $Q^\ell = \hat{Q}$. In view of Theorem 9.2.3 and
Corollary 9.2.5 the efficient vertex set of Q consists of strictly positive vertices of
$Q^\ell = \hat{Q}$.   □

Fig. 9.8 The third
approximation of $\hat{Q}$

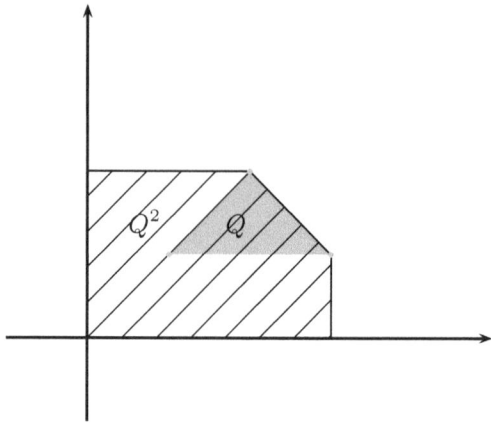

Computing the weakly efficient value set of (MOLP)

The outcome space algorithm allows us to obtain not only the efficient vertex value set of (MOLP), but it provides also scalarizing vectors for generating the weakly efficient value set of (MOLP). Indeed, at the final iteration we have $\hat{Q} = Q^\ell$. The vectors $\lambda^i, i = 0, \cdots, \ell - 1$ collected in Step 2 through the ℓ iterations determine all weakly faces of $\hat{Q}$. Then, the weakly efficient faces of Q are given by

$$\text{WMax}(Q) = \bigcup_{i=0}^{\ell-1} \{Cx \in \mathbb{R}^k : \langle \lambda^i, Cx \rangle = \langle \lambda^i, w^i \rangle, Ax = b, x \geq 0\}.$$

It is the union of all vectors $y \in \mathbb{R}^k$ for which there is some $x \in \mathbb{R}^n$ such that (x, y) is an optimal solution of the following linear problem

$$\begin{aligned}
\text{maximize} \quad & \langle \lambda^i, y \rangle \\
\text{subject to} \quad & y = Cx \\
& Ax = b, x \geq 0.
\end{aligned}$$

for $i = 0, \cdots, \ell - 1$.

When the hypothesis $Q \subseteq e + \mathbb{R}^k_+$ is not satisfied

As we have already discussed when the value set Q of (MOLP) is bounded, but the hypothesis $Q \subseteq e + \mathbb{R}^k_+$ is not guaranteed, one may shift Q to Q' to meet this requirement and performs the outcome space algorithm for Q'. Namely, one solves the linear programs

$$\text{maximize} \qquad \langle e^i, y \rangle$$
$$\text{subject to} \qquad y = Cx$$
$$Ax = b, x \geq 0.$$

for $i = 1, \cdots, k$ to find the infimum y^0 of Q that is the vector of optimal values of these programs. The set Q shifted by the vector $-y^0 + e$ is given by

$$Q' = Q - y^0 + e$$
$$= \{y \in \mathbb{R}^k : y = Cx - y^0 + e \text{ for some } x \in X\}.$$

The simplex Q^0 in Step 0 and the free disposal hull $\hat{Q}$ are shifted by $-y^0 + e$. The program in Step 1 is modified to

$$\text{maximize} \qquad t$$
$$\text{subject to} \quad q + t(y - q) \leqq Cx - y^0 + e$$
$$Ax = b$$
$$x \geqq 0, \ t \geqq 0.$$

There is no change in the dual program of Step 2. At the final iteration the set of efficient vertices of Q is equal to the set of efficient vertices of Q^ℓ shifted by the vector $-y^0 + e$.

9.5 Exercises

9.5.1 Let F be a face of X in (MOLP) such that $x = 0$ when $Cx = 0$. Prove that $dim C(F) = dim F$.

9.5.2 Prove that the conclusion of 9.1.4 remains true when Q is unbounded, but has a vertex.

9.5.3 Let Q be the value set of problem (MOLP) considered in this chapter such that $Q \subseteq e + \mathbb{R}^k_+$ and $\hat{Q}$ is its free disposal hull. Let q be an interior point of $e + \mathbb{R}^k_+$ and $y \in \mathbb{R}^k_+ \setminus \hat{Q}$. Prove that a point w is an efficient point of Q if and only if $w = q + t_0(y - q)$ where t_0 is an optimal value of the problem

$$\text{maximize} \qquad t$$
$$\text{subject to} \quad q + t(y - q) \geqq Cx$$
$$Ax = b$$
$$x \geqq 0,$$

and for any two optimal solutions (t_0, x^1) and (t_0, x^2) one has $Cx^1 = Cx^2$.

9.5.4 *Consider the value set Q of problem (MOLP) that satisfies $Q \subseteq e + \mathbb{R}^k_+$ and its free disposal set $\hat{Q}$. Let w be a weakly efficient element of $\hat{Q}$. Prove that a k-vector $\lambda \geq 0$ determines a face of $\hat{Q}$ that contains w if and only if there is a vector $\mu \in \mathbb{R}^m$ such that (λ, μ) is a solution to the system*

$$\lambda^T C - \mu^T A \leq 0$$
$$\langle v, \lambda \rangle - \langle b, \mu \rangle = 0$$
$$\lambda \geq 0.$$

9.5.5 *Consider the outcome space algorithm described in Sect. 9.4. Assume that at the iteration $i \geq 0$, one obtains a strictly positive vector v that is a vertex of Q^i.*

(i) *Show that v is an efficient element of Q^i.*
(ii) *Deduce that if $v \in \hat{Q}$, then v is also an efficient element of Q and the following system is consistent:*

$$\lambda^T C - \mu^T A \leq 0$$
$$\langle v, \lambda \rangle - \langle b, \mu \rangle = 0$$
$$\lambda \geq e.$$

9.5.6 *We consider the outcome algorithm for program (MOLP) having values in $\mathbb{R}^2$ which terminates after ℓ iterations.*

(i) *Show that among the vectors $\lambda^i, i = 0, \cdots, \ell - 1$ obtained in Step 2 there are at most two vectors that are not strictly positive.*
(ii) *Prove that if there are strictly positive vectors among the vectors $\lambda^i, i = 0, \cdots, \ell - 1$, then*

$$\text{Max}(Q) = \bigcup_{\lambda^i > 0, i \in \{0, \cdots, \ell-1\}} \left\{ y \in \mathbb{R}^2 : \begin{matrix} \langle \lambda^i, y \rangle = \langle \lambda^i, w^i \rangle, \\ y = Cx \text{ for some } x \in X \end{matrix} \right\}$$
$$\bigcup \left\{ y \in \mathbb{R}^2 : \langle e, y \rangle = \alpha, y = Cx \text{ for some } x \in X \right\}.$$

9.5.7 *Solve the following problem by the outcome space method*

$$\text{Maximize} \quad \begin{pmatrix} 3x_1 + 2x_3 \\ 3x_2 + 2x_3 \end{pmatrix}$$
$$\text{subject to} \quad x_1 + x_2 + x_3 = 1$$
$$x_1, x_2, x_3 \geq 0.$$

9.5.8 *Find the vertices of the efficient value set of the following problem by using the outcome space method*

$$\text{Minimize} \quad \begin{pmatrix} 1 & 0 & 0 \\ 0 & 1 & -1 \\ 1 & 1 & 1 \end{pmatrix} \begin{pmatrix} x_1 \\ x_2 \\ x_3 \end{pmatrix}$$

$$\text{subject to} \quad \begin{pmatrix} 2 & 1 & 0 \\ 1 & 0 & 3 \\ -1 & 0 & 1 \end{pmatrix} \begin{pmatrix} x_1 \\ x_2 \\ x_3 \end{pmatrix} \leq \begin{pmatrix} 4 \\ 4 \\ 6 \end{pmatrix}$$

$$x_1, x_2, x_3 \geq 0.$$

Bibliographical Notes

Chapter 1: Introduction.
This short introduction to the subject of multiobjective optimization is partly taken from the Chap. 7 [140]. Many real-word problems that motivate the research of multiobjective programming problems are described in [67, 197, 231]. See also the references given in Further Reading "Applications of multiobjective linear programming".

Chapter 2: Convex Polyhedra.
Basic notions from convex analysis are presented with emphasis on convex polyhedral sets that are defined by a finite system of linear inequalities. Farkas' alternative theorem and a version of it on separation of two disjoint convex polyhedra are key tools in the study of mathematical programming problems. They are found in textbooks on convex analysis and linear programming (see for instance [49, 141, 176]).

Chapter 3: Linear Programming.
This is a brief introduction to linear programming. Two topics of particular interest for further developments about duality programs and simplex algorithm are given in details. There are a lot of textbooks on linear programming with many exercises. We refer to the well known classic "Linear Programming and Extensions" by Dantzig [49] and two excellent books [83, 217] among many others on this topic.

Chapter 4: Pareto Optimality.
Basic concepts of Pareto optimal solutions are originally given in [131, 163]. For deep understandings of this topic the reader is referred to a number of classic books [41, 67, 87, 106, 136, 151, 184, 197, 231]. The first monograph on multiple objective linear programming (MOLP) is [233]. The book [197] contains a lot of practical problems that motivate multiobjective linear programming study and detailed methods to solve them.

Chapter 5: Duality.
Despite the fact that the duality constructions of linear programming presented in Chap. 3 are not suitable for MOLP, there exist different approaches to duality in multiobjective optimization. Unfortunately, general approaches developed for vector

© Springer International Publishing Switzerland 2016
D.T. Luc, *Multiobjective Linear Programming*,
DOI 10.1007/978-3-319-21091-9

problems involve set-valued functions (see [34, 70, 136, 139, 184]) can be applied, but the objective function of dual problems are set-valued and difficult to handle with. The papers [44, 105, 123, 128, 177] are the first works that provide dual problems for MOLP. Further investigations are [18, 19, 34, 50, 78, 81, 88, 94–96, 138, 177]. The first approach to duality in this chapter is based on dual sets which are complements to the value set of the primal problems. The second approach is a known method of Lagrangian function. The results of this chapter are mainly taken from [88, 138].

Chapter 6: Sensitivity and stability.
Although a large number of investigations have been carried out on stability of vector optimization problems in general setting (see [87, 136] and the references given in these), there exist quite few works specifically devoted to stability of multiobjective linear problems (see [21, 58, 62, 77, 91, 144, 161, 219, 222]. Recent contributions to tolerance study and robustness are given in [83, 85, 93, 97–101, 104, 115, 133, 175, 191–193, 211]. The results of the two first sections are essentially from [137, 211]. Theory of error bounds of linear systems is well known in [17, 102]. The result on weak sharp maxima in Exercise 6.5.6 is due to [57]. The proof given by this exercise is rather simple. The results of Sect. 6.4 on post-optimal analysis are mainly taken from [85].

Chapter 7: Multiobjective simplex method.
The popular simplex method of linear programming is adapted to MOLP. Detailed descriptions of this method was given in [233] and also in [67, 68, 231, 232]. Using this method to find all maximal vertex solutions is studied by a number of works [14, 15, 63, 64, 66, 104, 185]. The last exercise of this chapter is taken from [233].

Chapter 8: Normal cone method.
This method was developed in [117, 118] for finding the set of efficient and weakly efficient solutions of MOLP. It makes use of normal directions to a convex polyhedral set to determine efficient and weakly efficient faces. Related approaches are given in [109, 213, 214].

Chapter 9: Outcome space method.
Since in many practical problems the number of criteria is relatively much smaller than the number of variables, the idea of working in the image space is quite helpful in computing efficient vertices of the value set. Analysis of efficient sets in the image space of MOLP was initiated in [23, 51, 54, 55, 80, 172]. The method described in this chapter was developed in [25]. Most of the theoretical results are found in [23] for which we give shorter proofs. Further developments of this method are given in [22, 24, 26, 29, 30, 52, 53, 55, 69, 116, 119, 130, 146].

Further Reading.
There are some topics of MOLP that are not covered by this book. The interested readers are asked to read the corresponding references, which are far from being exhausted.

• Multiobjective stochastic linear programming:

[1, 2, 153, 158, 227].

• Multiobjective fuzzy linear programming:
[12, 13, 33, 36, 39, 48, 103, 108, 113, 114, 134, 135, 147–149, 155, 164, 169, 178–183, 194, 219, 220, 225, 230, 235].

• Multiobjective integer linear programming:
[31, 56, 112, 121, 122, 147, 154, 207, 208, 218].

• Interactive and non-deterministic methods:
[4–8, 10, 11, 14–16, 20–23, 27, 28, 32, 35, 37, 38, 42, 45, 46, 60, 61, 63–66, 81, 82, 84, 90, 104, 109, 110, 120, 124–127, 129, 143, 162, 165–168, 170, 173, 174, 185, 186, 189, 190, 195–198, 200–206, 209, 212, 215, 221, 223, 228, 229, 236].

• Applications of multiobjective linear programming:
[3, 9, 40, 43, 56, 59, 62, 71–76, 79, 86, 89, 107, 111, 132, 142, 145, 147, 148, 150, 152, 156, 157, 159, 160, 171, 188, 199, 201, 210, 216, 224, 226, 237].

References

1. F.B. Abdelaziz, P. Lang, R. Nadeau, Distributional efficiency in multiobjective stochastic linear programming. Eur. J. Oper. Res. **85**, 399–415 (1995)
2. F.B. Abdelaziz, H. Masri, A compromise solution for the multiobjective stochastic linear programming under partial uncertainty. Eur J. Oper. Res. **202**, 55–59 (2010)
3. A.K. Aggarwal, E.R. Clayton, T.R. Rakes, J.R. Baker, A problem identification taxonomy for classification and automated formulation of multiple objective linear programming models. Ann. Oper. Res. **38**, 1–16 (1992)
4. B. Aghezzaf, T. Ouaderhman, An interactive interior point algorithm for multiobjective linear programming problems. Oper. Res. Lett. **29**(4), 163–170 (2001)
5. M.J. Alves, J. Climaco, An interactive reference point approach for multiobjective mixed-integer programming using branch-and-bound. Eur. J. Oper. Res. **124**, 478–494 (2000)
6. M.J. Alves, J.P. Costa, An exact method for computing the nadir values in multiple objective linear programming. Eur. J. Oper. Res. **198**(2), 637–646 (2009)
7. R.K. Anderson, M. Dror, O. Liu-Sheng, Interactive multiobjective linear programming: the issue of consistency. Found. Comput. Decis. Sci. **20**(4), 255–274 (1995)
8. R.K. Anderson, M. Dror, An interactive graphic presentation for multiobjective linear programming. Appl. Math. Comput. **123**(2), 229–248 (2001)
9. C.H. Antunes, A.G. Martins, I.S. Brito, A multiple objective mixed integer linear programming model for power generation expansion planning. Energy **29**, 613–627 (2004)
10. A. Arbel, Fundamentals of interior multiple objective linear programming algorithms. *Multicriteria Decision Making*. International Series in Operations Research and Management Science, vol. 21 (1999), pp. 367–396
11. A. Arbel, P. Korhonen, Using objective values to start multiple objective linear programming algorithms. Eur. J. Oper. Res. **128**(3), 587–596 (2001)
12. M. Arenas, A. Bilbao, M.V. Rodríguez, M. Jiménez, Fuzzy parameters multiobjective linear programming problem: solution analysis. *Trends in Multicriteria Decision Making* (Cape Town, 1997). Lecture Notes in Economics and Mathematical Systems, vol. 465 (Springer, Berlin, 1998), pp. 128–137
13. F. Arikan, Z. Gungor, A two-phase approach for multi-objective programming problems with fuzzy coefficients. Inf. Sci. **177**, 5191–5202 (2007)
14. P. Armand, Finding all maximal efficient faces in multiobjective linear programming. Math. Program. **61**, 357–375 (1993)
15. P. Armand, C. Malivert, Determination of the efficient set in multiobjective linear programming. J. Optim. Theory Appl. **70**(3), 467–489 (1991)

© Springer International Publishing Switzerland 2016
D.T. Luc, *Multiobjective Linear Programming*,
DOI 10.1007/978-3-319-21091-9

16. A. Aurovillian, H. Zhang, M.M. Wiecek, A bookkeeping strategy for multiple objective linear programs. Comput. Oper. Res. **24**(11), 1033–1041 (1997)
17. D. Aze, A survey on error bounds for lower semicontinuous functions. Proceedings of 2003 MODE-SMAI Conference, ESAIM Proceedings, vol. 13 (EDP Sciences, Les Ulis, 2003), pp. 1–17
18. A. Balbas, A. Heras, Duality theory for infinite-dimensional multiobjective linear programming. Eur. J. Oper. Res. **68**, 379–388 (1993)
19. A. Balbas, P. Jimenez, A. Heras, Duality theory and slackness conditions in multiobjective linear programming. Comput. Math. Appl. **37**(4–5), 101–109 (1999)
20. S.M. Belenson, K.C. Kapur, An algorithm for solving multicriterion linear programming problems with examples. Oper. Res. Q. **24** (1970–1977), **1**, 65–77 (1973)
21. H.P. Benson, Multiple objective linear programming with parametric criteria coefficients. Manag. Sci. **31**(4), 461–474 (1985)
22. H.P. Benson, Complete efficiency and the initialization of algorithms for multiple objective programming. Oper. Res. Lett. **10**, 481–487 (1991)
23. H.P. Benson, A geometrical analysis of the efficient outcome set in multiple objective convex programs with linear criterion functions. J. Glob. Optim. **6**, 231–251 (1995)
24. H.P. Benson, Further analysis of an outcome set-based algorithm for multiple-objective linear programming. J. Optim. Theory Appl. **97**(1), 1–10 (1998)
25. H.P. Benson, An outer approximation algorithm for generating all efficient extreme points in the outcome set of a multiple objective linear programming problem. J. Glob. Optim. **13**(1), 1–24 (1998)
26. H.P. Benson, Hybrid approach for solving multiple-objective linear programs in outcome space. J. Optim. Theory Appl. **98**(1), 17–35 (1998)
27. H.P. Benson, Y. Aksoy, Using efficient feasible directions in interactive multiple objective linear programming. Oper. Res. Lett. **10**(4), 203–209 (1991)
28. H.P. Benson, D. Lee, J.P. McClure, Global optimization in practice: an application to interactive multiple objective linear programming. J. Glob. Optim. **12**(4), 353–372 (1998)
29. H.P. Benson, E. Sun, Outcome space partition of the weight set in multiobjective linear programming. J. Optim. Theory Appl. **105**(1), 17–36 (2000)
30. H.P. Benson, E. Sun, A weight set decomposition algorithm for finding all efficient extreme points in the outcome set of a multiple objective linear program. Eur. J. Oper. Res. **139**, 26–41 (2002)
31. V. Blanco, J. Puerto, A new complexity result on multiobjective linear integer programming using short rational generating functions. Optim. Lett. **6**, 537–543 (2012)
32. V. Blanco, J. Puerto, S.E.H.B. Ali, A semidefinite programming approach for solving multiobjective linear programming. J. Glob. Optim. **58**(3), 465–480 (2014)
33. A.R. Borges, C.H. Antunes, A weight space-based approach to fuzzy multiple-objective linear programming. Decis. Support Syst. **34**, 427–443 (2002)
34. R.I. Bot, S.M. Grad, G. Wanka, *Duality in Vector Optimization* (Springer, Berlin, 2009)
35. S.T. Breslawski, S. Zionts, A simulation based study of modifications to the Zionts-Wallenius algorithm for multiple objective linear programming. Comput. Oper. Res. **21**(7), 757–768 (1994)
36. J.J. Buckley, Multiobjective possibilistic linear programming. Fuzzy Sets Syst. **35**, 23–28 (1990)
37. R. Caballero, M. Luque, J. Molina, F. Ruiz, MOPEN: a computational package for linear multiobjective and goal programming problems. Decis. Support Syst. **41**, 160–175 (2005)
38. L. Cayton, R. Herring, A. Holder, J. Holzer, C. Nightingale, T. Stohs, Asymptotic sign-solvability, multiple objective linear programming, and the nonsubstitution theorem. Math. Methods Oper. Res. **64**(3), 541–555 (2006)
39. S. Chanas, Fuzzy programming in multiobjective linear programming a parametric approach. Fuzzy Sets Syst. **29**(3), 303–313 (1989)
40. N.B. Chang, C.G. Wen, S.L. Wu, Optimal management of environmental and land resources in a reservoir watershed by multiobjective programming. J. Environ. Manag. **44**, 145–161 (1995)

41. V. Chankong, Y.Y. Haimes, *Multiobjective Decision Making: Theory and Methodology*. North-Holland Series in System Science and Engineering (North-Holland Publishing Co., New York, 1983)
42. H.K. Chen, H.W. Chou, Solving multiobjective linear programming problems-generic approach. Fuzzy Sets Syst. **82**(1), 35–38 (1996)
43. F. Christensen, M. Lind, J. Tind, On the nucleolus of NTU-Games defined by multiple objective linear programs. Math. Methods Oper. Res. **43**, 337–352 (1996)
44. H.W. Corley, Duality theory for the matrix linear programming problem. J. Math. Anal. Appl. **104**, 47–52 (1984)
45. J.P. Costa, M.J. Alves, A reference point technique to compute nondominated solutions in MOLP. J. Math. Sci. **161**(6), 820–831 (2009)
46. S. Csikai, M. Kovács, A multiobjective linear programming algorithm based on the Dempster-Shafer composition rule. *Multiple Objective and Goal Programming*. Advances Software Computing, vol. 12 (2002), pp. 31–38
47. J. Current, M. Marsh, Multiobjective transportation network design and routing problems: taxonomy and annotation. Eur. J. Oper. Res. **65**, 4–19 (1993)
48. R. Dadashzadeh, S.B. Nimse, On stability in multiobjective linear programming problems with fuzzy parameters. J. Approx. Theory Appl. **2**(1), 49–56 (2006)
49. G.B. Dantzig, *Linear Programming and Extensions*, Reprint of the 1968 corrected edition, Princeton Landmarks in Mathematics (Princeton University Press, Princeton, 1998)
50. L.N. Das, S. Nanda, Symmetrie dual multiobjective programming. Eur. J. Oper. Res. **97**, 167–171 (1997)
51. J.P. Dauer, Analysis of the objective space in multiple objective linear programming. J. Math. Anal. Appl. **126**(2), 579–593 (1987)
52. J.P. Dauer, O.A. Saleh, Constructing the set of efficient objective values in multiple objective linear programs. Eur. J. Oper. Res. **46**, 358–365 (1990)
53. J.P. Dauer, Y.H. Liu, Solving multiple objective linear programs in objective space. Eur. J. Oper. Res. **46**, 350–357 (1990)
54. J.P. Dauer, O.A. Saleh, A representation of the set of feasible objectives in multiple objective linear programs. Linear Algebra Appl. **166**, 261–275 (1992)
55. J.P. Dauer, On degeneracy and collapsing in the construction of the set of objective values in a multiple objective linear program. Ann. Oper. Res. **47**, 279–292 (1993)
56. R.F. Deckro, J.E. Hebert, A multiple objective programming framework for tradeoffs in project scheduling. Eng. Costs Prod. Econ. **18**, 255–264 (1990)
57. S. Deng, X.Q. Yang, Weak sharp minima in multicriteria linear programming. SIAM J. Optim. **15**, 456–460 (2004)
58. D.V. Deshpande, S. Zionts, Sensitivity analysis in multiple objective linear programming: changes in the objective function matrix. *Multiple Criteria Decision Making Theory and Application*. Proceedings of Third Conference, Hagen, Knigswinter, 1979. Lecture Notes in Economics and Mathematical Systems, vol. 177 (Springer, Berlin, 1980) pp. 26–39
59. D.K. Despotis, Y. Sisko, Agricultural management using the ADELAIS multiobjective linear programming software: a case application. Theory Decis. **32**, 113–131 (1992)
60. C.E. Downing, J.L. Ringuest, An experimental evaluation of the efficacy of four multiobjective linear programming algorithms. Eur. J. Oper. Res. **104**, 549–558 (1998)
61. M. Dror, M.F. Shakun, Bifurcation and adaptation in evolutionary interactive multiobjective linear programming. Eur. J. Oper. Res. **93**, 602–610 (1996)
62. E.C. Duesing, Multiple objective linear programming and the theory of the firm: I. Substitution and sensitivity analysis. *Essays and Surveys on Multiple Criteria Decision Making* (Mons, 1982). Lecture Notes in Economics and Mathematical Systems, vol. 209 (Springer, Berlin, 1983)
63. J.G. Ecker, N.S. Hegner, I.A. Kouada, Generating all maximal efficient faces for multiple objective linear programs. J. Optim. Theory Appl. **30**(3), 353–381 (1980)
64. J.G. Ecker, I.A. Kouada, Finding all efficient extreme points for multiple objective linear programs. Math. Program. **14**, 249–261 (1978)

65. J.G. Ecker, N.E. Shoemaker, Multiple objective linear programming and the tradeoff-compromise set. *Multiple Criteria Decision Making Theory and Application*. Proceedings of Third Conference, Hagen, Knigswinter, 1979. Lecture Notes in Economics and Mathematical Systems, vol. 177 (Springer, Berlin, 1980)

66. J.G. Ecker, N.E. Shoemaker, Selecting subsets from the set of nondominated vectors in multiple objective linear programming. SIAM J. Control Optim. **19**(4), 505–515 (1981)

67. M. Ehrgott, *Multicriteria Optimization*, 2nd edn. (Springer, Berlin, 2005)

68. M. Ehrgott, J. Puerto, A.M. Rodríguez-Chía, Primal-dual simplex method for multiobjective linear programming. J. Optim. Theory Appl. **134**(3), 483–497 (2007)

69. M. Ehrgott, A. Lohne, L. Shao, A dual variant of Benson's "outer approximation algorithm" for multiple objective linear programming. J. Glob. Optim. **52**(4), 757–778 (2012)

70. G. Eichfelder, *Variable Ordering Structures in Vector Optimization* (Springer, Berlin, 2014)

71. J.H. Ellis, Multiobjective mathematical programming models for acid rain control. Eur. J. Oper. Res. **35**, 365–377 (1988)

72. T. Emam, Method of centers algorithm for multi-objective programming problems. Acta Math. Sci. **29**(5), 1128–1142 (2009)

73. C.E. Eseobar-Toledo, Industrial petrochemical production planning and expansion: a multiobjective linear programming approach. Sociedad de Estadística e Investigación Operativa, Top **9**(1), 77–89 (2001)

74. M.P. Estellita Lins, L. Angulo-Meza, A.C. Moreira da Silva, A multi-objective approach to determine alternative targets in data envelopment analysis. J. Oper. Res. Soc. **55**, 1090–1101 (2004)

75. P. Fiala, Data envelopment analysis by multiobjective linear programming methods. *Multiple Objective and Goal Programming*. Advances in Software Computing, vol. 12 (Physica, Heidelberg, 2002), pp. 39–45

76. J. Fritzsch, S. Wegener, G. Buchenrieder, J. Curtiss, S.G.Y Paloma, Is there a future for semi-subsistence farm households in Central and southeastern Europe? A multiobjective linear programming approach. J. Policy Model. **33**, 70–91 (2011)

77. R. Fuller, M. Fedrizzi, Stability in multiobjective possibilistic linear programs. Eur. J. Oper. Res. **74**, 179–187 (1994)

78. E.A. Galperin, P.J. Guerra, Duality of nonscalarized multiobjective linear programs: dual balance, level sets, and dual clusters of optimal vectors. J. Optim. Theory Appl. **108**(1), 109–137 (2001)

79. R.J. Gallagher, O.A. Saleh, Constructing the set of efficient objective values in linear multiple objective transportation problems. Eur. J. Oper. Res. **73**, 150–163 (1994)

80. R.J. Gallagher, O.A. Saleh, A representation of an efficiency equivalent polyhedron for the objective set of a multiple objective linear program. Eur. J. Oper. Res. **80**, 204–212 (1995)

81. E. Galperin, J. Jimerez Guerra, Duality of nonscalarized multiobjective linear programs: dual balance, level sets and dual clusters of optimal vectors. J. Optim. Theory Appl. **108**, 109–137 (2001)

82. L.R. Gardiner, R.E. Steuer, Unified interactive multiple objective programming. Eur. J. Oper. Res. **74**, 391–406 (1994)

83. S.I. Gass, *Linear Programming: Methods and Applications*, Corrected reprint of the 1995 fifth edition (Dover Publications Inc., Mineola, 2003)

84. S.I. Gass, P.G. Roy, The compromise hypersphere for multiobjective linear programming. Eur. J. Oper. Res. **144**(3), 459–479 (2003)

85. P.G. Georgiev, D.T. Luc, P.M. Pardalos, Robust aspects of solutions in deterministic multiple objective linear programming. Eur. J. Oper. Res. **229**(1), 29–36 (2013)

86. J. Glackin, J.G. Ecker, M. Kupferschmid, Solving bilevel linear programs using multiple objective linear programming. J. Optim. Theory Appl. **140**(2), 197–212 (2009)

87. A. Gopfert, H. Riahi, C. Tammer, C. Zalinescu, *Variational Methods in Partially Ordered Spaces*. CMS Books in Mathematics/Ouvrages de Mathematiques de la SMC, vol. 17 (Springer, New York, 2003)

88. D. Gourion, D.T. Luc, Saddle points and scalarizing sets in multiple objective linear programming. Math. Methods Oper. Res. **80**(1), 1–27 (2014)
89. F. Glover, F. Martinson, Multiple-use land planning and conflict resolution by multiple objective linear programming. Eur. J. Oper. Res. **28**, 343–350 (1987)
90. B.L. Gorissen, D.D. Hertog, Approximating the Pareto set of multiobjective linear programs via robust optimization. Oper. Res. Lett. **40**, 319–324 (2012)
91. P. Hansen, M. Labbé, R.E. Wendell, Sensitivity analysis in multiple objective linear programming: the tolerance approach. Eur. J. Oper. Res. **38**(1), 63–69 (1989)
92. F. Hassanzadeh, H. Nemati, M. Sun, Robust optimization for multiobjective programming problems with imprecise information. Procedia Comput. Sci. **17**, 357–364 (2013)
93. F. Hassanzadeh, H. Nemati, M. Sun, Robust optimization for interactive multiobjective programming with imprecise information applied to R&D project portfolio selection. Eur. J. Oper. Res. **238**(1), 41–53 (2014)
94. A. Hamel, F. Heyde, A. Lohne, C. Tammer, K. Winkler, Closing the duality gap in linear vector optimization. J. Convex Anal. **11**, 163–178 (2004)
95. F. Heyde, A. Lohne, Geometric duality in multiple objective linear programming. SIAM J. Optim. **19**, 836–845 (2008)
96. F. Heyde, A. Lohne, C. Tammer, Set-valued duality theory for multiple objective linear programs and application to mathematical finance. Math. Methods Oper. Res. **69**, 159–179 (2009)
97. M. Hladik, Additive and multiplicative tolerance in multiobjective linear programming. Oper. Res. Lett. **36**(3), 393–396 (2008)
98. M. Hladik, Computing the tolerances in multiobjective linear programming. Optim. Methods Softw. **23**(5), 731–739 (2008)
99. M. Hladik, On the separation of parametric convex polyhedral sets with application in MOLP. Appl. Math. **55**(4), 269–289 (2010)
100. M. Hladik, Complexity of necessary efficiency in interval linear programming and multiobjective linear programming. Optim. Lett. **6**(5), 893–899 (2012)
101. M. Hladik, S. Sitarz, Maximal and supremal tolerances in multiobjective linear programming. Eur. J. Oper. Res. **228**(1), 93–101 (2013)
102. A.J. Hoffman, On approximate solutions of systems of linear inequalities. J. Res. Natl. Bur. Stand. **49**, 263–265 (1952)
103. M.L. Hussein, M.A. Abo-Sinna, Decomposition of multiobjective programming problems by hybrid fuzzy dynamic programming. Fuzzy Sets Syst. **60**, 25–32 (1993)
104. M. Ida, Efficient solution generation for multiple objective linear programming based on extreme ray generation method. Eur. J. Oper. Res. **160**(1), 242–251 (2005)
105. S.H. Isermann, On some relations between a dual pair of multiple objective linear programs. Zeitschrift für Operations Research **22**(1), 33–41 (1978)
106. J. Jahn, *Vector Optimization: Theory, Applications, and Extensions* (Springer, Berlin, 2004)
107. B. Javadia, M.S. Mehrabadb, A. Hajic, I. Mahdavia, F. Jolaid, N. Mahdavi-Amiri, No-wait flow shop scheduling using fuzzy multi-objective linear programming. J. Franklin Inst. **345**, 452–467 (2008)
108. M. Jimeneza, A. Bilbao, Pareto-optimal solutions in fuzzy multi-objective linear programming. Fuzzy Sets Syst. **160**, 2714–2721 (2009)
109. J.M. Jorge, Maximal descriptor set characterizations of efficient faces in multiple objective linear programming. Oper. Res. Lett. **31**(2), 124–128 (2003)
110. J.M. Jorge, A bilinear algorithm for optimizing a linear function over the efficient set of a multiple objective linear programming problem. J. Glob. Optim. **31**, 1–16 (2005)
111. T. Joro, P. Korhonen, J. Wallenius, Structural comparison of data envelopment analysis and multiple objective linear programming. Manag. Sci. **44**, 962–970 (1998)
112. J. Karaivanova, S. Narula, V. Vassilev, An interactive procedure for multiple objective integer linear programming problems. Eur. J. Oper. Res. **68**, 344–351 (1993)
113. E.E. Karsak, O. Kuzgunkaya, A fuzzy multiple objective programming approach for the selection of a flexible manufacturing system. Int. J. Prod. Econ. **79**, 101–111 (2002)

114. H. Katagiri, M. Sakawa, K. Kato, I. Nishizaki, Interactive multiobjective fuzzy random linear programming: maximization of possibility and probability. Eur. J. Oper. Res. **188**, 530–539 (2008)

115. E. Khorram, M. Zarepisheh, B.A. Ghaznavi-ghosoni, Sensitivity analysis on the priority of the objective functions in lexicographic multiple objective linear programs. Eur. J. Oper. Res. **207**, 1162–1168 (2010)

116. N.T.B. Kim, Efficiency equivalent polyhedra for the feasible set of multiple objective linear programming. Acta Mathematica Vietnamica **27**(1), 77–85 (2002)

117. N.T.B. Kim, D.T. Luc, Normal cones to a polyhedral convex set and generating efficient faces in linear multiobjective programming. Acta Mathematica Vietnamica **25**(1), 101–124 (2000)

118. N.T.B. Kim, D.T. Luc, The normal cone method in solving linear multiobjective problems, generalized convexity, generalized monotonicity, optimality conditions and duality in scalar and vector optimization. J. Stat. Manag. Syst. **5**(1–3), 341–358 (2002)

119. N.T.B. Kim, N.T. Thien, L.Q. Thuy, Generating all efficient extreme solutions in multiple objective linear programming problem and its application to multiplicative programming. East-West J. Math. **10**(1), 1–14 (2008)

120. S.H. Kim, A new interactive algorithm for multiobjective linear programming using maximally changeable dominance cone. Eur. J. Oper. Res. **64**, 126–137 (1993)

121. K. Klamroth, J. Tind, S. Zust, Integer programming duality in multiple objective programming. J. Glob. Optim. **29**, 1–18 (2004)

122. D. Klein, E. Hannan, An algorithm for the multiple objective integer linear programming problem. Eur. J. Oper. Res. **9**, 378–385 (1982)

123. J. Kolumbán, Dualität bei Optimierungsaufgaben. In *Proceedings of the Conference on Constructive Theory of Functions*, Akademiai Kiado, Budapest (1969), pp. 261–265

124. P. Korhonen, Reference direction approach to multiple objective linear programming: historical overview. *Essays in Decision Making* (Springer, Berlin, 1997), pp. 74–92

125. P. Korhonen, J. Wallenius, A multiple objective linear programming decision support system. Decis. Support Syst. **6**, 243–251 (1990)

126. P. Korhonen, J. Wallenius, Using qualitative data in multiple objective linear programming. Eur. J. Oper. Res. **48**, 81–87 (1990)

127. P. Korhonen, J. Wallenius, A computer graphics-based Decision Support System for multiple objective linear programming. Eur. J. Oper. Res. **60**, 280–286 (1992)

128. J.S.H. Kornbluth, Duality, indifference and sensitivity analysis in multiple objective linear programming. Oper. Res. Q. **25**, 599–614 (1974)

129. S. Krichen, H. Masri, A. Guitouni, Adjacency based method for generating maximal efficient faces in multiobjective linear programming. Appl. Math. Model. **36**(12), 6301–6311 (2012)

130. K.H. Kufer, On the asymptotic average number of efficient vertices in multiple objective linear programming. J. Complex. **14**(3), 333–377 (1998)

131. H.W. Kuhn, A.W. Tucker, Nonlinear programming. In *Proceedings of the Second Berkeley Symposium on Mathematical Statistics and Probability*, University of California Press, Berkeley and Los Angeles (1951), pp. 481–492

132. P.L. Kunsch, J. Teghem Jr, Nuclear fuel cycle optimization using multi-objective stochastic linear programming. Eu. J. Oper. Res. **31**, 240–249 (1987)

133. D. Kuchta, A concept of a robust solution of a multicriterial linear programming problem. Cent. Eur. J. Oper. Res. **19**, 605–613 (2011)

134. H. Kuwano, On the fuzzy multi-objective linear programming problem: Goal programming approach. Fuzzy Sets Syst. **82**, 57–64 (1996)

135. T.F. Liang, Distribution planning decisions using interactive fuzzy multi-objective linear programming. Fuzzy Sets Syst. **157**, 1303–1316 (2006)

136. D.T. Luc, *Theory of Vector Optimization*. Lecture Notes in Economics and Mathematical System, vol. 319 (Springer, Berlin, 1989)

137. D.T. Luc, Smooth representation of a parametric polyhedral convex set with application to sensitivity in optimization. Proc. Am. Math. Soc. **125**(2), 555–567 (1997)

138. D.T. Luc, On duality in multiple objective linear programming. Eur. J. Oper. Res. **210**, 158–168 (2011)
139. D.T. Luc, J. Jahn, Axiomatic approach to duality in optimization. Numer. Funct. Anal. Optim. **13**(3–4), 305–326 (1992)
140. D.T. Luc, A. Ratiu, Vector optimization: basic concepts and solution methods. *Fixed Point Theory, Variational Analysis, and Optimization* (CRC Press, A Chapman and Hall Book, 2014), pp. 247–305
141. D.G. Luenberger, Y. Ye, *Linear and Nonlinear Programming*. International Series in Operations Research and Management Science, vol. 116, 3rd edn. (Springer, New York, 2008)
142. M. Maino, J. Berdegue, T. Rivas, Multiple objective programming—an application for analysis and evaluation of peasant economy of the VIIIth Region of Chile. Agric. Syst. **41**, 387–397 (1993)
143. B. Malakooti, An exact interactive method for exploring the efficient facets of multiple objective linear programming problems with quasi-concave utility functions. IEEE Trans. Syst. Man Cybern. **18**(5), 787–801 (1988)
144. A.M. Marmol, J. Puerto, Special cases of the tolerance approach in multiobjective linear programming. Eur. J. Oper. Res. **98**, 610–616 (1997)
145. A.G. Martins, D. Coelho, C.H. Antunes, J. Climacio, A multiple objective linear programming approach to power generation planning with demand-side management (DSM). Int. Trans. Oper. Res. **3**(3–4), 305–317 (1996)
146. H. Masri, S. Krichen, A. Guitouni, Generating efficient faces for multiobjective linear programming problems. Int. J. Oper. Res. **15**(1), 1–15 (2012)
147. G. Mavrotas, D. Diakoulaki, L. Papayannakis, An energy planning approach based on mixed 0–1 multiple objective linear programming. Int. Trans. Oper. Res. **6**(2), 231–244 (1999)
148. G.A. Mendoza, G.E. Campbell, G.L. Rolfe, Multiple objective programming: an approach to planning and evaluation of agroforestry systems-Part 1: model description and development. Agric. Syst. **22**, 243–253 (1986)
149. G.A. Mendoza, B.B. Bare, Z. Zhou, A fuzzy multiple objective linear programming approach to forest planning under uncertainty. Agric. Syst. **41**, 257–274 (1993)
150. D.P. Memtsas, Multiobjective programming methods in the reserve selection problem. Eur. J. Oper. Res. **150**, 640–652 (2003)
151. K. Miettinen, *Nonlinear Multiobjective Optimization*. International Series in Operations Research and Management, vol. 12 (Kluwer, Boston, 1999)
152. R.S. Moreno, F. Szidarovszky, A.R. Aguilar, I.L. Cruz, Multiobjective linear model optimize-water distribution in Mexican valley. J. Optim. Theory Appl. **144**, 557–573 (2010)
153. R. Nadeau, B. Urli, L.N. Kiss, PROMISE: a DSS for multiple objective stochastic linear programming problems. Ann. Oper. Res. **51**, 45–59 (1994)
154. S.C. Narula, V. Vassilev, An interactive algorithm for solving multiple objective integer linear programming problems. Eur. J. Oper. Res. **79**, 443–450 (1994)
155. H. Nehi, H.R. Maleki, M. Mashinchi, Multiobjective linear programming with fuzzy variable. Far East J. Math. Sci. **5**(2), 155–172 (2002)
156. I. Nishizaki, M. Sakawa, Stackelberg solutions to multiobjective two-level linear programming problems. J. Optim. Theory Appl. **103**(1), 161–182 (1999)
157. I. Nishizaki, M. Sakawa, On computational methods for solutions of multiobjective linear production programming games. Eur. J. Oper. Res. **129**, 386–413 (2001)
158. D.C. Novak, C.T. Ragsdale, A decision support methodology for stochastic multi-criteria linear programming using spreadsheets. Decis. Support Syst. **36**, 99–116 (2003)
159. W. Ogryczak, Multiple criteria linear programming model for portfolio selection. Ann. Oper. Res. **97**, 143–162 (2000)
160. V. Okoruwa, M.A. Jabbarht, J.A. Akinwumi, Crop-livestock competition in the West African derived savanna: application of a multi-objective programming model. Agric. Syst. **52**, 439–453 (1996)
161. C. Oliveira, C.H. Antunes, Multiple objective linear programming models with interval coefficients—an illustrated overview. Eur. J. Oper. Res. **181**, 1434–1463 (2007)

162. D.L. Olson, Tchebycheff norms in multi-objective linear programming. Math. Comput. Model. **17**(1), 113–124 (1993)
163. V. Pareto, Statistique et Economie Mathématique. *Oevres Complètes de Vilfredo Pareto VIII. Travaux de Droit, d'Economie, de Sciences Politiques, de Sociologie et d'Anthropologie*, vol. 48 (Librairie Droz, Geneva, 1989)
164. M.A. Parra, A.B. Terol, M.V.R. Uria, Solving the multiobjective possibilistic linear programming problem. Eur. J. Oper. Res. **117**, 175–182 (1999)
165. C.O. Pieume, L.P. Fotso, P. Siarry, An approach for finding efficient points in multiobjective linear programming. J. Inf. Optim. Sci. **29**(2), 203–216 (2008)
166. J. Ch. Pomerol, T. Trabelsi, An adaptation of PRIAM to multiobjective linear programming. Eur. J. Oper. Res. **31**, 335–341 (1987)
167. L. Pourkarimi, M.A. Yaghoobi, M. Mashinchi, Determining maximal efficient faces in multiobjective linear programming problem. J. Math. Anal. Appl. **354**(1), 234–248 (2009)
168. M.A. Quaddus, K.L. Poh, A comparative evaluation of multiple objective linear programming methods. Optimization **1–2**, 317–328 (1992)
169. K.S. Raju, L. Duckstein, Multiobjective fuzzy linear programming for sustainable irrigation planning: an Indian case study. Soft Comput. **7**, 412–418 (2003)
170. G.R. Reeves, L.S. Franz, A simplified interactive multiple objective linear programming procedure. Comput. Oper. Res. **12**(6), 589–501 (1985)
171. G.R. Reeves, K.D. Lawrence, S.M. Lawrence, J.B. Guerard Jr, Combining earnings forecasts using multiple objective linear programming. Comput. Oper. Res. **15**(6), 551–559 (1988)
172. R. Rhode, R. Weber, The range of the efficient frontier in multiple objective linear programming. Math. Program. **28**(1), 84–95 (1984)
173. J.L. Ringuest, D.B. Rinks, Interactive solutions for the linear multiobjective transportation problem. Eur. J. Oper. Res. **32**, 96–106 (1987)
174. J.L. Ringuest, T.R. Gulledge Jr, A preemptive value function method approach for multiobjective linear programming problems. Decis. Sci. **14**(1), 76–86 (1983)
175. S. Rivaz, M.A. Yaghoobi, Minimax regret solution to multiobjective linear programming problems with interval objective functions coefficients. Cent. Eur. J. Oper. Res. **21**(3), 625–649 (2013)
176. R.T. Rockafellar, *Convex Analysis*. Princeton Mathematical Series, vol. 28 (Princeton University Press, Princeton, 1970)
177. W. Rodder, A generalized saddlepoint theory: Its application to duality theory for linear vector optimum problems. Eur. J. Oper. Res. **1**(1), 55–59 (1977)
178. O.M. Saad, Stability on multiobjective linear programming problems with fuzzy parameters. Fuzzy Sets Syst. **74**(2), 207–215 (1995)
179. O.M. Saad, An iterative goal programming approach for solving fuzzy multiobjective integer linear programming problems. Appl. Math. Comput. **170**, 216–225 (2005)
180. B. Sahoo, A.K. Lohani, R.K. Sahu, Fuzzy multiobjective and linear programming based management models for optimal land-water-crop system planning. Water Resour. Manag. **20**, 931–948 (2006)
181. M. Sakawa, Interactive fuzzy decision making for multiobjective linear programming problems and its application. *Fuzzy Information, Knowledge Representation and Decision Analysis* (Marseille, 1983). IFAC Proceeding Series, vol. 6. (IFAC, Laxenburg, 1984), pp. 295–300
182. M. Sakawa, H. Yano, I. Nishizaki, *Linear and Multiobjective Programming with Fuzzy Stochastic Extensions*. International Series in Operations Research and Management Science, vol. 203 (Springer, New York, 2013)
183. M. Sasaki, M. Gen, An extension of interactive method for solving multiple objective linear programming with fuzzy parameters. Comput. Ind. Eng. **25**(1–4), 9–12 (1993)
184. Y. Sawaragi, H. Nakayama, T. Tanino, *Theory of Multiobjective Optimization*. Mathematical Science and Engineering, vol. 176 (Academic Press Inc., Orlando, 1985)
185. S. Sayin, An algorithm based on facial decomposition for finding the efficient set in multiple objective linear programming. Oper. Res. Lett. **19**(2), 87–94 (1996)

186. S. Sayin, Measuring the quality of discrete representations of efficient sets in multiple objective mathematical programming. Math. Program. Ser. A **87**, 543–560 (2000)
187. D. Schweigert, P. Neumayer, A reduction algorithm for integer multiple objective linear programs. Eur. J. Oper. Res. **99**, 459–462 (1997)
188. L. Shao, M. Ehrgott, Approximately solving multiobjective linear programmes in objective space and an application in radiotherapy treatment planning. Math. Methods Oper. Res. **68**, 257–276 (2008)
189. L. Shao, M. Ehrgott, Approximating the nondominated set of an MOLP by approximately solving its dual problem. Math. Methods Oper. Res. **68**, 469–492 (2008)
190. J. Siskos, D.K. Despotis, A DSS oriented method for multiobjective linear programming problems. Decis. Support Syst. **5**, 47–55 (1989)
191. S. Sitarz, Postoptimal analysis in multicriteria linear programming. Eur. J. Oper. Res. **191**, 7–18 (2008)
192. S. Sitarz, Standard sensitivity analysis and additive tolerance approach in MOLP. Ann. Oper. Res. **181**, 219–232 (2010)
193. S. Sitarz, Sensitivity analysis of weak efficiency in multiple objective linear programming. Asia-Pacific J. Oper. Res. **28**, 445–455 (2011)
194. R. Slowinski, A multicriteria fuzzy linear programming method for water supply system development planning. Fuzzy Sets Syst. **19**, 217–237 (1986)
195. R.S. Solanki, P.A. Appino, J.L. Cohon, Approximating the noninferior set in multiobjective linear programming problems. Eur. J. Oper. Res. **68**, 356–373 (1993)
196. R.E. Steuer, Multiple objective linear programming with interval criterion weights. Manag. Sci. **23**(3), 305–316 (1976/1977)
197. R.E. Steuer, *Multiple Criteria Optimization: Theory, Computation, and Application* (Krieger Publishing Co., Malabar, 1989)
198. R.E. Steuer, Random problem generation and the computation of efficient extreme points in multiple objective linear programming. Comput. Optim. Appl. **3**(4), 333–347 (1994)
199. R.E. Steuer, R.L. Olivier, An application of multiple objective linear programming to media selection. OMEGA **4**(4), 455–462 (1976)
200. R.E. Steuer, A.A. Piercy, A regression study of the number of efficient extreme points in multiple objective linear programming. Eur. J. Oper. Res. **162**(2), 484–496 (2005)
201. R.E. Steuer, A.T. Schuler, An interactive multiple-objective linear programming approach to a problem in forest management. Forest Ecol. Manag. **2**, 191–205 (1979)
202. T.J. Stewart, A combined logistic regression and Zionts-Wallenius methodology for multiple criteria linear programming. Eur. J. Oper. Res. **24**, 295–304 (1986)
203. T.J. Stewart, An interactive multiple objective linear programming method based on piecewise-linear additive value functions. IEEE Trans. Syst. Man Cybern. **17**(5), 799–805 (1987)
204. M. Sun, Some issues in measuring and reporting solution quality of interactive multiple objective programming procedures. Eur. J. Oper. Res. **162**, 468–483 (2005)
205. M. Sun, A. Stam, R.E. Steuer, Interactive multiple objective programming using Tchebycheff programs and artificial neural networks. Comput. Oper. Res. **27**, 601–620 (2000)
206. H. Suprajitno, Solving multiobjective linear programming problem using interval arithmetic. Appl. Math. Sci. **6**(77–80), 3959–3968 (2012)
207. J. Sylva, A. Crema, A method for finding the set of non-dominated vectors for multiple objective integer linear programs. Eur. J. Oper. Res. **158**, 46–55 (2004)
208. J. Sylva, A. Crema, A method for finding well-dispersed subsets of non-dominated vectors for multiple objective mixed integer linear programs. Eur. J. Oper. Res. **180**, 1011–1027 (2007)
209. S.F. Tantawy, Detecting non-dominated extreme points for multiple objective linear programming. J. Math. Stat. **3**(3), 77–79 (2007)
210. Z. Tao, J. Xu, A class of rough multiple objective programming and its application to solid transportation problem. Inf. Sci. **188**, 215–235 (2012)
211. L.V. Thuan, D.T. Luc, On sensitivity in linear multiobjective programming. J. Optim. Theory Appl. **107**(3), 615–626 (2000)

212. T.B. Trafalis, R.M. Alkahtani, An interactive analytic center trade-off cutting plane algorithm for multiobjective linear programming. Comput. Ind. Eng. **37**, 649–669 (1999)
213. T.V. Tu, Optimization over the efficient set of a parametric multiple objective linear programming problem. Eur. J. Oper. Res. **122**(3), 570–583 (2000)
214. T.V. Tu, A common formula to compute the efficient sets of a class of multiple objective linear programming problems. Optimization (2014). doi:10.1080/02331934.2014.926357
215. B. Urli, R. Nadeau, PROMISE/scenarios: an interactive method for multiobjective stochastic linear programming under partial uncertainty. Eur. J. Oper. Res. **155**, 361–372 (2004)
216. J. Vada, O. Slupphaug, T.A. Johansen, Optimal prioritized infeasibility handling in model predictive control: parametric preemptive multiobjective linear programming approach. J. Optim. Theory Appl. **109**(2), 385–413 (2001)
217. R.J. Vanderbei, *Linear Programming: Foundations and Extensions*. International Series in Operations Research and Management Science, vol. 37, 2nd edn. (Kluwer, Boston, 2001)
218. T. Vincent, F. Seipp, S. Ruzika, A. Przybylski, X. Gandibleux, Multiple objective branch and bound for mixed 0–1 linear programming: corrections and improvements for the biobjective case. Comput. Oper. Res. **40**, 498–509 (2013)
219. H.F. Wang, M.L. Wang, Inter-parametric tolerance analysis of a multiobjective linear programming. J. Math. Anal. Appl. **179**(1), 275–296 (1993)
220. R.C. Wang, T.F. Liang, Application of fuzzy multi-objective linear programming to aggregate production planning. Comput. Ind. Eng. **46**, 17–41 (2004)
221. E. Werczberger, Multi-objective linear programming with partial ranking of objectives. Socio Econ. Plan. Sci. **15**(6), 331–339 (1981)
222. D.J. White, Duality, indifference and sensitivity analysis in multiple objective linear programming. Oper. Res. Q. **25**, 599–614 (1974) by J.S.H. Kornbluth. With a reply by Kornbluth. Oper. Res. Q. **26**(3, part 2), 659–662 (1975)
223. M.M. Wiecek, H. Zhang, A parallel algorithm for multiple objective linear programs. Comput. Optim. Appl. **8**, 41–56 (1997)
224. P.A. Wojtkowski, G.H. Brister, F.W. Cubbage, Using multiple objective linear programming to evaluate multi-participant agroforestry systems. Agrofor. Syst. **7**, 185–195 (1988)
225. F. Wu, L. Lu, G. Zhang, A new approximate algorithm for solving multiple objective linear programming problems with fuzzy parameters. Appl. Math. Comput. **174**(1), 524–544 (2006)
226. W. Xiao, Z. Liu, M. Jiang, Y. Shi, Multiobjective linear programming model on injection oilfield recovery system. Comput. Math. Appl. **36**(5), 127–135 (1998)
227. J. Xu, L. Yao, A class of multiobjective linear programming models with random rough coefficients. Math. Comput. Model. **49**(1–2), 189–206 (2009)
228. H. Yan, Q. Wei, J. Wang, Constructing efficient solutions structure of multiobjective linear programming. J. Math. Anal. Appl. **307**(2), 504–523 (2005)
229. J.B. Yang, J. Chen, Z.J. Zhang, The interactive decomposition method for multiobjective linear programming and its applications. Inf. Decis. Technol. **14**(4), 275–288 (1988)
230. H. Yano, K. Matsui, Hierarchical multiobjective fuzzy random linear programming problems. Procedia Comp. Sci. **22**, 162–171 (2013)
231. P.L. Yu, *Multiple-Criteria Decision Making: Concepts, Techniques, and Extensions*. ed. by Y.R. Lee, A. Stam. Mathematical Methods in Engineering and Science, vol. 30 (Plenum Press, New York, 1985)
232. P.L. Yu, M. Zeleny, Linear multiparametric programming by multicriteria simplex method. Manag. Sci. **23**(2), 159–170 (1976)
233. M. Zeleny, *Linear Multiobjective Programming* (Springer, Berlin, 1974)
234. X. Zeng, S. Kang, F. Li, L. Zhang, P. Guo, Fuzzy multi-objective linear programming applying to crop area planning. Agric. Water Manag. **98**, 134–142 (2010)
235. X. Zhang, G.H. Huang, C.W. Chan, Z. Liu, Q. Lin, A fuzzy-robust stochastic multiobjective programming approach for petroleum waste management planning. Appl. Math. Model. **34**, 2778–2788 (2010)
236. S. Zionts, J. Wallenius, An interactive multiple objective linear programming method for a class of underlying nonlinear utility functions. Manag. Sci. **29**(5), 519–529 (1983)
237. C. Zopounidis, D.K. Despotis, I. Kamaratou, Portfolio selection using the ADELAIS multi-objective linear programming system. Comput. Econ. **11**, 189–204 (1998)

Index

© Springer International Publishing Switzerland 2016
D.T. Luc, *Multiobjective Linear Programming*,
DOI 10.1007/978-3-319-21091-9

MIX

Papier | Fördert
gute Waldnutzung

FSC® C083411

Zeitfracht Medien GmbH
Ferdinand-Jühlke-Straße 7
99095 Erfurt, Deutschland
produktsicherheit@kolibri360.de